T0387970

Cultivation for Climate Change Resilience

Volume 1

Tropical Fruit Trees

Editors

Adel A. Abul-Soad

Tropical Fruit Research Department
Horticulture Research Institute, Agricultural Research Center
Giza, Egypt

Jameel M. Al-Khayri

Department of Agricultural Biotechnology
College of Agriculture and Food Sciences
King Faisal University
Saudi Arabia

CRC Press is an imprint of the
Taylor & Francis Group, an **informa** business
A SCIENCE PUBLISHERS BOOK

First edition published 2023
by CRC Press
6000 Broken Sound Parkway NW, Suite 300, Boca Raton, FL 33487-2742

and by CRC Press
4 Park Square, Milton Park, Abingdon, Oxon, OX14 4RN

CRC Press is an imprint of Taylor & Francis Group, LLC

Library of Congress Cataloging-in-Publication Data (applied for)

ISBN: 978-0-367-15189-8 (hbk)
ISBN: 978-1-032-37116-0 (pbk)
ISBN: 978-0-429-05558-4 (ebk)

DOI: 10.1201/9780429055584

Typeset in Palatino
by Radiant Productions

Preface

Awareness of the adverse impact associated with the global climate change on the future of agriculture, researchers are devoting efforts for finding solutions to mitigate undesired effects based on intelligent predictions and improved utilization of available genetic resources. This book highlights the contemporary knowledge of the impacts of abiotic and biotic stresses inflected by climate changes on the production, horticultural practices and physiological processes of various fruit tree species. Moreover, it describes the adaptation of innovative approaches to mitigate the climatic adverse effects and enhance resilience characteristics of fruit crops.

This book consists of 2-volume set. This Volume 1 subtitled Tropical Fruit Trees contains 12 chapters grouped in 3 parts. Part I consists of 2 chapters addressing conceptual framework emphasizing climate change impact on agriculture and food security and mitigation strategies. Part II describes the impact of climate change on some of rain forest tropical fruits species including banana, coconut palm, oil palm, papaya, and passion fruit. Part III addresses sub-tropical fruit species including avocado, guava, jujube, Indian jujube and snake fruit.

Each chapter addresses one fruit crop and updates available information in relation to the various concepts. Topics discussed include a general introduction on climatic requirements of a fruit crop, significant symptoms of climate change impacts, crop management under changed climate conditions, natural adaptation of genetic resources, mitigation strategies against biotic and abiotic stresses, remote sensing and environmental certification, and concludes with future prospects and literature.

This book is essential for researchers and students concerned with improving the productivity and quality of fruit crops to achieve sustainable fruit cultivation and conservation of this important nutritional food source for future generations. This book is geared toward a variety of general readers, fruit farmers, research scholars and scientists interested in learning about the impact of climate changes on fruit cultivation. It is a valuable source of information for students to increase their awareness of fruit cultivation under climate change conditions.

Our sincere gratitude is presented to the contributing authors for their generous cooperation and much appreciation to the CRC Press for the opportunity to publish this book.

Adel A. Abul-Soad
Giza, Egypt

Jameel M. Al-Khayri
Al-Ahsa, Saudi Arabia

Contents

Part III: Sub-Tropical Fruits

Part I
Conceptual Framework

1

Implications of Climate Change on Agriculture and Food Security

Adam Ahmed,[1,2,*] *Azharia Elbushra*[1,3] and *Mutasim Elrasheed*[1]

1. Introduction

1.1 Importance of Agriculture and Fruit Cultivation

Agricultural sector plays an important role in livelihood, food security and economic development for the majority of the world. This contribution is well documented in the literature. According to the Food and Agriculture Organization of the United Nation (FAO, 2008), about 36% of the total world labor force depends primarily for its livelihood on the state agricultural sector. However, contribution of workforce varies according to region and population density. For instance, it constitutes about 4.2% in Europe (EUROSTAT, 2018); 20% in Arab countries (World Bank, 2020); more than one-third in Southeast Asia (Zhai and Zhuang, 2009); 40–50% in highly populated countries of Asia and the Pacific (FAO, 2008); 48% for the total African population (NEPAD, 2013) and up to 66% in sub-Saharan Africa (FAO, 2008). Furthermore, a number of studies have addressed the

[1] Department of Agribusiness and Consumer Sciences, College of Agricultural and Food Sciences, King Faisal University, P.O. Box 400, Al-Ahsa 31982, Saudi Arabia.
[2] Department of Agricultural Economics, Faculty of Agriculture, University of Khartoum, P.O. Box 32, Postal Code 13314, Khartoum North, Sudan.
[3] Department of Agricultural Economics, College of Agriculture, University of Bahri, 12327, Postal Code 111111, Khartoum North, Sudan.
Emails: aaali@kfu.edu.sa; melrasheed@kfu.edu.sa
* Corresponding author: ayassin@kfu.edu.sa; adamelhag2002@yahoo.com

contribution of agriculture to labor force in different countries. According to EUROSTAT (2018) statistics, the contribution of agricultural sector to the total labor employment accounts for 23% in Romania, 17.5% in Bulgaria, 10.7% in Greece and 10.1% in Poland, and about 1% in Canada and Germany (World Bank, 2020). While the share of agriculture in the total employment in the African countries showed a wide spectrum, it is only one-fourth in Egypt, about half in Kenya, two-third in Ethiopia, two-fifth in Sudan and about three-quarter in Uganda. Similarly, it constitutes about 25, 29 and 37% in China, Indonesia and Pakistan, respectively (World Bank, 2020).

A report produced by the Food and Agricultural Organization (FAO) of the United Nations estimated that 2.5 billion out the 3 billion, who lived in the rural areas, depend on agriculture as the main source of livelihood (FAO, 2013). It is worth mentioning that the demand for agricultural products has increased three times from 1960 to date to meet the needs of the growing world population. Edame et al. (2011) stated that agricultural production needs to be increased by more than two-third to satisfy the food demands of the growing population by 2050.

Evidence-based research revealed that the percentage share of the agriculture sector to the gross domestic product (GDP) of the higher income countries is very small, indicating an inverse relation between national income and the percentage share of agriculture in GDP (FAO, 2019). As depicted in Table 1, the average agricultural contribution to GDP at the global level represented 3.4% in 2017 whereas it contributed about one-third in the developing countries (FAO, 2013).

The contribution of the agricultural sector to GDP is stable at around 4% since 2000 at the global level (FAO, 2021) and varies considerably among different countries and regions in the world. It amounted to 5.4% in USA (USDA, 2019); 15% in Africa, where it ranged from 3% in Botswana and South Africa to 50% in Chad (OECD-FAO, 2016); more than

Table 1. Percentage contribution of agricultural sector value added to the total Gross Domestic Product at global and regional level (2017–2018).

Regions	Year	
	2017	**2018**
Arab world	5.4	5.2
Europe and Central Asia	2.0	1.9
European Union	1.7	1.7
Middle East and North Africa	5.1	4.0
Sub-Saharan Africa	15.8	15.7
World	3.4	-

Source: World Bank, 2020

10% in Southeast Asia (Zhai and Zhuang, 2009); and less than 2% in Gulf countries (Woertz, 2017).

The main objective of this chapter is to describe the implications of climate change on agriculture and food security. It particularly reviews the effect of greenhouse gas (GHG) emissions, temperature, precipitation and the associated weather phenomenon, such as drought, dry and wet spells, floods and heat waves on food systems and food security pillars. Moreover, fruits' contributions to kilocalories, protein and fat supply as well as fruits' footprint of land use, fresh water withdrawal, greenhouse gases and eutrophying emissions are covered. In this context, Ritchie and Roser (2020) define eutrophying emissions as the 'runoff of excess nutrients into the surrounding environment and waterways, which affect and pollute ecosystems with nutrient imbalances and they are measured in kilograms of phosphate equivalents.'

This chapter is divided into five sections: Section one is concerned with the introduction, which highlights the importance of the agricultural sector and the basic concepts of food security and climate change issues; Section two covers interlinks among climate change, food systems and food security; the third section is devoted to fruits, climate change and food security, while climate change adaptation and mitigation measures are presented in Section four. Lastly, Section five is devoted to the conclusions drawn.

1.2 Concept and Causes of Climate Change

The concept of climate change has been addressed by many researchers and concerned agencies. According to a study conducted by FAO (2008), the numbers of climate change definitions have been stated. For instance, the World Metrological Organization (WMO) defines climate change as 'long-term changes in average weather conditions', while Global Climate Observing System (GCOS) refers to it as 'all changes in the climate system, including the drivers of change, the changes themselves and their effects' and the United Nation Framework Convention on Climate Change (UNFCCC) defines it as 'only human-induced changes in the climate system'. Contrarily to UNFCCC, the Intergovernmental Panel on Climate Change (IPCC) refers to climate change as not only changes induced by human activity but also by natural variability (IPCC, 2007a). It is worth noting that most of the experts stress the importance of differentiation between climate change and climate variability based on the time horizon. Accordingly, climate change is defined as a significant long-term variation in weather patterns, including global temperature, precipitation, wind, rainfall and snowfall in a certain place or a planet as whole (FAO, 2012; UCDAVIS, 2019; Takepart, 2020), while climate variability refers to short-term variation (daily, seasonal, inter-annual, several years) in the weather patterns (FAO, 2012).

On the one hand, climate change can be attributed to natural factors and human activities. The natural causes of climate change include among others, volcanic eruptions, changes in solar energy (EPA, 2016a), earth orbit and CO_2 level in the atmosphere (FAO, 2012). The United States Environmental Protection Agency (EPA) summarizes the main causes of climate change as the energy balance (between energy entering and leaving the earth); greenhouse effects, changes in the sun's energy and reflectivity (EPA, 2016a). On the other hand, human activities, including GHG emissions from burning fossil fuel, usually result in global warming (FAO, 2012). Relevant researches in the area indicate that burning of oil, gas and coal results in release of CO_2 into the air and hence, the rising planet temperature (Met. Office, 2020). For example, human activities increased the global greenhouse gas emissions by 70% during the period from 1970 to 2004 (IPCC, 2007c).

1.3 *Concepts of Food Security and Food Security Pillars*

The FAO (1996) defines food security as a 'situation that exists when all people, at all times, have physical, social, and economic access to sufficient, safe and nutritious food that meets their dietary needs and food preferences for an active and healthy life.' Whereas IPCC (2013) defines food security as 'a state that prevails when people have secure access to sufficient amounts of safe and nutritious food for normal growth, development, and an active and healthy life.' Additionally, FAO (2003) stated that 'food security depends more on socio-economic conditions than on agroclimatic ones, and on access to food rather than the production or physical availability of food.' It also stated that to evaluate the potential impacts of climate change on food security, 'it is not enough to assess the impacts on domestic production in food-insecure countries. One also needs to (a) assess climate change impacts on foreign exchange earnings; (b) determine the ability of food surplus countries to increase their commercial exports or food aid; and (c) analyze how the incomes of the poor will be affected by climate change.'

Based on the available literature, food security is decomposed into four pillars, namely: availability, accessibility, utilization and stability.

Food availability refers to the existence of food in a particular place at a particular time (Brown et al., 2015), which means that each individual in the country has a sufficient amount of food in terms of quantity and quality from the country's own production, or via imports, food aid and donation (Stamoulis and Zezza, 2003; Scialabba, 2011). It also indicates that there is adequate food for everyone at the household, community, regional, state and international levels (Clay, 2003). Food availability reflects the supply side of food security (Brown et al., 2015). The components of food security supply side include food production, distribution, exchange (Armstrong-Mensah, 2017) as well as stored and processed food (FAO, 2008).

Food accessibility refers to the individual's economical and physical ability to access appropriate and nutritious food (Schmidhuber and Tubiello, 2007; Brown et al., 2015). Furthermore, Gericke (2003) decomposed food accessibility into four dimensions: physical, economic, social and technological access. The physical access dimension is concerned with infrastructure, such as market, transportation and distribution facilities, while economic access refers to the ability of individuals to produce their own food and/or have enough money to purchase it. Furthermore, the culturally accepted food by the community reflects the social access dimensions, while technological access refers to the technological resources owned by the households for food preparation and/or preservation. The many factors that affect food accessibility and stability are: food price, market infrastructure, distance from market, income, consumer's preferences, self-sufficiency, transportation costs, socio-economic condition and technological facilities. Moreover, according to Brown and others (2015), food accessibility and stability are highly affected by changing climate variables.

Food utilization refers to how efficiently available and accessible food is utilized in order to obtain the adequate nutritional value required for individual needs. Food utilization is influenced by many factors, such as food safety, quality, consumption patterns and trends (Brown et al., 2015), food processing, storage and preservation facilities. In this context, some researchers emphasize the importance of incorporating non-food inputs, such as health when dealing with the concept of food utilization (Stamoulis and Zezza, 2003). Lastly, food stability refers to the sustainability of all or any one of the other food security dimensions (availability, accessibility and utilization) over time (Schmidhuber and Tubiello, 2007).

1.4 *Concepts of Food Systems*

Researchers have defined food systems as complex and connected activities including production, processing, transportation, consumption and food waste (FAO, 2008; Oxford University Press, 2017; FAO, 2018). According to Grubinger et al. (2010), the food system could be defined as 'an interconnected web of activities, resources and people that extends across all domains involved in providing human nourishment and sustaining health, including production, processing, packaging, distribution, marketing, consumption and disposal of food. The organization of food systems reflects and responds to social, cultural, political, economic, health and environmental conditions and can be identified at multiple scales, from a household kitchen to a city, county, state or nation.'

Each activity of the food system is further disaggregated into assets (FAO, 2008) and subsystems (FAO, 2018), which interact with each other to affect the food system activities and hence food security pillars. Further, breakdown of the food system activities, elements of food security pillars as well as outcomes and their interaction are well captured in the Global Environmental Change and Food Systems' (GECAFS) definition. They state that 'food systems encompass (i) activities related to the production, processing, distribution, preparation and consumption of food; and (ii) the outcomes of these activities contributing to food security (food availability, with elements related to production, distribution and exchange; food access, with elements related to affordability, allocation and preference; and food use, with elements related to nutritional value, social value and food safety). The outcomes also contribute to environmental and other securities (e.g., income). Interactions between and within biogeophysical and human environments influence both the activities and the outcomes' (GECAFS, 1994).

The FAO (2008) differentiates between food systems and food chain. The former tackles the issue from a holistic approach consisting of simultaneously interacting food activities, while, the latter refers to linear sequential activities that happen for people to get their food.

To fulfil the purpose of this chapter, climate change and food security, a framework document is used to highlight the impact of climatic change factors (environmental) on food systems and their effect on food security pillars.

2. Interlinks Among Climate Change, Food Systems and Food Security

Sustainable agriculture and food systems are important in achieving food security. However, agriculture and food systems are threatened by climate change. Accordingly, mitigation and adaptation strategies that are used to address the impact of the climate change (MDGs 13) are essential for achieving the Millennium Development Goals (MDGS): no poverty (MDGs 1), zero hunger (MDGs 2) and clean water and sanitation (MDGS 6). Available evidence declared that 924 million peoples were severe food insecure in 2020 (FAO, IFAD, UNICEF, WFP and WHO, 2021), while during 2012–14, approximately 805 million people were food insecure at the globe, where 98% of them lived in the developing countries (FAO, IFAD and WFP, 2014), and at least 2 billion live with inadequate nutritious food (Pinstrup-Andersen, 2009). Overweight and obesity reached 2.1 billion in 2013 compared to only 857 thousand in 1980 (Ng et al., 2014).

Agriculture is the main sector for providing food to the growing population in the world. Agricultural production is generally governed by the availability of resources, such as arable land, water, production inputs, technologies, human resources and the environment. Interlink between agriculture production and climate change is widely acknowledged in the literature. Climate change risk factors, such as extreme changes in temperature and precipitation, wind and other extreme weather events and their associated phenomenon; negatively influence agricultural production and hence food availability. Moreover, these environmental factors exert influence on food accessibility, utilization and their stability over time. These weather events-associated phenomena include among others, heat waves, dry spells, shifting seasons, biodiversity, drought, deforestation, desertification, soil fertility, erosion, flood, rising sea level, melting glacier, water scarcity, pollution and land scarcity.

The impact of climate change on food production, processing, distribution, consumption and waste disposable could be summarized in a wide range of effects at the local, national, regional and global levels. Its impact on production activity is manifested in agricultural yield (crops, livestock, forestry, poultry and fisheries) and labor productivity fluctuation; changing cropping patterns and irrigation water requirements; and soil types. It is worth stating that most agricultural products are characterized by seasonality in production. This necessitates performing storage and processing functions in order to sustain food security all through the year. The effect of climate change on storage and processing activity is illustrated in terms of needs for energy and chemical requirement as well as the availability of sufficient and hygienic water. A report released by FAO (2008) declared that the impact of climate change on global food systems could be reflected not only via influencing market price of food processing, storing, transporting and production, but also through food price.

The distribution activity requires the presence of facilities in terms of vehicles, paved roads, railways, cargo airplane, ships to deliver agricultural products from production to consumption centers. Any disruption in the distribution activity channels as a consequence of climate change might result in increase of agricultural products' marketing costs and hence negatively influence the food security situation. In this regard, McGuirk and others (2009) stated that efficiency and effectiveness of land, water and air transportation facilities are prone to climate change adverse events and are ultimately reflected in higher transportation costs.

The effect of changing temperature, precipitation, as well as other extreme weather events affect food consumption activity in different ways. For instance, high temperature may result in reducing the shelf-life of perishable agricultural products, increasing water consumption and energy requirements for storage facilities. Moreover, rainfall fluctuation might hamper agricultural production and in turn reduce the availability of diversified food to satisfy the consumer preference for acquiring healthy and nutritious food.

Different argumentations might be initiated with reference to food systems and food security determinants other than climate change factors. Some researchers summarized these factors into technological and structural changes in food system's activities; population growth, socioeconomic and demographic changes; changes in food consumption patterns; disasters and changes in both energy availability and usage (Brown et al., 2015); conflict and economic downturn (FAO, IFAD, UNICEF, WFP and WHO, 2021) as well as the incidence of pandemics, such as Coronavirus.

2.1 *Food Systems and Greenhouse Gas Emissions*

The contribution of the agriculture sector to the global emissions from human activities, during the period from 2007 to 2016, were estimated to be 13, 44 and 82% of carbon dioxide, methane and nitrous oxide, respectively. These contributions collectively accounted for 23% of total net anthropogenic GHG emissions (IPCC, 2019). In a study focused on trend in global GHG emissions conducted by Olivier and his colleagues (2005), it was found that agricultural contribution to the global man-made methane emissions constitutes about 43%, out of which, animal fermentation; rice farming and animal waste and Savanah burning contributed about 25, 12 and 6%, respectively. Moreover, they confirm that agriculture is the main source of origin of global anthropogenic nitrous oxide emissions. It contributes about 38% from animal waste, grazing and manure; 13% from crop cultivation; 12% fertilizers and 19% from indirect agricultural sources.

Based on FAO report that focused on climate change, agriculture and food security, changing land use systems and land degradation contributed significantly to the total GHG emissions (FAO, 2016). One example of it is that about 5 billion metric tons of CO_2eq to the atmosphere was produced because of deforestation of peat-lands during the periods 2005–2017. In this context, only 20 countries accounted for more than 66% of the total agricultural GHG emissions, out of which, only four countries (China, India, Brazil and the United States of America) shared more than 50% of the total agricultural GHG emissions, with Asia being the biggest contributor (FAO, 2019).

Worldwide food systems accounted for about more than one-third (34%) of the GHG emissions (Crippa et al., 2021). According to IPCC, the contribution of the food systems to GHG emissions in term of CO_2 and non-CO_2 gases was from different agriculture sources. These sources are:

within the farm gate (10–12%); land use and land use change dynamics (8–10%); and food supply chains activities beyond farm gate (5–10%), ending up with a total food system share of 25–30% (IPCC, 2019; Poore and Nemecek, 2018). Moreover, scientists stated that food systems, including crop and livestock production, are responsible for 19–29% of the global anthropogenic GHG emissions, out of which the contribution of agricultural production might reach up to 86% (Sonja et al., 2012).

In this context, Table 2 summarizes the global agricultural sector emissions of CO_2eq (CO_2eq), from CH_4 and (CO_2eq) from NO_2 (in million Giga grams) during the period 1961–2017. Asia contributes the highest

Table 2. Global agricultural emissions of CO_2eq, CH_4 and NO_2 in million Giga grams (1961–2017).

Regions	Years											
	1961–70		1971–80		1981–90		1991–2000		2001–2010		2011–2017	
	value	%	Value	%	value	%	value	%	value	%	value	%
Emissions (CO_2eq)												
Africa	2.6	8.6	3.1	8.7	3.8	9.4	6.2	13.5	7.6	15.5	6.2	16.8
Americas	7.7	25.7	9.5	26.4	10.4	25.5	11.4	24.8	12.6	25.8	9.2	25.1
Asia	10.9	36.1	12.8	35.6	15.3	37.5	19.0	41.3	20.9	42.6	16.0	43.4
Caribbean	0.2	0.7	0.2	0.6	0.2	0.6	0.2	0.5	0.2	0.5	0.2	0.5
Europe	7.6	25.2	9.0	25.0	9.7	23.9	7.3	15.9	5.9	12.1	4.0	10.9
Oceania	1.3	4.4	1.5	4.3	1.5	3.7	2.1	4.5	2.0	4.1	1.4	3.9
Emissions (CO_2eq) from CH_4												
Africa	1.5	7.4	1.8	7.8	2.1	8.6	3.2	12.0	3.9	14.3	3.3	16.1
Americas	5.0	24.6	5.9	25.5	6.3	25.4	6.7	25.4	7.4	26.8	5.3	25.8
Asia	0.1	0.7	0.1	0.6	0.1	0.6	0.1	0.6	0.1	0.5	0.1	0.5
Caribbean	8.3	40.5	9.2	39.8	10.1	40.5	11.5	43.4	12.2	44.1	9.1	44.8
Europe	4.8	23.5	5.3	22.9	5.5	22.0	4.0	15.0	3.0	11.0	2.0	9.7
Oceania	0.8	4.0	0.9	4.0	0.9	3.5	1.1	4.1	1.1	3.8	0.7	3.6
Emissions (CO_2eq) from NO_2												
Africa	1.1	11.1	1.3	10.3	1.7	10.6	3.0	15.4	3.6	17.0	2.9	17.7
Americas	2.7	27.8	3.6	27.8	4.0	25.4	4.7	23.9	5.2	24.4	4.0	24.1
Asia	0.1	0.7	0.1	0.6	0.1	0.6	0.1	0.4	0.1	0.4	0.1	0.4
Caribbean	2.6	26.6	3.6	27.8	5.2	32.7	7.5	38.3	8.7	40.4	6.8	41.4
Europe	2.8	28.4	3.7	28.7	4.3	26.8	3.3	17.0	2.9	13.4	2.0	12.3
Oceania	0.5	5.4	0.6	4.7	0.6	3.9	1.0	5.0	0.9	4.4	0.7	4.2

Source: Calculated from FAOSTAT, 2020

share with an increasing trend of CO_2 emissions in the world as it produced 36% during the sixties and reaching up to 43.4% during 2011–2017. Similarly, Africa's share of the global CO_2 emissions show an increasing trend as it was almost double during the period 1961–2017. Americas are ranked second in terms of CO_2 emissions; however, its contribution is almost constant during the entire period, ranging from 24.8 to 26.4%. In contrast, Europe's share to the global CO_2 emissions show a declining trend as the emissions decreased from 25% in the sixties to less than 11% in 2011–2017.

On the other hand, although Caribbean shows the lowest share of CO_2eq emissions in the world, it had the highest share of CO_2eq emissions from methane (CH_4), ranging from 39.8 to 44.8% for the period 1961–2017. Americas ranked second in CO_2eq emissions from methane followed by Europe. However, a constant trend of CO_2eq emissions from methane is noticed for Americas and a decreasing one for Europe, where it decreased from 23.5% in 1961–70 to only 9.7% in 2011–17 of the global emissions.

Pertaining to CO_2eq emissions of NO_2, Europe, Americas and Caribbean collectively represented more than three-quarters of the global emissions. It was noticed that Americas and Europe showed a decreasing trend in the percentage share of global CO_2eq emissions from NO_2 over time (1961–2017). However, Caribbean showed an increasing trend during the same period (Table 2).

On the other hand, the negative impacts of the elevating CO_2 level and its consequences in global warming can be manifested in increasing the potential effect of insects, weeds and pests in damaging agricultural production (U.S. Global Change Research Program, 2009a) and reducing crops absorption of some essential nutrients such as zinc and iron (Myers et al., 2014). In contrast, the positive impact can be depicted in flourishing biomass and improving crops yields (FAO, 2008; Cho, 2018).

2.2 *Food Systems and Temperature*

The impact of rising temperature on climate change is well documented in the literature. Globally, rise in temperature will result in significant economic losses (Gurdeep et al., 2021). The rise or decline in the mean temperature influenced agricultural production, food processing and consumption (FAO, 2008) and consequently food security. Its impact on production can be witnessed in the decline of plants' and animals' productivities; change in crop and livestock pattern; lower labor productivity and increase in pest and diseases infestation thresholds. These impacts are widely tackled by researchers, for instance, EPA (2016a) declared that temperature plays a crucial role in the determination of plant and animal types that survive in certain locations on the planet. In this sense, FAO (2016) reported that rising temperatures and frequent occurrences of extremely dry and wet years in sub-Saharan Africa were expected to have a negative effect on crop and livestock production.

On the same lines, some researchers mentioned that the occurrence of heat waves adversely affected livestock fertility; dairy cows' productivity (Cho, 2018); and increase in production and capital expenditure (U.S. Global Change Research Program, 2009b). Moreover, livestock become prone to parasite infestations and diseases (Cho, 2018).

Temperature could have direct and indirect impacts on agricultural production. The direct effect is noticed in yield reduction as response to temperature changes beyond the optimal level. For instance, IPCC (2007b) modelling studies indicated that slight increase in mean temperature ranging between 1–3°C accompanied with an increase in the atmospheric carbon dioxide (CO_2) and changing rainfall are expected to improve crop productivity in the temperate zones. Furthermore, in low-latitude areas, moderate increases in the mean temperature between 1–2°Centigrade are likely to cause decline in the major cereal crops, where warming of more than 3°Centigrade negatively affects all regions (IPCC, 2007b). According to the National Research Council (2011), any global warming by 1°Celsius may result in 5–10% changes in both precipitation across many regions and stream flow of rivers; 5–15% decrease in crop productivities; and increase in the severity of summers and ocean acidity.

Labor plays an important factor in the agricultural sector, particularly in developing countries, where agriculture is a labor-intensive activity. Agricultural labor productivity is expected to be influenced by economic, social and environmental factors. The economic factors include all factors that affect agricultural labor efficiency, such as wages, labor availability, skills and assets ownerships. On the other hand, social factors include demographic characteristics and dependency ratio. The environmental factors influencing agricultural labor are weather conditions, such as temperature, precipitation and other extreme events (wind patterns, snow, floods). The impact of extreme changes in temperature is expected to seriously decrease labor productivity. Thus, the consequences of reduction in labor productivity might have a severe effect on production costs, food availability and soaring food priced, ending up with reduction in access to food.

Regarding food processing and storage, any increase in temperature may cause an additional cost in food storage, distribution and transportation. This could be attributed to the additional need of energy requirements for cool transportation and storage. Absence of appropriate storage facilities and high temperature-related effects might not only result in drastic increase in postharvest losses due to storage affected by pests' attacks, but also lowering in food quality. Ultimately, due to the seasonality of many agricultural products, particularly vegetables and fruits, well-equipped and proper processing and storage facilities might improve food quality preservation, food safety, longer shelf-life and lower food disposal, thus playing a significant role in food security pillars.

One of the main challenges of achieving global food and nutrition security is the provision of enough supply of nutritive, healthy and accessible food. Food processing and associated marketing functions play an important role in achieving food security. Food processing is the process of transformation of raw agricultural products into food ready for consumption (Lund, 2003). Agro-processed food improves foods' nutritional ingredients and quality. Moreover, agro-processing reduces post-harvest losses and thus conserves healthy food for the consumer.

Based on the available literature, a number of studies addressed the impact of rising temperatures and changing weather events on food systems. McGuirk and others (2009) state one example—increasing temperature and the occurrence of extreme weather events usually influence food distribution systems at different levels. Others state that the delivery of safe and quality food necessitates the use of proper food processing, storing, packaging and transporting facilities. They also added that any obvious increase in temperature as a consequence of climate change would lead to food poisoning and spoilage, which require improved cold storage facilities (James and James, 2010).

In discussing the concept of food system, food consumption is defined as the processes that relates to food utilization and consumption. It is broadly concerned with food processing; household's decision-making regarding food distribution practices and choices; as well as individual access to health care, sanitation and knowledge (FAO, 1997). The exposure to extreme temperature alters an individual's food consumption and increases incidences of heat-related diseases, such as heat stroke and dehydration particularly for agricultural outdoor labor, thereby resulting in lower labor productivity.

2.3 *Food Systems and Altered Precipitation Patterns*

Precipitation means any liquid or frozen water that constituted in the atmosphere and falls back to the Earth in different shape such as rain, snow and sleet (Graham et al., 2010). Rainfall is one of the main environmental factors that influence agricultural production, particularly rain-fed crops, where about 60 and 90% of the stable food are produced under rain-fed globally and in sub-Saharan Africa, respectively (Savenije, 2001). Furthermore, globally, rain-fed agriculture constitute about 83% of all cultivated land and three fifth of all food (FAO, 2002). Shortage and/or extreme rainfall affect agriculture and livestock production through different phenomenon such as the occurrence of droughts, dry and wet spells, heavy precipitations and floods. The frequency, severity and duration of these phenomena determine the expected impact level on food security and the livelihoods of the population. Drought and floods adversely affect access to clean water for drinking and agricultural uses in different ways. For instance, drought might generate water stress problem due to water scarcity, while flood might lead to water contamination.

Drought could be defined as the situation in which the lack of precipitation causes a prolonged period of dry weather, which result in water shortages (EPA, 2016b). While the United Nations Convention to Combat Desertification (UNCCD, 1994) define it as 'drought means the naturally occurring phenomenon that exists when precipitation has been significantly below normal recorded levels, causing serious hydrological imbalances that adversely affect land resource production systems.'

The risk of drought varies across the countries depending on the degree of severity (moderate, severe, extreme and exceptional drought), thus most countries are susceptible to different levels of damages and losses in crops, pastures, and water sources. On the other hand, dry spell is defined as the deficit of precipitation for a short period, while, the occurrence of intensive periods of precipitation is known as wet spell (EPA, 2016b). It is worth mentioning that, the definition of dry spell required the determination of different threshold values for the successive dry days relative to location, crop types and researcher's aims. Whereas, heavy precipitation is the situation in which the amount of precipitation in a certain place considerably exceeds the normal level (EPA, 2016b). Lastly, flood can be defined as a situation in which dry land is changed into wetland or open water resulting from climate change in form of heavy rains and/or increasing sea water level (EPA, 2016b).

The expected impact of drought, dry & wet spells, heavy precipitations and flood could be depicted in soil moisture contents, soil fertility, cropping pattern, length of the growing season, pasture, biodiversity, and pest & disease infestations. Consequently, this might significantly influence food production, storing, processing and consumption activities and eventually food security.

On the production side, the occurrence of precipitation phenomenon such as heavy precipitation, wet and dry spells and drought might change cropping pattern, interrupt crops growth and increase uncertainty of crops and livestock production. However, it is worth mentioning that, shortages of rainfall might necessitate the need for supplementary irrigation as water become scarcer to fulfil crops water requirements.

Drought and dry spell might negatively affect food availability through affecting crops' yields and quality, soil fertility, and might even lead to complete crop loss. Moreover, their impact on livestock is manifested in overgrazing, water stress, decrease in livestock yield and may even cause animal death. Some researchers argued that incidences of drought and dry spell cause changes in forests and other ecosystems, resulting in deforestation and desertification (Olagunju, 2015). In this context, a study investigating the impact of drought on crop production losses, during the period 1983–2009, found that about 75% of the harvested areas from maize, rice, soybeans, and wheat worldwide are subject to yield loss amounting to about USD 166 billion due to drought (Kim et al., 2019). Moreover, a study focused on climate change and grazing herds in Ethiopia concluded that the intensity and durations of drought not only affect animal stocking rates, but also forage production (Godde et al., 2019). Drought also usually affects rural farmers' income generated from agricultural production as they resort to felling of trees for charcoal and wood as a source of income, thereby aggravating the severity of desertification.

The impact of climate change events, such as heavy precipitation, drought and dry and wet spells, might be extended to not only affecting agricultural and livestock production activities but also disrupting international food trade and increasing the number of households to face food insecurity and malnourished; and hence increase the need for food aid.

On the other side, flooding, which results in the conversion of arable lands that is used for food production and as livestock pasture into wet land or open water, will negatively impact agricultural activities, food prices and hence, food security. The situation is more obvious in countries located near the oceans or in locations facing heavy precipitation. For instance, South Asian countries, such as Bangladesh, India and Pakistan are more prone to severe, extreme and frequent floods due to climate changes, such as rise in temperatures and sea levels (Mirza, 2010). Moreover, the pasture carrying capacity of non-flooded grazing land gets seriously affected by overgrazing. This situation of overgrazing is usually accompanied with widespread livestock diseases and pest infestations which in turn lower the quality and quantity of livestock products, thereby affecting the contribution of animal-origin products to the total dietary energy supply (DES) of the majority of the population in the affected areas. Moreover, overgrazing accompanied with the occurrence of drought results in a high livestock mortality rate in Mongolia (Nandintsetseg, 2018).

Another related adverse effect of drought and dry spells is obvious as it causes agricultural labor to migrate to urban centers, resulting in skilled agricultural labor resorting to small trading and marginal activities. This is in line with the findings of Nawrotzki et al. (2017) who reported that the probability of rural-urban migration would increase by 3.6 per cent because of a monthly increase of drought.

Drought, dry and wet spells and precipitation might impair farmers' assets as well as public infrastructure concerned with food production, storage, possessing and distribution facilities, thereby exacerbating the problem of food availability and accessibility. Moreover, in such circumstances, consumers may shift their food consumption patterns into less preferable and non-diversified food as a coping mechanism to climate change and ultimately influence food utilization. A study conducted by Carpena (2019) to assess the effect of drought on food spending and macronutrient consumption in rural India revealed a shift of household consumption pattern from highly nutritive food into lower ones. Further, he concluded that household incomes generated during drought is the main cause of consuming lower-quality food. Another study concerned with droughts and famine in Kenya attributed the drought-triggered food insecurity factors to not only rainfall but also assets and labor constraints; poor policy implementation and poverty (Speranza, 2008).

2.4 Food System and Extreme Weather Events

Climate change is sometimes associated with the occurrence of extreme weather events, such as storms, high winds and frequent floods. The impact of these events varies across locations according to their degree of frequency and intensity. Extreme weather events influence households' human, social, financial, physical and natural assets. For instance, high winds and storms may destroy the physical productive assets, such as crops, livestock, buildings, stores and equipment. The implication of these events would be reflected in the reduction of food availability; diversity and quality; rise in food prices; increase in water pollution; increase in pest infestations and diseases; disruption in food supply chains; destruction of infrastructure facilities; decline in soil fertility; deterioration of human health and nutritional status; changing land-use pattern; threat to household's income sources and hence negatively affect food availability, accessibility and utilization (FAO, 2008).

In this regard, agriculture, rangelands and livestock in the Near East are expected to be seriously affected by climate change in terms of rising temperatures and the associated events, such as drought, floods and soil degradation. The impact varies across countries; for instance, the national production of rice and soya beans in Egypt will decrease by 11% and 28%, by the year 2050, respectively, relative to the current situation (FAO, 2014). In the same vein, the occurrence of severe heat waves in Russia 2010–11 resulted in huge grain crop loss, amounting to 30% of the total production. Likewise, the severe floods that happened in the 1980's led to a considerable reduction in the cultivated land area (14%) (Yadav, 2019).

3. Fruits, Climate Change and Food Security

3.1 Role of Fruits in Enhancing Food Security

Fruits and vegetables is one of the major five food groups that plays a crucial role in food security. Their consumption is considered as an important source of vitamins, minerals and fibers, which may reduce the risk of heart disease and certain types of cancer (Pérez, 2002), lower occurrences of cardiovascular diseases and obesity (Slavin and Lloyd, 2012). Moreover, the collective contribution of fruits, nuts and vegetables as sources of vitamins in the U.S. diet amounts to approximately more than 90% of vitamin C, about half of vitamin A, more than one-quarter of vitamin B6, and less than one-fifth each of thiamine and niacin (Kader, 2001).

Worldwide, the average fruit supply amounted to 75 kilograms per capita per year during the period 2014–2017. However, it varied across regions, ranging from 64 kg/capita/year in Africa to 166 kg/capita/year in Caribbean (Table 3). The recommended amount of fruits and vegetables consumption per capita/year is estimated to be 146 kg/year (WHO/FAO, 2003). Table 3 revealed that fruit consumption in Africa was less than the world average, while Asia, Oceania and Europe consumes an amount that was almost similar to the world average. However, Caribbean and Americas consumed 2.2 and 1.3-fold of world average, respectively. Similarly, the percentage share of fruit to food supply (kcal/capita/day), protein and fat supply (gram/capita/day) showed slight variability among all regions except Caribbean, where fruit contributed about 8% of food supply, 3.5 and 3.8% of protein and fat supply, respectively.

Table 3. Fruit Food Supply (FS) (kg/capita/yr) and macronutrient contribution at global and regional levels (2014–2017).

	FS(kg/capita/yr)	**FS (kcal/c/d)**		**Protein (g/c/d)**		**Fat (g/c/d)**	
		Value	**%**	**Value**	**%**	**Value**	**%**
World	75	99	3.4	1.2	1.4	0.7	0.8
Africa	64	105	4.0	1.2	1.8	0.5	1.0
Americas	100	129	3.9	1.5	1.6	1.2	1.0
Caribbean	166	229	8.2	2.4	3.5	2.8	3.8
Asia	71	89	3.2	1.0	1.3	0.6	0.8
Europe	78	108	3.2	1.3	1.2	0.8	0.6
Oceania	72	105	3.2	1.1	1.1	1.1	0.8

Source: FAO, 2020

3.2 *Impact of Climate Change on Fruit Crops*

A recent study that addresses the environmental impact of food production stated that agriculture represents about half of the global habitable land, out of which, more than three-quarters (77%) is used for meat and dairy production, while crops occupy less than one-quarter. Although livestock used more than three-quarters of the agricultural land, it contributed less than one-fifth (18%) and lesser than two-fifth (37%) of the global calorie and protein supply, respectively. Moreover, agriculture used more than two-third (70%) of fresh water and produced more than three-quarters of the global ocean and fresh-water pollutants (Ritchie and Roser, 2020).

Table 4 illustrated the footprints of land use, freshwater withdrawal, GHG and eutrophying emissions for select crops, mainly fruits, cereal and nuts. For comparison purposes, the footprints were measured in kilograms of food products, 100 grams of protein and 1000 Kilocalories for each of the selected crops.

With reference to the land-use footprint, all fruits used less land per square meters to produce 1 kg of food products as compared to nuts and cereals crops. Contrarily, in general, fruits used more area for supplying both 100 grams of proteins and 1000 Kcal compared to cereals and nuts.

On the other hand, fresh water withdrawal footprint refers to the amount of fresh water in liters required to produce one kg of food products or for supplying 100 grams of proteins and 1000 Kcal. It is worth mentioning that fruits, such as citrus, banana, apple, berries and grapes withdraw less amounts of water to produce one kg of food products compared to nuts, rice and wheat. The highest amount of water required for supplying 100 grams of proteins was reported for banana, followed by berries and grapes, with least water withdrawal for wheat. On the other hand, berries and grapes ranked first with reference to the amount of fresh water withdrawal for supplying 1000 Kcal which is relatively similar to nuts and rice.

In the same vein, GHG emission footprint refers to the amount of GHG emissions in Kg of CO_2eq required to produce 1 kg of product, 100 grams of proteins and 1000 Kcal. Table 4 shows that GHG emissions from producing 1 kg of rice outnumbered all the selected crops. In other words, GHG emissions generated from producing 1 kg of rice is equivalent to that produced from 5.7 kg banana, 10 kg apple and 13.3 kg citrus fruits.

According to Ritchie and Roser (2020), eutrophying emissions is defined as 'runoff of excess nutrients into the surrounding environment and waterways, which affect and pollute ecosystems. They are measured in grams of phosphate equivalents (PO_4eq).' The ecosystem pollutants in grams of PO_4eq generated from the production of one kg of rice corresponds to the pollutants resulting from the production of 24 kg of citrus, 16 kg of apple, 11 kg of banana and 6 kg of berries and grapes. In fact, fruits are perishable products, which require proper handling at pre- and post-harvest processes. Accordingly, they are highly susceptible to climate change. For instance, high temperature is associated with additional costs in transportation and storages facilities. According to Moretti and others

Table 4. Footprint for land use, freshwater withdrawal, and GHG and eutrophying emissions of fruit species compared to field crops in 2018.

Crop	Land use (m^2)			Fresh water (Liter)			GHG emissions (Kg Co_2eq)			Eutrophying emissions (g)		
	Kg product	100 g protein	1000 kcal	kg of product	100 protein	1000 kcal	Kg product	100 protein	1000 kcal	KG of product	100 g protein	1000 kcal
Citrus	0.9	14.3	2.7	83	1378	258	0.3	6.5	1.2	2.2	37.3	7.0
Banana	1.9	21.4	3.2	115	1272	191	0.7	9.6	1.4	3.3	36.6	5.5
Nuts	13.0	7.9	2.1	4134	2531	672	0.3	0.3	0.1	19.2	11.7	3.1
Apple	0.7	21	1.3	180	6003	375	0.4	14.3	0.9	1.5	48.3	3.0
Berries and grapes	2.4	24.1	4.2	420	4196	736	-	15.3	2.7	6.1	61.2	10.7
Wheat and rye	3.9	3.2	1.4	648	531	242	1.4	1.3	0.6	7.2	5.9	2.7
Rice	2.8	3.9	0.8	2248	3167	610	4.0	6.3	1.2	35.1	49.4	9.5
Maize	2.9	3.1	0.7	216	93	48	1.0	1.8	0.4	4.0	4.2	0.9

Source: Ritchie and Roser (2020)

(2010), high levels of temperature, CO_2 and O_3 have direct and indirect impact on both quantity and quality of fruits and vegetables.

4. Adaptation and Mitigation to Climate Changes

Scientists stress the importance of adaptation and mitigation and differentiate between them as strategies for addressing climate change. They define any strategy concerned with the reduction of sources and amount of GHG emissions as mitigation, whereas adaptation is referred to any 'adjustment in natural or human systems in response to actual or expected climatic stimuli or their effects, which moderates harm or exploits beneficial opportunities' (IPCC, 2001). Moreover, Tol (2005) distinguished between them according to spatial (global vs. district and region) and temporal (long-term vs. current and short-terms) scales and economic sectors (energy, transportation vs. urban planning, water, agriculture). Researchers differentiate between them according to their causal association; hence, mitigation tackles the root causes of climate change, whereas adaptation addresses the effects of climate change (CIFOR, 2011; Duguma et al., 2014).

Climate change effect is widely noticed at global, regional, country, sector, local and household levels. Accordingly, different measures for adaptation and mitigation are necessary. These measures vary according to the frequency, intensity, duration and severity of the climatic change events. In this sense, different adaptation and mitigation strategies and policies are used to mitigate the impact of climate change on agriculture and food security. In general, these might include, among others, (a) food supply, demand and any disruption in food chain; (b) access to services, such as credit, insurance and extension services; (c) soil and water management strategies; and (d) livelihood and agricultural and non-agricultural income activities.

Some of the climate change mitigation and adaptation strategies are usually used to improve the efficiency of agricultural products through use of different measures. These include adoption of improved and high-yielding varieties that are tolerant to climate change events; use of modern technologies; use of clean energy sources, such as solar energy; activation of agricultural extension agents; and encouragement to agricultural research and development. Moreover, provision of agricultural services in term of credit and insurance would increase and sustain agricultural production and lighten the adverse effect of climate change. Additionally, soil management strategies might be applied to reduce the climate change's negative effect on soil fertility, soil moisture contents, soil erosion and waterlogged land.

The adaptation and mitigation strategies not only cover the agricultural production side but also extend to cover other food system activities (processing, storing, distributions, consumption and disposal) to elevate the adverse effects of climate change and hence realize sustainable food security (Brown et al., 2015). Moreover, different countries adopted different adaptation and mitigation strategies to modify food demand and these might include loss and waste reduction and recycling; taxes and subsidies; awareness campaign.

In addition to this, additional adaptation and mitigation measures for achieving sustainable fruits production could be highlighted, based on the fruit's long life cycle which makes them more prone to environmental changes relative to short-duration crops. These measures include among others, diversification of cropping pattern; selection of crops varieties that suit the environment (Sarkar et al., 2021); and implementing precise strategies for water management (Mukhopadhyay and Mandal, 2021).

Fruits' long life cycles make them more prone to environmental changes relative to short-duration crops. This necessitates additional adaptation and mitigation measures for achieving sustainable fruit production. The measures include, among others, diversification of cropping pattern; selection of crops varieties that suit the environment (Sarkar et al., 2021); and implementation of precise strategies for water management (Mukhopadhyay and Mandal, 2021).

5. Conclusion and Prospects

Climate change risk factors, such as extreme changes in GHG emissions, temperature, precipitation, wind and other extreme weather events and their associated phenomena negatively influence the agricultural sector, food systems and hence, food security. The agricultural sector contributes significantly to global warming as it generates about one-fourth of the total net anthropogenic GHG emissions. Analysis of the global agricultural emissions of CO_2eq during the period 1961–2017 revealed:

- Asia has the highest share (36–43%) with increasing trends.
- Americas ranked second with almost constant trends (24.8–26.4%).
- Africa displays an increasing trend (8.6–16.8%).
- Europe shows a declining trend (25–11%).

Rising temperatures in the tropical zone negatively influence the food systems. The impact of changing temperatures on agricultural production can be manifested in the decline of plants' and animals' productivities; change in crop and livestock patterns; lower labor productivity and increase in pest infestations and diseases. According to the National Research Council (2011), any global warming by 1°Celsius may result in 5–10% changes in both precipitation across many regions and stream flow of rivers; 5–15% decrease in crop productivities; and increase in the severity of summers and ocean acidity. On the other hand, food systems are also affected by precipitation events, such as heavy precipitation, drought, flood and dry and wet spells. These effects could lead to disruption in international food trade, impairment of farmers' assets, changes in food consumption patterns, and eventually rendering the number of households insecure and malnourished by way of food.

With reference to land use, fresh water withdrawal and GHG emissions footprints:

- All fruits used less land in square meters to produce 1 kg of food products compared to nuts and cereals crops. Contrarily, they used more area for supplying both 100 grams of proteins and 1000 Kcal compared to cereals and nuts.
- Some fruits, such as citrus, banana, apple, berries and grapes withdraw less amounts of water to produce one kg of food products as compared to nuts, rice and wheat.
- GHG emissions generated from producing 1 kg of rice are equivalent to that produced from 5.7 kg banana, 10 kg apple and 13.3 kg citrus fruits.

Based on the aforementioned conclusion, adaptation and mitigation measures should be implemented to reduce the adverse implications of climate change on food security.

References

Armstrong-Mensah, E.A. (2017). *Global Health: Issues, Challenges, and Global Action*. John Wiley & Sons. https://www.wiley.com/enus/Global+Health%3A+Issues%2C+Challenges%2C+and+Global+Action-p-9781119110217.

Brown, M.E., Antle, J.M., Backlund, P., Carr, E.R., Easterling, W.E., Walsh, M.K., Ammann, C., Attavanich, W., Barrett, C.B., Bellemare, M.F., Dancheck, Funk, V.C., Grace, K., Ingram, J.S.I., Jiang, H., Maletta, H., Mata, T., Murray, A., Ngugi, M., Ojima, D., O'Neill, B. and Tebaldi, C. (2015). *Climate Change, Global Food Security and the U.S. Food System*. 146 pp. http://www.usda.gov/oce/climate_change/FoodSecurity2015Assessment/FullAssessment.pdf.

Carpena, F. (2019). How do droughts impact household food consumption and nutritional intake? A study of rural India. *World Development*, 122: 349–369. https://doi.org/10.1016/j.worlddev.2019.06.005.

Cho, R. (2018). *Agriculture and Climate: How Climate Change will Alter Our Food, State of the Planet*. Earth Institute, Colombia University. https://blogs.ei.columbia.edu/2018/07/25/climate-change-food-agriculture/.

Clay, E. (2003). Chapter 2. *Food Security: Concepts and Measurement in Food Reforms and Food Security: Conceptualizing the Linkages Expert Consultation*. FAO. http://www.fao.org/3/y4671e/y4671e06.htm#bm06.

Crippa, M., Solazzo, E., Guizzardi, D. et al. (2021). Food systems are responsible for a third of global anthropogenic GHG emissions. *Nat. Food*, 2: 198–209. https://doi.org/10.1038/s43016-021-00225-9.

Duguma, L.A., Wambugu, S.W., Minang, P.A. and Noordwijk, M.V. (2014). A systematic analysis of enabling conditions for synergy between climate change mitigation and adaptation measures in developing countries. *Environmental Science & Policy*, 42: 138–148. http://dx.doi.org/10.1016/j.envsci.2014.06.003.

Edame, G.E., Anam, B.E., Fonta, W.M. and Duru, E. (2011). Climate change, food security and agricultural productivity in Africa: Issues and policy directions. *International Journal of Humanities and Social Science*, 1(21): 205–223. https://www.ijhssnet.com/journals/Vol_1_No_21_Special_Issue_December_2011/21.pdf.

EPA. (2016a). *Causes of Climate Change*. United States Environmental Protection Agency. https://19january2017snapshot.epa.gov/climate-change-science/causes-climate-change_.html.

EPA. (2016b). *Climate Change Indicators in the United States: Drought*. United States Environmental Protection Agency. https://www.epa.gov/sites/production/files/2016-08/documents/print_drought-2016.pdf.

EUROSTAT. (2018). *Agriculture, Forestry and Fishery Statistics*. Luxembourg: Publications Office of the European Union. https://ec.europa.eu/eurostat/documents/3217494/9455154/KS-FK-18-001-EN-N.pdf/a9ddd7db-c40c-48c9-8ed5-a8a90f4faa3f.

FAO. (1996). *Technical Background Document Executive Summary*. World Food Summit, 13–17 Nov. 1996, Rome, Italy. Available at: http://www.fao.org/3/w2612e/w2612e00.htm.

FAO. (2002). Crops and Drops: Making the Best Use of Water for Agriculture. Rome, Italy. *In*: Brown, M.E., Antle, J.M., Backlund, P., Carr, E.R., Easterling, W.E., Walsh, M.K., Ammann, C., Attavanich, W., Barrett, C.B., Bellemare, M.F., Dancheck, V., Funk, C., Grace, K., Ingram, J.S.I., Jiang, H., Maletta, Mata, H.,T., Murray, A., Ngugi, M., Ojima, D., O'Neill, B. and Tebaldi, C. (2015). *Climate Change, Global Food Security, and the U.S. Food System*. Available online at http://www.usda.gov/oce/climate_change/FoodSecurity2015Assessment/FullAssessment.pdf.

FAO. (2003). World agriculture: Toward 2015/2030, Chapter 13, Rome, *Earthscan*. http://www.fao.org/3/a-y4252e.pdf.

FAO. (2008). *Climate Change and Food Security: A Framework Document*. http://www.fao.org/docrep/010/k2595e/k2595e00.htm.

FAO. (2012). *Climate Change and Food Security, Food and Agricultural Organization of the United Nations e-Learning Courses*. This course is funded by the European Union's Food Security Thematic Programme and implemented by the Food and Agriculture Organization of the United Nations. Available online: https://elearning.fao.org/course/view.php?id=143.

FAO. (2013). *FAO Statistical Yearbook*. World Food and Agriculture Organization. http://www.fao.org/3/i3107e/i3107e.pdf.

FAO. (2014). *World Food and Agriculture – Statistical Year Book 2014*. Rome. http://www.fao.org/3/i3591e/i3591e.pdf.

FAO. (2016). *The State of Food and Agriculture, Climate Change, Agriculture and Food Security Report*. Available at: http://www.fao.org/3/a-i6030e.pdf; accessed 2 June 2020.

FAO. (2018). *A Sustainable Food System Conceptual and Framework*. Food and Agriculture Organization, Rome. http://www.fao.org/3/ca2079en/CA2079EN.pdf.

FAO. (2019). *Statistical Pocketbook 2019*. World Food and Agriculture Organization, Rome. http://www.fao.org/3/ca6463en/CA6463EN.pdf.

FAO. (2021). *Statistical Yearbook 2021*. Food and Agriculture Organization of the United Nations, Rome. https://www.fao.org/3/cb4477en/online/cb4477en.html#chapter-1.

FAO, IFAD and WFP. (2014). *The State of Food Insecurity in the World: Strengthening the Enabling Environment for Food Security and Nutrition*. Rome, Italy. Retrieved from http://www.fao. org/3/a-i4030e.pdf.

FAO, IFAD, UNICEF, WFP and WHO. (2021). *The State of Food Security and Nutrition in the World: Transforming Food Systems for Food Security, Improved Nutrition and Affordable Healthy Diets for All*. Rome, FAO. https://doi.org/10.4060/cb4474en.

FAOSTAT. (2020). http://www.fao.org/faostat/en/#data/FBS. Accessed 2 June 2020.

GECAFS. (1994). About GECAFS. *In*: *FAO 2008: Climate Change and Food Security: A Framework Document*. http://www.fao.org/docrep/010/k2595e/k2595e00.htm.

Gericke, G.J. (2003). Better eating for better health: principles and practices of planning a healthful diet. Schönfeldt, H.C. (ed.). *Fundamentals of Nutrition Security in Rural Development: Graduate Readings*, vol. 3, University of Pretoria, Pretoria, 7 pp.

Godde, C., Dizyee, K., Ash, A., Thornton, P., Sloat, L., Roura, E., Henderson, B. and Herrero, M. (2019). Climate change and variability impacts on grazing herds: Insights from a system dynamics approach for semi-arid Australian rangelands. *Glob. Change Biol.*, 25: 3091–3109. Doi: 10.1111/gcb.14669.

Graham, S., Parkinson, C. and Chahine, M. (2010). *The Water Cycle*. NASA Earth Observatory. Accessed on 19 June 2020. https://earthobservatory.nasa.gov/features/Water.

Grubinger, V., Berlin, L., Berman, E., Fukagawa, N., Kolodinsky, J., Neher. D., Parsons, B., Trubek, A. and Wallin, K. (2010). University of Vermont Transdisciplinary Research Initiative Spire of Excellence Proposal: Food Systems. Proposal, Burlington, University of Vermont. *In:* Chase, L.P and Grubinger, V. (2014). *Introduction to Food Systems*. In the book, Chase, L. (ed.). *Food, Farms and Community: Exploring Food Systems*. University Press of New England. https://muse.jhu.edu/book/36007; https://www.cifor.org/fileadmin/fileupload/cobam/ENGLISH-Definitions%26ConceptualFramework.pdf.

Gurdeep, S., Manpreet, M. and Prashant K. (2021). Impact of climate change on agriculture and its mitigation strategies: a review. *Sustainability*, 13(3): 1318. https://doi.org/10.3390/su13031318.

IPCC. (2001). Climate Change 2001, Synthesis Report, Cambridge University Press. *In*: CIFOR (2011). *Climate Change and Forests in the Congo Basin Synergies between Adaptation and Mitigation in a Nutshell*. Center for International Forestry Research. Retrieved in 30 May 2020, 19:30 GMT.

IPCC. (2007a). *Climate Change 2007 - The Physical Science Basis Contribution of Working Group I to the Fourth Assessment Report of the IPCC*. ISBN 978 0521 88009-1 Hardback; 978 0521 70596-7 Paperback. https://www.ipcc.ch/site/assets/uploads/2018/05/ar4_wg1_full_report-1.pdf.

IPCC. (2007b). *Climate Change 2007: Impacts, Adaptation and Vulnerability*. Contribution of Working Group II to the Fourth Assessment Report of the IPCC. https://www.ipcc.ch/site/assets/uploads/2018/03/ar4_wg2_full_report.pdf.

IPCC. (2007c). *Climate Change 2007 – Mitigation of Climate Change*. Contribution of Working Group III to the Fourth Assessment Report of the IPCC. https://www.ipcc.ch/site/assets/uploads/2018/03/ar4_wg3_full_report-1.pdf.

IPCC. (2013). Glossary. *In: Climate Change 2014: Impacts, Adaptation, and Vulnerability*. Annex II, Contribution of Working Group II to the Fifth Assessment Report of the Intergovernmental Panel on Climate Change, Cambridge University Press, Cambridge and New York, NY, p.13. Available at www.ipcc.ch/report/ar5/wg2/.

IPCC. (2019). *IPCC Special Report on Climate Change, Desertification, Land Degradation, Sustainable Land Management, Food Security, and Greenhouse Gas Fluxes in Terrestrial Ecosystems*. IPCC Intergovernmental Panel for Climate Change. https://www150.statcan.gc.ca/n1/en/pub/82-003-x/2001003/article/6103-eng.pdf?st=qASktFSv.

James, S.J. and James, C. (2010). The food cold-chain and climate change. *Food Res. Int.*, 43: 1944–1956. https://www.sciencedirect.com/science/article/abs/pii/S0963996910000566. doi:10.1016/j.foodres.2010.02.001.

Kader, A. (2001). Importance of fruits, nuts, and vegetables in human nutrition and health, Department of Pomology, UC Davis, No. 106. *Perishables Handling Quarterly*. https://ucanr.edu/datastoreFiles/234-104.pdf.

Kim, W., Iizumi, T. and Nishimori, M. (2019). Global patterns of crop production losses associated with droughts from 1983 to 2009. *Journal of Applied Meteorology and Climatology*, 58: 1233–2144. Doi: 10.1175/JAMC-D-18-0174.1.

Lund, D. (2003). Predicting the impact of food processing on food constituents. *Journal of Food Engineering*, 56: 113–117. https://dokumen.tips/download/link/predicting-the-impact-of-food-processing-on-food-constituents.

McGuirk, M., Shuford, S., Peterson, T.C. and Pisano, P. (2009). Weather and climate change implications for surface transportation in the USA. *WMO Bulletin*, 58(2): 84–93. https://public.wmo.int/en/bulletin/weather-and-climate-change-implications-surface-transportation-usa.

Met. Office. (2020). *Weather and Environment, Causes of Climate Change*. https://www.metoffice.gov.uk/weather/climate-change/causes-of-climate-change; accessed 19 May 2020.

Mirza, M.M.Q. (2011). Climate change, flooding in South Asia and implications. *Reg. Environ Change*, 11: 95–107. https://doi.org/10.1007/s10113-010-0184-7.

Moretti, C.L., Mattos, L.M., Calbo, A.G. and Sargent, S.A. (2010). Climate changes and potential impacts on postharvest quality of fruit and vegetable crops: A review. *Food Research International*, 43: 1824–1832. Doi:10.1016/j.foodres.2009.10.013.

Mukhopadhyay, S. and Mandal, A.K. (2021). Impact of climate change on groundwater resource of India: A geographical appraisal. pp. 125–154. *In*: Md. N. Islam and Amstel, A. (eds.). *India: Climate Change Impacts, Mitigation and Adaptation in Developing Countries*. Springer Nature, Switzerland AG, 2021. https://doi.org/10.1007/978-3-030-67865-4.

Myers, S.S., Zanobetti, A., Kloog, I. et al. (2014). Increasing CO_2 threatens human nutrition. Nature, 510: 139–142. *In*: John Hopkins Center of a Livable Future (undated), *Food System Premier – Food and Climate Change*. Retrieved on 30 May 2020; 16:30 GMT. http://www.foodsystemprimer.org/food-production/food-and-climate-change/.

Nandintsetseg, B., Shinoda, M. and Erdenetsetseg, B. (2018). Contributions of multiple climate hazards and overgrazing to the 2009/2010 winter disaster in Mongolia. *Nat. Hazards*, 92: 109–126. https://doi.org/10.1007/s11069-017-2954-8.

National Research Council. (2011). *Climate Stabilization Targets: Emissions, Concentrations, and Impacts over Decades to Millennia*. Washington, DC: The National Academies Press. https://doi.org/10.17226/12877.

Nawrotzki, R.J., DeWaard, J., Bakhtsiyarava, M. and Ha, J.T. (2017). Climate shocks and rural-urban migration in Mexico: Exploring nonlinearities and thresholds. *Clim. Change*, 140(2): 243–258. Doi:10.1007/s10584-016-1849-0.

NEPAD. (2013). *Agriculture in Africa, Transformation and Outlook*. NEBAD transforming Africa. https://www.un.org/en/africa/osaa/pdf/pubs/2013africanagricultures.pdf.

Ng, M., Fleming, T., Robinson, M., Thomson, B., Graetz, N., Margono, C., and Gakidou, E. (2014). Global, regional, and national prevalence of overweight and obesity in children and adults during 1980–2013: A systematic analysis for the Global Burden of Disease Study 2013. *Lancet*, 384(9945): 766–781. https://www.thelancet.com/action/showPdf?pii=S0140-6736%2814%2960460-8.

OECD/FAO. (2016). *OECD-FAO Agricultural Outlook 2016–2025*. OECD Publishing, Paris. http://dx.doi.org/10.1787/agr_outlook-2016-en. http://www.fao.org/3/a-i5778e.pdf.

Olagunju, T.E. (2015). Drought, desertification and the Nigerian environment: A review. *Journal of Ecology and the Natural Environment*, 7(7): 196–209. Doi: 10.5897/JENE2015.0523.

Olivier, J.G.J., Aardenne, J.A.V., Dentener, F.J., Pagliari, V., Ganzeveld, L.N. and Peters, J.A.H.W. (2005). Recent trends in global greenhouse gas emissions: Regional trends

1970–2000 and spatial distribution of key sources in 2000. *Environmental Sciences*, 2: 2–3, 81–99. DOI: 10.1080/15693430500400345.

Oxford University Press. (2015). *What is the Food System?* Oxford Martin Programme on the future of food. Available on line at https://www.futureoffood.ox.ac.uk/what-food-system; last access 15 June 2020.

Pérez, C.E. (2002). Fruit and vegetable consumption. *Health Reports*, Statistics Canada, Catalogue No. 82-003, 13(3). https://www150.statcan.gc.ca/n1/en/pub/82-003-x/2001003/article/6103-eng.pdf?st=qASktFSv.

Pinstrup-Andersen, P. (2009). Food security: Definition and measurement. *Food Security*, 1: 5–7. https://link.springer.com/content/pdf/10.1007/s12571-008-0002-y.pdf.

Poore, J. and Nemecek, T. (2018). Reducing food's environmental impacts through producers and consumers. *Science*, 360(6392): 987--992. *In*: Ritchie, H. and Roser, M. (2020). *Environmental Impacts of Food Production*. published online at OurWorldInData.org. https://ourworldindata.org/environmental-impacts-of-food.

Ritchie, H. and Roser, M. (2020). *Environmental Impacts of Food Production*. published online at Our World in Data.org. Retrieved from: https://ourworldindata.org/environmental-impacts-of-food.

Sarkar, T., Roy, A., Choudhary, S.M. and Sarkar, S.K. (2021). Impact of climate change and adaptation strategies for fruit crops. pp. 79–98. *In*: Md. N. Islam and Amstel, A. (eds.). *India: Climate Change Impacts, Mitigation and Adaptation in Developing Countries*. Springer Nature, Switzerland AG, 2021. https://doi.org/10.1007/978-3-030-67865-4.

Savenije, H.H.G. (2001). The Role of Green Water in Food Production in sub-Saharan Africa. FAO, the Netherlands. *In*: Alam, M., Toriman, M., Siwar, C. and Talib, B. (2010). Rainfall variation and changing pattern of agricultural cycle. *American Journal of Environmental Sciences*, 7(1): 82–89. https://www.researchgate.net/publication/279407992_Rainfall_Variation_and_Changing_Pattern_of_Agricultural_Cycle.

Schmidhuber, J. and Tubiello, F.N. (2007). Global food security under climate change. *Proceedings of the National Academy of Sciences (PNAS) of the United States of America*, PNAS, December 11, 2007, 104(50): 19703–19708. https://doi.org/10.1073/pnas.0701976104.

Scialabba, N.E. (2011). *Food Availability and Natural Resource Use FAO/OECD*. Expert Meeting on Greening the Economy with Agriculture Paris, 5–7 September Natural Resources Management and Environment Department, FAO. http://www.fao.org/fileadmin/user_upload/suistainability/Presentations/Availability.pdf.

Slavin, J.L. and Lloyd, B. (2012). Health benefits of fruits and vegetables. *Advances in Nutrition*, **3**(4): 506–516. https://doi.org/10.3945/an.112.002154.

Sonja, J.V., Campbell, B.M. and Ingram, J.S.I. (2012). Climate change and food systems. *Annu. Rev. Environ. Resour.*, 37: 195–222. Doi:10.1146/annurev-environ-020411-130608.

Speranza, C., Kiteme, B. and Wiesmann, U. (2008). Droughts and famines: The underlying factors and the causal links among agro-pastoral households in semi-arid Makueni district, Kenya. *Global Environmental Change*, 18(1): 220–233. https://doi.org/10.1016/j.gloenvcha.2007.05.001.

Stamoulis, K. and Zezza, A. (2003). A Conceptual Framework for National Agricultural, Rural Development and Food Security Strategies and Policies, *ESA Working Paper*, No. 03–17 November 2003, FAO. Available at http://www.fao.org/3/ae050e/ae050e00.pdf. accessed 2 June 2020.

Takepart. (2020). *What is Climate Change?* http://www.takepart.com/flashcards/what-is-climate-change/ accessed 19 May 2020.

Tol, R.S.J. (2005). Adaptation and mitigation: Trade-offs in substance and methods. *Environmental Science and Policy*, 8(6): 572–578. *In*: CIFOR (2011). *Climate Change and Forests in the Congo Basin Synergies between Adaptation and Mitigation in a Nutshell*. Center for International Forestry Research. Retrieved in 30 May 2020, 19:30 GMT. https://

www.cifor.org/fileadmin/fileupload/cobam/ENGLISH-Definitions%26ConceptualFramework.pdf.

UCDAVIS. (2019). *Science and Climate*. Available online: https://climatechange.ucdavis.edu/science/climate-change-definitions/.

UNCCD. (1994). *United Nations Convention to Combat Desertification in Countries Experiencing Serious Drought and/or Desertification, particularly in Africa (UNCCD)*. https://www.jus.uio.no/english/services/library/treaties/06/6-02/combat-desertification.xml#:~:text=%22drought%22%20means%20the%20naturally%20occurring,d.

USAD. (2019). Economics Research Service, U.S. Department of Agriculture. https://www.ers.usda.gov/data-products/ag-and-food-statistics-charting-the-essentials/ag-and-food-sectors-and-the-economy.aspx.

U.S. Global Change Research Program. (2009a). Global Climate Change Impacts in the United States, Cambridge, New York, Melbourne, Madrid, Cape Town, Singapore, São Paulo, Delhi: Cambridge University Press. *In*: John Hopkins Center of a Livable Future (2020). *Food System Premier – Food and Climate Change*. Retrieved on 30 May 2020; 16:30 GMT. http://www.foodsystemprimer.org/food-production/food-and-climate-change/.

U.S. Global Change Research Program. (2009b). *Global Climate Change Impacts in the United States*, Karl, T.R., Melillo, J.M. and Peterson, T.C. (eds.). Cambridge University Press, 2009. http://www.iooc.us/wp-content/uploads/2010/09/Global-Climate-Change-Impacts-in-the-United-States.pdf.

WHO/FAO. (2003). *Expert Report on Diet, Nutrition and the Prevention of Chronic Diseases*, Technical Report Series 916, https://apps.who.int/iris/bitstream/handle/10665/42665/WHO_TRS_916.pdf;jsessionid=D2551BA86C8E19B8A0A03AB71FDAD003?sequence=1.

Woertz, E. (2017). Agriculture and development in the wake of the arab spring. *International Development Policy, Revue Internationale de Politique de développement* [Online], 8.1 | 2017, Online since 12 February 2017, connection on 23 February 2017. URL: http://poldev.revues.org/2274. Doi: 10.4000/poldev.2274.

World Bank. (2020). *Agriculture, Forestry, and Fishing, Value Added (% of GDP)*. https://data.worldbank.org/indicator/NV.AGR.TOTL.ZS. Accessed 2 June 2020.

Yadav, S.S., Hegde, V.S., Habibi, A.B., Dia, M. and Verma, S. (2019). Climate change, agriculture and food security. pp 1–24. *In*: Yadav, S.S., Redden, R.J., Hatfield, J.L., Ebert, A.W. and Hunter, D. (eds.). *Food Security and Climate Change*. first edition, 2019, John Wiley & Sons Ltd., published 2019 by John Wiley & Sons Ltd. https://www.researchgate.net/publication/329870678_Climate_Change_Agriculture_and_Food_Security.

Zhai, F. and Zhuang, J. (2009). *Agricultural Impact of Climate Change: A General Equilibrium Analysis with Special Reference to Southeast Asia*, ADBI Working Paper 131, Tokyo: Asian Development Bank Institute. Available: https://www.adb.org/sites/default/files/publication/155986/adbi-wp131.pdf.

2

Mitigation Strategies of the Impact of Global Climate Change on Fruit Crops

Rinny Swain,[1] *Jannela Praveena*[2] and *Gyana Ranjan Rout*[1,*]

1. Introduction

Climate change is the chief concern and an inevitable threat to sustainable development of agriculture worldwide. As per the Intergovernmental Panel on Climate Change (IPCC, 2019) report, by 2015 the mean land surface air temperature had increased by 1.53°C more than the global mean surface (land and ocean) temperature (GMST), which is 0.87°C. In 2015, after the Paris Agreement convention, all governments worldwide agreed to limit global warming below 2°C by putting in efforts to limit GMST to 1.5°C, thereby enhancing adaptation to the adverse impacts of climate change. According to the projected Intergovernmental Panel on Climate Change report (IPCC), there is a high probability of 10–40% loss in crop production by 2080–2100 due to global warming in India. Many of the commercial tropical, sub-tropical and temperate fruits will perform poorly and unpredictably due to climate change. With an increase

[1] Department of Agricultural Biotechnology, College of Agriculture, Odisha University of Agriculture & Technology, Bhubaneswar- 751003, Odisha, India; presently working at Crop Improvement Division, School of Agriculture, GIET University, Gunupur-765022, Rayagada, Odisha, India.

[2] Department of Fruit Science and Horticulture Technology, College of Agriculture, Odisha University of Agriculture & Technology, Bhubaneswar-751003, Odisha, India.

Emails: swain.rinny12@gmail.com; jannela.praveena17@gmail.com

* Corresponding author: grrout@rediffmail.com

in average global temperature, several commercial fruit varieties have changed or shifted to new latitudinal belts with favorable climatic zones. Therefore, the crops that are used to be productive in one particular area may no longer be so, or the other way round. Climate change is believed to be a variation of climate over a long period of time and is credited directly or indirectly to human activities that have altered the total composition of the global atmosphere. Climate change continues to create challenges to life and livelihoods worldwide, causing changes in water quantity and quality, and shifts in geographical areas, seasonal activities, migration patterns, changes in species abundance and interactions between many terrestrial as well as freshwater or marine species with more negative than positive impacts on the yields of most crops (IPPC, 2021). Climatic variability has occurred mostly due to enormous emissions of greenhouse gases rapidly since the 1900s. The change in the atmosphere has severely affected agriculture by causing changes in crop physiology, biochemistry, and floral biology, biotic and abiotic stresses. This has eventually resulted in the reduction of yield and quality of crop plants. The climate change has led to loss of vigor, early or late flowering, low fruit-bearing capability, reduction in fruit size, reduction in juice content, less color, reduced shelf-life, and escalating pest attacks, consequentially resulting in low production and poor quality of fruit crops. An economic literature was also published recently, exploring the impacts of climate change on farm yields in several parts of the world (Chandio et al., 2020). Mitigation strategies are essential measures to reduce the devastating consequences of climate change. Also, the negative impacts of climate change can be controlled through implementation of adaptation strategies that are relevant, robust and effortlessly operated by scientists and farmers. In view of the rising concern at worldwide climate change, this chapter focuses on the imminent impact of changing climate on the horticultural crops, especially fruit crops and strategies to lessen the risk of crop failure and minimize monetary loss involved in crop production.

1.1 Global Climate Change

The rise in global mean surface temperature (GMST) is mostly attributed to the alarming increase in the concentration of greenhouse gases (GHGs), like carbon dioxide (CO_2), methane (CH_4), nitrous oxide (N_2O), sulfur dioxide (SO_2) and chlorofluorocarbons (CFC) primarily due to accelerated growth of industrialization worldwide (Singh, 2010). The increase in GHGs has occurred as a result of manmade interventions, like coolants in industrial enterprises, excessive automobiles, production of energy-intensive agro-chemicals including fertilizers that has led to ozone depletion. The global GHG emissions reached 31.2% and above in 2016 when compared with the GHG level, at an average annual increase of

0.9% since 2010 (UNFCCC, 2019). As per the United Nations Framework Convention on Climate Change (UNFCC), sectors that contributed a major share in GHG emissions during 2016 are energy (34%), industry (22%), and transport (14%). In addition, about 23% of net emissions of GHG were from anthropogenic activities like agriculture, forestry and other land use accounting to 13% of CO_2, 44% of CH_4 and 82% of N_2O emissions globally during 2007–2016. After fossil fuel, land-use change, forest deforestation and degradation are the biggest emitters of CO_2 (Baumert et al., 2009). The CO_2 emission is primarily accountable for 77% of global warming over the period of past 100-years and hence is the most significant aspect of GHG emissions (Climate Analysis Indicators Tool, 2011). Khan et al. (2021) reported that energy use, CO_2 emissions, labor force, harvested land and fruit crop growth have a long-term co-integrated relationship. The long-term CO_2 emission and rural population have negative influence on fruit crop production. They suggested that 1% rise in rural population and CO_2 emissions will reduce fruit crop production by –0.59% and –1.97% in the long term.

The International Kyoto Protocol, set out and adopted in 1997 for binding emission reduction commitments to industrialized countries, came into force in 2005. In 2012, a second commitment to the Doha Amendment was adopted and pertained to a new period up to 2020. Finally, in December 2015, the UNFCC adopted the Paris Agreement that came into force on November 2016 with an outcome to develop a legal instrument applicable to all parties involved in reducing GHG emissions and which would be implemented from 2020. It was decided to build climate resilience and foster climate finance flows consistent with a pathway towards low emission of GHGs. The increase in GMT in relation to industrial development involved the threat of desertification, water scarcity and land degradation. All these events can ultimately cause instability between crop yield and food supply, leading to food insecurities. Such an adverse climate change will also aggravate land degradation processes through an increase in rainfall, flooding, frequent droughts, heat stress, dry spells, strong wind, rise in sea level and high tides, and permafrost thaw with gravely devastating outcomes. The enhanced land surface air temperature, evapotranspiration and decreased rainfall in interaction with climate variability and anthropogenic activities are mainly accountable for increased desertification in some dry land areas, like Sub-Saharan Africa, parts of East and Central Asia, and Australia (UNFCCC, 2019). To combat desertification, activities like climate change adaptation with mitigation as well as halting biodiversity loss through sustainable development are being carried out. Avoidance and reduction of desertification would enhance soil fertility, increase carbon storage in soil and biomass, and provide benefits to agricultural productivity and

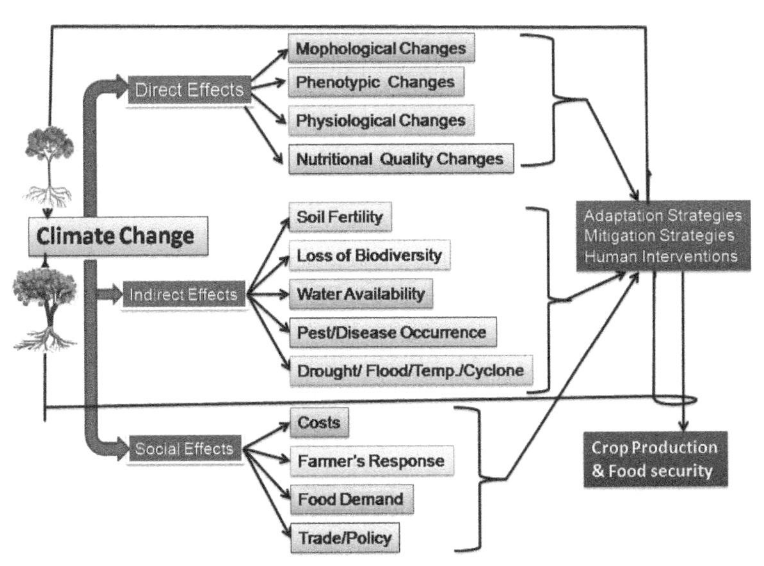

Fig. 1. Diagrammatic representation of climate change impact on agriculture.

food security. It is preferable to attempt to restore degraded land for better utilization and to reduce pressure on land. On the other hand, IPCC has estimated that the sea level in 2100 will be 40 cm higher than what it is today and nearly 80 million residents in coastal areas of Asia will be flooded. Most flooded areas will lie in South Asia, particularly Bangladesh and India. Climate change has had three deleterious effects on agriculture and crop production (Fig. 1). Farmers have to adopt different strategies (adaptation, mitigation, and human interventions) to manage the climate change scenario and food security.

1.2 Climate Change Indian Contest

The changes in Indian climatic conditions are consistent with global trends over the last 100 years and thus represent a serious risk to human health, environment, agriculture as well as the economy of the country. India is considered to be one of the world's most vulnerable countries to climate change as it is exposed to sea-level rise, shifts in precipitation patterns, and extreme weather events. These events continuously pose a risk to livelihoods, human health, water availability, biodiversity, and ecosystem and food security. About 12% (40 million ha) of India is flood-prone, while 16% (51 million ha) is drought-prone (Chadha, 2015). According to IPCC (2007), Indian economy accounts for nearly 14% of GHG emissions, i.e., about 1331.6 Mt (Million tonnes/Metric tonnes) of carbon dioxide in the GHG emissions. Methane, nitrous oxide, and carbon dioxide are the three main GHGs emitted usually from different activities of agriculture and allied sectors (Puri et al., 2017). India ranks about 11th in GHG emission level as it improved its position from 14th as given in the previous report (CCPI, 2019). The most noteworthy improvement in its performance is in the renewable energy category. However, India is planning a new coal-fired power plant which poses the risk of offsetting positive development in the renewable energy sector. India has an overall high rating in the emission category because of comparatively low levels of per capita GHG emissions and a relatively determined mitigation target plan for 2030 (IPCC, 2009). In 2019, it was reported that states like Assam, Mizoram, and Jammu & Kashmir (J&K) were 'highly vulnerable' to climate change with little capacity to resist or cope up with the environment. In India, there is a high probability of loss estimated as 10–40% in crop production by 2080–2100 due to global warming (Parry et al., 2004; IPCC, 2014). The IPCC's fourth assessment report suggests that climate change will affect chiefly agricultural yields ranging up to 50% because of rain-based farming. In India, rain-fed agro-ecology accounts for 60% of the net sown area and therefore climate change is expected to cause severe difficulties for Indian farmers. On the other hand, the agriculture sector has a direct impact on the Indian economy.

In 2017, the agriculture sector contributed about 15.4% to Gross Domestic Product (GDP) in India (Indiastat, 2019). With this GDP contribution, the agriculture sector provided $375.61 billion, making India the second largest producer of agricultural products. The total global agricultural output accounts for 7.39% in India (Indiastat, 2019). The agricultural sector's contribution to the Indian economy is much higher than the agricultural contribution of 6.4% to the world's economy. Among the agricultural crops, the contribution of horticulture to Indian GDP is nearly 30% (Chadha, 2015). The horticulture growth rate is high due to the improvement in productivity of horticulture crops which was approximately 28% between 2001–2002 and 2011–2012. India is the second largest producer of fruits (10.9%) and vegetables (8.6%) in the world (FAOSTAT, 2019). It is the third largest producer of (1,396.8 Million nuts) coconuts worldwide (CDB, 2019). India is also the biggest producer, processor, consumer and exporter of cashew.

2. Impact of Climate Change on Fruit Crops

2.1 Impact on Dormancy and Chilling Requirements

The rapid climatic changes have altered the adaptability of many temperate fruit crops, causing low productivity problems. There are specific chilling requirements for temperate fruit crops, like apple, pear, peach, almond, cherry, plum, etc. for breaking dormancy and growth (Samish and Lavee, 1982). However, due to increase in temperature, temperate fruit crops receive insufficient chilling that leads to the development of symptoms like delayed foliation, reduced fruit set, increased buttoning, and reduced fruit quality. Lack of chilling during the mild winter results in abnormal patterns of bud-breakage and development in temperate fruit trees (Ameglio et al., 2000). The dormancy symptoms of fruits may get prolonged when winter is neither long nor cold enough to break the dormancy adequately; such a condition is referred to as 'delayed foliation' (Black, 1952; Ruck, 1975). For commercially successful cultivation of many temperate fruit trees, the fulfillment of winter chilling is very essential.

2.2 *Impact on Temperature and Crop Shift*

The majority of plant processes associated with maximum yield are high-temperature dependent and require an optimum temperature range for maximum growth and production. The optimum temperature corresponds to the optimum growth through active photosynthesis and other metabolic reactions. In the case of perennial crops, an increase in temperature which is more than optimum reduces photosynthesis, thereby affecting productivity. At higher temperatures, the biochemical processes and reactions in plants get accelerated up to a threshold where enzyme systems can destroy and die (Chmielewski et al., 2004). The rise in mean temperature from 1.45°C to 2.32°C over the last 20 years adversely affected vernalization of all temperate fruit crops (Ahmed et al., 2011). A study conducted by Fraga and Santos (2021) in Portugal was involved in the quantification of the impacts of climate change due to chilling and forcing on the main fresh fruit regions using bias-corrected computed data from several RCM-GCM model chains. They suggested that in future, chilling and forcing may lead to limitations and changes in phenological stages in the most important Portuguese temperate-fruit regions.

In India, the northern hilly zone showed an elevation in temperature approximately by 0.65°C to 2.3°C, while the north-western Himalayas showed about 0.5°C ever since the last century (Bhutiyani and Kale, 2002). A shift in the cultivation area of the temperate crops has been witnessed in these zones. Thus, it indicates that a crop that was productive in one area is now no longer so or it may be the other way round. The consequential effect of climate change is very evident in apple cultivation as the shift in a crop from lower elevations to higher altitudes has occurred in India (Rai et al., 2015). Due to the rise in temperature and decline in rainfall, fruits like apricots and cherries are rapidly disappearing from a few areas of Kashmir Valley, deteriorating both yield and quality of apples. However, the use of proper rootstocks provides a huge number of alternatives to the growers to enhance fruit quality and yield, attain early fruiting and uniform cropping, maintain the tree size, and to produce an opportunity for high-density crop planting, etc. The increase in frost incidence with change in climate is a foremost threat to fruit crops grown in the subtropical regions of north-western India. The occurrence of the cold waves from west to eastern India, during 2002–2003, caused a serious damage to fruit plantations. The level of damage reported in mango (40–100%) and litchi (30–80%) was still higher (Samra et al., 2003). In 2007–2008, due to 'western disturbances', the temperature dipped and frost occurred in many parts of India, causing frost and chilling damage to diverse fruit plants, predominantly in mango cultivation (Gill and Singh, 2012). The damage caused was significantly high in frost-prone regions of Punjab where a majority of mango (50–100%) plantations were damaged severely.

2.3 *Impact on Phenology*

The change in phenology, i.e., the timing of different physiological activities, is one of the main distinctive effects of climate change (Cleland et al., 2007). A 10-days prior date of the full bloom was observed in apple varieties of 'Boskoop', 'Cox's Orange Pippin' and 'Golden Delicious' on comparing the last 20 years with the earlier 30 years (Blanke and Kunz, 2011). The advancing trends in the blooming dates of many fruit crops indicate the response to climate change.

2.4 *Impact on Flowering, Fruiting and Fruit Dropping*

The most crucial events in all fruit crops are flowering and fruiting which are regulated by the climatic condition. In Himachal Pradesh, the warmer temperature during the dormant season induces early flowering in apple varieties of the 'Delicious' group that become vulnerable to spring frost (Bhatia, 2010). Similarly, the climatic shifts affect the fruit-bud differentiation pattern in subtropical fruit crops, which, in turn, affects the time of flowering and fruit production (Ravishanker and Rajan, 2011). The climatic variations disturb the flowering pattern and fruit set in many fruit crops by reducing pollination activity. Heavy rains during flowering generally wash out the pollen from the stigma of flowers, resulting in poor or no fruit set. Varu and Viradia (2015) reported that about 80–90% less mango production occurred due to unseasonal rains followed by heavy dew attacks during the flowering season. As a result, there was reduced fruit set, increased fruit drop at pea stage and greater incidence of sooty mold and powdery mildew in mango. During 2008–09, mango yield was drastically low as around 2°C higher temperature was recorded in Gujarat during the flower induction (Parmar et al., 2012). This seems to have a detrimental effect on mango due to poor flowering, eventually affecting the crop yield. High intensity of relative humidity during the flowering period also resulted in a higher infestation of mango hopper and powdery mildew. Transpiration rate and respiration also influence the fruit temperature, which later on influences the normal fruit-ripening processes and lead to the development of spongy tissues (Hemanth et al., 2008). In strawberries, the shift in flowering time was observed in comparison to the prevailing climate of the Baltic States as it typically occurs in the middle of June, instead of the middle of May (Bethere et al., 2016). In the case of mango, it was reported that higher temperature caused greater drying of shoots in fruiting plants in comparison to non-fruited ones; amongst different varieties, Dashehari mango (77.28%) showed maximum twig drying followed by var. Baneshan (32%) and var. Rajapari (33.93%), whereas lowest twig drying amounting to 19.5% was recorded in var. Langra (Reddy and Singh, 2011).

2.5 *Impact on Crop Duration*

Crop duration and maturity indices are important physiological processes in fruit crops as they affect the fruit quality and final yield. In lowland tropical areas, citrus fruits mature quickly due to high respiration rates in warm temperatures and thus do not have adequate time to accumulate high total soluble solids (TSS) (Zekri, 2011). The acidity of citrus declines rapidly, leading to increase in soluble solids or acid ratio and insipid taste and quick drying of the fruit. It has been reported that the effects of climate change are extremely visible at grape harvest time which starts two weeks earlier than ever before in India (*The Economic Times*, 2016). In Greece, the harvest dates of grapes were earlier than in India due to the change in maximum and minimum temperatures (George et al., 2014). In the case of bananas, a rise in temperature by 1–2°C beyond 25–30°C increased leaf production, thereby reducing crop duration and improved production (Chaddha and Kumar, 2011).

2.6 *Impact of Effect on Pollination*

Climate change even affected bees and natural pollinators at different levels, interfering with their pollinating efficiency (Reddy et al., 2012). Thus, the disturbing climate change scenario has contributed to a significant decrease in the population of pollinating insects. The temperature extremes, i.e., very low or very high are unfavorable to the pollinators, resulting in no fertilization, finally affecting the fruit set. The cross-pollinated fruits, such as walnuts and pistachios, show reduced pollination because of insufficient chilling, leading to reduced crop yields (Gradziel et al., 2007). In temperate fruits, like apple, pear, plum, cherry, etc., the optimum temperature for pollination and fertilization is between 20–25°C. The low temperature and foggy conditions had a detrimental effect during pollination in sour cherry in the USA (Zavalloni et al., 2008). A higher temperature during flowering reduces the effective time for pollination by the pollinators, causing poor pollinator activities and desiccation of pollen. Also, the higher temperatures, ranging from 33°C to 36°C during the microsporogenesis or gamete formation in pollen, decrease pollen viability up to 60–85% (Issarakraisila et al., 1993). Even cooler temperatures below 17°C would produce abnormal as well as non-viable pollen grains. The reproductive stage during microsporogenesis appears to be most sensitive to temperatures below 10°C. Hence, cooler conditions also negatively affect pollen germination and pollen-tube growth, which is completely repressed at temperatures below 15°C (Issarakraisila and Considine, 1994). The pre-monsoon showers are also very harmful to fruit plantations as they destroy complete crops of fruits, like grapes, mango and date plants. The rain washes away the pollen from the stigma of flowers, resulting in poor or no fruit set (Rajan et al., 2020).

2.7 *Impact on Post-Harvest Quality*

The temperature variations directly affect crop photosynthesis and thus have a major impact on the post-harvest quality of fruits by altering important quality parameters, like sugar content, organic acids, antioxidants, peel color and fruit firmness. The expression of anthocyanin biosynthetic genes is induced by low temperatures, so a higher temperature has a negative effect on its biosynthesis. Such a condition arises in red oranges (Lo Piero et al., 2005), apples (Ubi et al., 2006) and grapes (Yamane et al., 2006). In the case of grapes, higher temperatures (above 46°C) cause thick skin of berries. When grown under high temperature, grapes show higher sugar content and lower levels of tartaric acid (Kliewer and Lider, 1970). Similarly, higher night temperatures also reduce anthocyanin accumulation in berry skin because of the degradation of anthocyanin pigment as well as the inhibition of RNA transcript which is responsible for its biosynthesis (Mori et al., 2005). In ripening apples, anthocyanin is induced at temperatures below 10°C (Curry, 1997). However, the synthesis of anthocyanin takes place at higher irradiation at mild temperatures (20–27°C) in mature apples (Curry, 1997; Reay, 1999). High temperatures result in poor red color development in the apple peel and reduce its post-harvest quality (Wand et al., 2002, 2005). The fruit development and quality of climate change is presented in Table 1.

Table 1. Development of climate resilient fruit crops for food security.

Name of fruit crop	Climate variables	Salient observations	Quality development	References
Apple (*Malus domestica* Borkh.)	Temperature, humidity, and rainfall	Terpene and volatile contents influenced by the temperature, rainfall and humidity	Flavor development	Vallat et al. (2005)
	Temperature	High temperatures responsible for developing poor red color in apples	Post harvest quality	Wand et al. (2002, 2005)
Var. York Imperial	Drought	Better growth and yield	Growth and yield	NICRA, 2015
Bilberries (*Vaccinium myrtillus* L.)	Overall climate and thermal effect	Anthocyanin concentration in bilberries is influenced by climatic factors	Sensory quality; health-related benefits	Akerstrom et al. (2009)
Fig (*Ficus carica* Linn.)	Overall climate	Secondary compounds are variable due to the influence of climatic conditions	Sensory qualities	Darjazi and Larijani (2012)
Grapes (*Vitis vinifera* L.)	Temperature, solar radiation, rainfall	Presence of phenolic compounds and antioxidant properties are significantly developed	Sensory quality; benefit to human health	Xu et al. (2011)
	Levels of Carbon dioxide	Tartaric acid increase with a rise in carbon dioxide level	Sensory quality	Bindi et al. (2001)
	Temperature	Higher temperatures cause thick skin of berries	Processing and quality	Yamane et al., 2006
	Night temperatures	Higher night temperatures also reduce anthocyanin accumulation in berry skin because of factors such as anthocyanin degradation	Benefit to human health	Mori et al. (2005)

Table 1 contd. ...

...Table 1 contd.

Name of fruit crop	Climate variables	Salient observations	Quality development	References
Hops (*Humulus lupulus* L.)	Overall climate	Concentrations of key compounds depend on climatological conditions with highest levels in poorest weather conditions	Sensory qualities; benefit to human health	Keukeleire et al. (2007)
Pomegranate (*Punica granatum* L.)	Temperature	Seasonal temperature inversely correlated to anthocyanin accumulation	Sensory quality; benefit to human health	Borochov-Neori et al. (2011)
Strawberry (*Fragariax ananassa* Duch.)	Temperature	Higher antioxidant activity and flavonoids on cooler days	Sensory quality; benefit to human health	Wang and Zheng (2001)
Red oranges (*Citrus sinensis* (L.) Osbeck.) Var.'Blood orange')	Temperature	Expression of anthocyanin biosynthetic genes induced by low temperature	Sensory quality; benefit to human health	Lo Piero et al. (2005)
Citrus sinensis (L.) Osbeck.) Var. Mosambi	Drought	Sustain in high temperature and better growth	Better quality	NICRA (2015)
Apricot (*Prunus armeniaca* L.)	Drought stress	High yield	Better quality	NICRA (2015)
Banana (*Musa acuminata* L.) Var. Karpuravalli (ABB genome)	Drought	Sustain in drought condition and better growth	Sensory quality; benefit to human health	NICRA (2015)
Banana (*Musa acuminata* L.) Var. Shrimanti, Grand Naine	Temperature	Sustain in high temperature and better growth	Sensory quality; benefit to human health	NICRA (2015)

Table 1 contd. ...

...Table 1 contd.

Name of fruit crop	Climate variables	Salient observations	Quality development	References
Mango (*Mangifera indica* L.) Var. Arka,, Neelachal, Sinduri/Jalore seedless	Drought	Sustain in high temperature; better growth; ripening with flavor; high storage quality	Good quality; benefit to human health	NICRA (2015)
Guava (*Psidium guajava* L.) Var. Allahabad Safeda, Lucknow-49	Salinity stress	Better growth and storage quality; maintain the ascorbic acid content	Good quality; benefit to human health	NICRA (2015)

2.8 *Impact on Irrigation Water*

Agriculture always demands water for irrigation; it is considered more sensitive to climate change. The elevated temperature increases the demand for more irrigation, which is another restraint affecting the productivity of fruit crops. A change in the field-level climate may alter the need, amount and timing of irrigation, especially the fruit crops cultivated under rain-fed conditions as these are drastically affected. The fruit crop requirement of annual irrigation will increase, not because of higher evaporation, but because the trees develop at a faster rate during the annual period due to higher atmospheric CO_2 levels. About 80% of the reduction in apple yield was predictable due to water shortage for irrigation and 20% due to the high evaporation rate in the apple-growing areas of Himachal Pradesh, India (Singh et al., 2016). It is predicted that around 2025, all irrigated areas of India would need more irrigation water and also the global net irrigation requirements would increase, irrespective of climate change by 3.5–5% (Pathak et al., 2014).

2.9 *Impact on Physiological Disorders*

Furthermore, climate change has increased the occurrence of physiological disorders in plants. High moisture and temperature stress results in increased sunburn and cracking of apples, apricot, and cherries. Three types of sunburn were reported to be caused by heat and or light stress (Felicetti and Schrader, 2008). The first is 'sunburn necrosis' and is induced by heat burn that occurs when the temperature of the fruit surface reaches 52°C in about 10 min. This causes the death of cells in the peel followed by dark brown or black necrotic spots. In the second case, 'sunburn browning', which is the most common type, leads to yellow, brown, or dark tan spots on the sun-exposed side of the fruit. The third and final type of sunburn is induced when fruits are suddenly exposed to full sunlight, as they are not acclimated to sun exposure. The apples with sunburn necrosis show higher relative electrical conductivity than the fruits with no sunburn or with sunburn browning (Schrader et al., 2001). Increased relative electrical conductivity in fruits with sunburn necrosis showed that the membrane integrity was damaged, allowing the leakage of electrolytes. The increased temperature at the maturity level will lead to fruit cracking and burning, as in the case of litchi (Mark and Marin, 2016). In custard apple, the maximum black spots (35.63%) were reported on the skin of the fruit due to the high-speed wind was recorded in the same year (Varu et al., 2010).

2.10 Impact on Pest and Disease Incidence

The rise in temperature as a consequence of climate change has indirectly affected the pest and disease incidences in crop species. All stages of the insect life cycle are affected by climate, i.e., life span, molting, fecundity, dispersal, mortality and genetic adaption. The higher temperature in the spring season results in a faster reproduction rate of insects, thereby boosting the pest population (Patterson et al., 1999). The crop protection chemicals are also affected as their efficacy reduces due to changes in temperature and precipitation. In addition, there is a direct impact of climate change on apple productivity; it has also aggravated the susceptibility to various diseases and pest attacks, resulting in further losses in yield (Gautam et al., 2013). Increased summer precipitation, particularly heavy storms, increases the incidences of *Rhynchosporium* leaf blotch and Septoria leaf spot diseases (Royle et al., 1986). With increased temperature and elevated CO_2 it was suggested that pests, such as aphids and weevil larvae, will respond positively (Newman, 2004; Staley and Johnson, 2008). The change in climatic parameters also increased the threat of new incursions. In the case of mango, hot and dry climate reduces the risk of fungal diseases like anthracnose and powdery mildew because sunlight, low humidity and temperature extremes (below 18°C or more than 35°C) rapidly inactivate the spores (Alfonso and Brent, 2014). Correlation analysis of weather factors and inflorescence pests showed that relative humidity had a negative correlation with hopper population and a highly negative correlation with flower bug, thrips/leaf and thrips/inflorescence of mango. The minimum temperature and evening relative humidity had a significant negative correlation with flower bug and thrips/inflorescence population, while displaying a highly negative correlation with thrips/leaf population (Bhut and Jethva, 2015). Hence, climate change has a varied and detrimental effect on fruit crops. The rise in mean temperature, long spells of drought during the summer season, delay in the start of winter season and reduced rainfall or snowfall has condensed the total area supposed to be slightly suitable for apple and other temperate fruit cultivation. The critical chill units required for apple production have been showing a decline. A trend analysis study indicates that snowfall is decreasing at the rate of 82.7 mm/annum in Himachal Pradesh, India. Consequently, the apple cultivation region is moving further up in elevation due to the warmer climate (Gautam et al., 2014). In India, a total decrease of around 2–3% in apple yield had been reported in the districts of Shimla, Kullu, Lahul and Spiti during mid-2000s and a maximum decline of about 4% was witnessed in marginal farms (Bhagat et al., 2009). IPCC projected that the average air-temperature rise will be a maximum of 4°C during the 21st century. That's why it has become very urgent to develop effective and promising technologies to mitigate the severe risks induced by climate change in fruit crop production. Nowadays, climate change adaptation and mitigation strategies are being incorporated more deeply in governmental structures and policies with the rising outline of climate action in national political agendas. Several countries are appointing inter-ministerial committees to oversee climate action and comprehensive national systems to monitor, evaluate and report on progress in mitigating climate change.

3. Adaptation of Fruit Crop to Climate Change

Global climate change is an outcome of a complex phenomenon that seems to be beyond one's control. It is considered to be a challenge and prioritizes action against it. In addition, its a science is specifically specialized, but our approach to this effect has more or less been generalizing in nature. Adaptation to climate change is nothing but embracing the actions that will include adjusting practices, processes and making strategies for minimizing the risk of climate change on fruits crops in an integrated approach. Diversification in crop production is a key strategy applied to minimize risk and build resilience in the farming systems (Sthapit, 2012). Monocultures may produce bumper harvests during favorable weather as well as market conditions, but they also expose the producers to the risk of complete crop failure and loss. There is an enormous diversity in agricultural practices with respect to cultural, institutional and economic factors and their interactions because of the variable climatic and environmental conditions. This provides us various opportunities to develop a correspondingly large array of possible adaptations for the existing agricultural systems and often supports climate risk management (Rajatiya et al., 2018). An essential component of this approach is to implement adaptation frameworks that are very relevant, robust and can easily be operated by all stakeholders. Alternatively, agricultural production systems with high biodiversity bring high stability in yield and also limit pest and disease outbreaks while increasing resilience to disturbances of climate change

(Frison et al., 2011). Fruit trees add resilience to the farming systems as they can better withstand climate adversity than annual crop plants. Depending on the fruit crop species, they can also provide numerous use values as seen in the case of timber, fodder, firewood, nitrogen fixation in soil and protection from windbreaks. The reproductive stages of fruit trees are most susceptible to climate change due to implications for the quantity and quality of fruits produced (Ramos et al., 2011). They bring change or adaptation in varieties of longer-lived fruit trees and this is an emerging challenge because of the rapidly occurring climate changes. However, fruit trees have long productive lives, generally ranging from over two to four decades, and hence any change in variety can happen over this long period of time (Lobell et al., 2006).

Numerous strategies of adaptation and mitigation measures have been reported by various researchers to minimize the climate change effect (Li et al., 2002; Zhong, 2003; Cao and Sun, 2007; Wei et al., 2007; Duan et al., 2008; Xue et al., 2008; Sun et al., 2009; Chen, 2010). The traditional agronomical methods extensively adopted as adaptation strategies to address the adverse impacts of climate change on productivity and quality of crops are:

1) Raising awareness within the farming communities to adapt to the changing climate and providing better information on challenges and solutions.
2) Development of cost-effective and climate-resilient technologies suitable for farmers.
3) Changing varieties or cultivars and altering the planting and harvest dates to achieve an effective, low-cost option. However, it may increase the risk that the farmer's product will be put in a different market window with lower prices.
4) Adopting strategies like altering or shifting planting dates to combat the increasing temperature and water stress periods during the main crop-growing season.
5) Adopting crop-based adaptations by using climate-ready crops that could induce flowering at higher temperatures or rootstock having low vigor but a strong root system.
6) Introducing adaptation strategies through cropping patterns, i.e., cropping systems, intercropping, crop diversification and relocation of crops in alternative areas.
7) Adaptations based on using variable crop species or tolerant/resistant cultivars and rootstock against climate change, like drought, high temperature, etc.

8) Modifying the crop management practices, such as zero tillage or minimum tillage practices to improve soil drainage, using sustainable liquid fertilizer, changing inland use management practices, etc.
9) Weed management and maintaining crop residues in the field for efficient use of resources.
10) Improving existing irrigation systems or implementing new ones, like drip irrigation systems.
11) Providing irrigation during critical stages of the crop growth and thus conserving soil's moisture reserves.
12) Adopting new farming techniques and resource-conserving technologies, for example, bagging of fruits, fertigation, etc. In mango, bagging mango fruits at the marble stage with brown paper or securing bags give maximum fruit retention. Bagging with newspaper bags gave the highest fruit weight and also reduced the occurrence of spongy tissue (Haldankar et al., 2015). Again, bagging pomegranate fruits with prgmen bags reduces fruit cracking and sunburn-like physiological disorders (Mohamed, 2016).
13) Mulching of cultivation beds with reflective silver-color film is a commonly used method to improve the skin/peel coloring of apples as it increases sunlight reflection from the bottom. Mulching helps to conserve the soil moisture, soil micro-climate, improve microbial activities and soil health. The use of plastic mulch has increased the yield of papaya (64.24%), mango (45.23%), banana (33.95%), Indian jujube (ber-27.06%), guava (25.93%), pineapple (14.63%) and litchi (12.61%) when compared to the control of the plant (Patil et al., 2013).
14) Plastic mulching in combination with drip irrigation is a common practice to attain high-quality and high-yield production in citrus orchards.
15) In greenhouse cultivation of vegetables, a range of devices like efficient ventilator, shading, fogger cooling, heat pumps, photo-selective film, etc. are developed for practical use to minimize the rise in interior temperature.
16) Anti-traspirants, like chitosan, kaolin, etc. in agriculture are used to reflect the heat radiation from plant parts so that water losses through transpiration and temperature of fruit and leaf surface are reduced (Parashar and Ansari, 2012). The anti-transpirant kaolin is also an effective treatment for reducing sunburn in pomegranate fruit (Ehteshami et al., 2011).
17) Among the frost reduction approaches, the chemical called the Bordeaux mixture is extensively used for reduction of frost damages

on grapes grown in a moderately cold climate as compared to other frost-reduction approaches (Yadollahi, 2011).

18) Modifying the fertilizer application schemes to enhance nutrient availability and use of other soil amendments to improve soil fertility for enhancing nutrient uptake are other methods.
19) Wind breaks or shelterbelts should be prepared for modifying the microclimate of the orchard as they provide shelter to pollinating insects and protect orchards against natural disasters, like wind erosion. It was reported that the minimum mortality percentage of fruit plants affected by frost in an orchard surrounded by windbreaks was 2.97–30.81%, whereas in the absence of this barrier, maximum mortality reached up to 91.43% (Yadollahi, 2011).
20) Weather forecasting by use of GIS and crop insurance schemes should be introduced to the farmers.
21) In some cases, excessive soil moisture, because of heavy rains, becomes a major problem and such a condition can be overcome by growing crops on raised beds.
22) Water conservation measures need to be encouraged among the farmers to increase resilience to climate change and these activities should be adopted in a widespread manner.
23) Significant efforts at water conservation are necessary for regions where a major portion of the total water resource is used for agricultural purposes. Water-saving measures, such as rainwater harvesting, crop rotations make efficient use of available water and adjusting in sowing dates as per the temperature and rainfall patterns, use of suitable crop varieties for new weather conditions (e.g., crop varieties with shorter cycles, more resilient to water stress), the espousal of water conservation practices that favor in-filtration and the reuse of waters should be practiced on arable land that reduces water run-off and acts as a windbreaker.
24) Beyond the farm level, measures such as modernizing the irrigation infrastructure, can be applied.
25) Assessment of the 'vulnerability' of all the major regions and/or fruit commodities should be done to identify the current 'at-risk' production sites and/or industries.
26) Identification should be done of the processes and practices that in the long term will be a threat to horticultural regions and cropping systems as a consequence of climate change adaptation. So, developing adaptation strategies that are appropriate practically and economically is necessary, in consultation with framers and researchers.

27) Reviewing and/or developing, where necessary, Good Agriculture Practices (GAP) for fruit cultivation, including adaptation and mitigation components.
28) Adoption of 'conservation agriculture' by keeping the land covered with vegetation all round the year prevents soil erosion and rainwater runoff on soil developed on limestone.
29) Restoration of biodiversity and forest plantation in all barren lands.
30) River banks to be enclosed with agroforestry, with shrubs and trees, having high CO_2 demand.
31) High-density orchards and garden plant practices must be encouraged to keep the soil covered with shadow horticultural crops.
32) Propose legal bans against any construction or non-farming activities on productive land.
33) Promote the construction of highways and railway tracks along the river banks.
34) Topsoil restoration is a significant issue to be addressed in relation to the adjoining environment situations, including slope gradient, natural vegetation, effective soil depth, rainfall intensity, clay types, nature of land use and soil biodiversity.
35) Assessment of the economic benefits of silvi-horticulture as well as the benefits that might accrue through adaptation and mitigation.
36) Documentation of the effects of climate change on major overseas production regions, particularly in those countries that are chief competitors to India's production.

Along with the adoption of modified crop management practices, the challenges posed by climate change could be tackled by developing new climate-resilient tolerant varieties. Numerous institutions have produced hybrid varieties which are tolerant of heat, salinity and drought-stress conditions. Such hybrids must be planted efficiently to combat the consequence of climate change. Sufficient efforts should be focused on the development of new varieties well suited to diverse agro-ecological regions under changing climatic conditions. In contrast to annual crops, where the application of adaptation strategies can be visualized comparatively quickly by using a broad range of cultivars and altering the planting dates or season; planting and reorganizing orchards require contemplation of more long-term planning for climate change. Hence, before resorting to any adaptation strategy, detailed research on the impact of climate change on perennial crops is very essential.

4. Mitigation Strategies of Fruit Crop to Climate Change

Mitigation strategies are considered human interventions to reduce the effect or impact of human activities on the climate system (Fig. 2). The mitigation tools are more frequently used in fruits and vegetables as the climate-related issues are increasing, whereas the adaptive strategies could be considered for more long-term stress management. It predominately includes strategies that reduce GHG sources or emissions and hence enhance GHG sinks. 'Climate mitigation' primarily comprises steps to reduce the emissions or enhance the sinks of GHGs which are responsible for climate change with the help of conscious practices, which can permanently eradicate or reduce the long-term risk and hazard of climate change to humans life. So, mitigation includes the process in which the emission of GHGs is reduced or sequestered. The improved agronomic practices for enhanced nutrient use efficiency, water use efficiency, reduction of GHGs and eco-friendly disease and pest management strategies are also parts of mitigation. Application of the right amount of fertilizer, use of more biological control measures, an extension of water-saving technology like dripping irrigation or sparkling irrigation, water-fertilizer coupling effect technologies could be helpful in decreasing GHGs emissions. Several mechanical, as well as chemical efforts have been made to minimize CO_2, CH_4 and N_2O emissions. Planting trees and enhancing agroforestry for high CO_2 demand is one of the efforts that will reduce GHG's effects. A research by Forster et al. (2021) calculated the potential GHG mitigation in the UK by national planting strategy of 30,000 ha yr^{-1} from 2020 to 2050 using a dynamic life cycle assessment and recorded that the commercial forest could mitigate 1.64 Pg CO_2e by 2120. They found that forest growth rate was most important for determining cumulative mitigation, irrespective of whether trees were harvested by 2120. Greenhouse gas (GHGs) emissions from burning fossil, liquid, solid and gaseous fuels including other sources being chemically sequestered in some other forms. Minimizing the load on petroleum and coal, fossil fuel sources are being used through different biofuel extraction methods from *Jatropha* and *Pongamia* species and their utilization. The use of biofuels—fuels with lower carbon content, i.e., natural gas, CNG, cow-dung gas—will improve the efficiency of electricity generation, transmission and distribution throughout the world. Many architectural manipulations were employed by developing greenery over buildings through lawns and similar constructions. The reduction in CO_2 and CH_4 emissions were taken into consideration throughout the whole agriculture process. So, there is a need to quantify the GHG emissions from all improved agronomic practices before cataloging or tagging any technology as 'green' or 'low-carbon' technology. In processing technology, biogas production is extremely useful as it reuses or recycles various farm wastes and can be

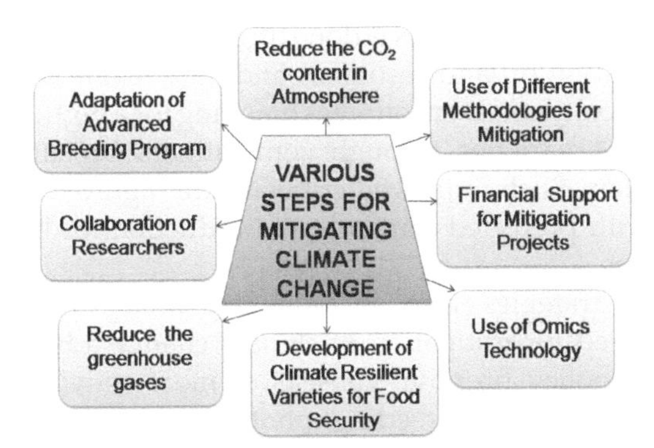

Fig. 2. Different mitigation strategies to cope with climate change and to sustain food security.

used together with possible renewable energies, like hydropower and wind energy. The biogas produced can also be used in transportation activities or storage facilities, thus reducing dependence on fossil fuels and reducing CO_2 emissions into the environment.

Intercropping in orchards or introduction of green manure crops or cover crops in orchards facilitates the carbon sink function in different fruit crops and their related vegetation in ecosystems. So, carbon sequestration is being encouraged through mechanical, chemical, biological and pedogenic manipulations. Sequestration of carbon refers to CO_2 removal from the atmosphere and its storage or accumulation in soil, biomass and harvested products, which are protected or preserved to avoid CO_2 to be released back into the atmosphere. This process of carbon accumulation is referred to as carbon stores or carbon sinks. Fruit trees are also an essential part of perennial-based solutions for climate change mitigation. In 2012, it was reported that perennial crops can sequester around 320–1,100 kg of soil carbon per hectare when compared to 0–450 kg of carbon in annual crops and hence are more likely to provide better yields than annual crops at higher temperatures (Glover et al., 2007; Sthapit and Scherr, 2013). The horticultural crops are perennial, such as plantations, fruit trees and tree spices which are potential candidates for carbon sequestration. Though currently most of these do not fall under carbon trade, there is a lot of scope for these tree species to be used in carbon sequestration and climate regulation system. Similar studies conducted on coconut plantations suggested that annual carbon sequestration in coconut above the ground biomass varies between 15–35 Mg CO_2 ha^{-1} yr^{-1} depending on cultivar, agro-climatic zone, soil type and management (Kumar, 2013). Cocoa-arecanut intercropping also is a good system for carbon sequestration with the potential to sequester 5–7 Mg CO_2 ha^{-1} yr^{-1} (Kumar, 2009; Balasimha and Kumar, 2010). The mitigation of climate change damage to various tropical fruits is done through zoning of crop suitability for proper allocation of land resources and an amiable environment for compatible fruit cultivars. It encourages breeding for a high and wide range of stress-resistant varieties through conventional breeding techniques and biotechnology. Such improved cultivars are highly recommended or incorporated with protected horticulture production for better and safer economic returns.

Mitigation strategies also include the policy and institutional capability required for mitigation of climate change adversities (Fig. 2). The adaptation policy proposal consists of more activities, i.e., further research focusing on plant breeding, photosynthetic capacity, biological nitrogen fixation rate, level of stress resistance, protected horticultural crops, and precision horticulture for augmentation of crop adaptability to climate change. Development of climate monitoring, forecasting and pre-warning capability in all provinces will contribute to adaptability to climate change. Therefore, improvement in public awareness of climate change problems and the need for betterment of adaptation and mitigation strategies is highly recommended. As a result, there is a critical need for a decision support system (DSS) to facilitate improved management of agriculture at the farm level and for sustainable and climate-resilient crop production. A farmer needs to answer several questions related to crop production and marketing before deciding on the technology best suited to the farm's condition. The crop models are of immense use in the climate change perspective as well as from the crop management point of view. The crop models developed also need to be calibrated and validated for the study purpose before conducting a large-scale analysis. Linking the models to GIS and remote-sensing data also offers enormous scope for regional estimations of GHG emissions and for significantly reducing the yield loss. The major sectors for application of crop simulation models include crop management, agro-ecological zoning, estimating potential production, yield gap analysis and developing breeding strategies. Strategic and anticipatory decision-making in the system also helps in finding crop potential zones for land use planning and hi-tech horticulture. Crop insurance and weather-based horti-advisory are also dependent on crop simulation models. Crop modeling is considered an important tool in the research and development of perennial plantation crops because it takes a lot of time for conducting research experiments. Similar to any other approach, simulation modeling also has limitations and uncertainties attached to it.

5. Research Development for Climate Resilience in Fruit Crop

Harmful effects of climate change include increased temperature followed by changes in time, intensity and pattern of rainfall, which in turn lead to a higher frequency and extent of natural calamities, like floods, cyclones, droughts and augmented soil salinity. Thus, the hazardous damage caused by climate change inevitably brings about a shift in the optimum planting time of various crops, particularly in rain-fed agricultural areas. The critical impacts of climate change (associated with high temperature, frost, GHGs emissions, etc.) can be easily visualized in the physiological processes of fruit crops, geographic shift of production areas, changes in cropping practices, changes in the frequency of disease and pest spread, crop production and yield, product quality, etc. This enhances the urgent need to initiate the development of potential cultural practices and other countermeasures to cope with climate change.

To alleviate the adverse effects of climate change on the productivity and quality of fruits produced, several strategies and technological countermeasures have been planned and developed. The adaptation technologies against increased temperature and other climatic abnormalities also include breeding and biotechnological intervention for increased production. The breeding interventions include measures like systematic breeding or phenotyping of fruit crop genetic wealth against temperature increase, moisture stress and genetic enhancement for tolerance towards biotic and abiotic stress. Marker-assisted selection, molecular characterization and development of transgenic or climate-resilient crops having resistance to various traits in relation to biotic and abiotic stress are some common measures adopted in biotechnological interventions for improving climate resilience. Gene pyramiding against stress and *in-vitro* conservation of rare and useful species can be used as the future thrust areas of research for climate change. So, there is a critical need right now to conduct a focused and precise research to generate adequate information on the impacts of climate change and thus derive suitable adaptation and mitigation strategies for a particular region. The literature review based on of PRISMA-P (Preferred Reporting Items for Systematic Review and Meta-Analysis Protocols) showed that small-scale farmers in the past 30 years have adopted climate-resilient crops and varieties to counter abiotic stresses, such as drought, heat, flooding and salinity in lower- and middle-income countries (Acevedo et al., 2020). On the basis of the collected literature, the use of climate-resilient varieties has been always recommended as an excellent way for farmers to cope with or adapt to climate change. Various researchers have reported a series of pathways and interventions that can contribute to ever higher adoption rates of climate-resilient crops.

6. Conclusion and Prospects

Agricultural production is highly affected due to climate change. Extreme weather conditions, such as heat waves, droughts, cyclones and floods lead in reduction of agricultural food production and poverty, especially in rural communities. Climate change significantly leads to yield reduction to the tune of about 30% in most crops apart from lower productivity and failure of the crop. The increase in the global population and change in food habits in developing countries will highly impact the natural resources and lead to food insecurity. To cope with climate change, farmers need to modify farm management practices including change in planting time, supplementing irrigation, intercropping, adopting conservation agriculture, accessing short- and long-term crops and planting more climate-resilient crop varieties. On the basis of the literature, inclusion of a series of pathways and interventions can contribute to higher adoption rates of climate-resilient crops. Climate resiliency at the farm level is of utmost importance to achieve food security and improve livelihood support to the rural communities, especially the communities that depend on local agricultural produce to ensure household income and achieve daily adequate caloric intake and balanced nutrition. Farmers are adopting various strategies under climate-smart agriculture to build highly resilient and sustainable agricultural systems. Environment friendly and ecologically sound chemicals must be used for breaking the rest period or dormancy. In tropical climate, dormancy can be induced artificially by defoliating after harvesting (Griesbach, 2007). Also, the application of sprays of hydrogen cyanamide has been effective in breaking dormancy, thereby promoting blooming (Erez et al., 2008; Ashebir et al., 2010; Chabchoub et al., 2010). Weather-based monitoring strategies must be adopted for rapid and effective diagnosis of insects, pests and diseases using the GIS system. Recent achievements in the development of newer technologies and crop genetic improvement that include variety adaptation, crop diversification, biodiversity identification and underutilized crops manipulation have identified cultivars that can adapt to high temperature, drought, and floods.

References

Acevedo, M., Pixley, K., Zinyengere, N., Meng, S., Tufan, H., Cichy, K., Bizikova, L., Isaacs, K., Ghezzi-Kopel, K. and Porciello, J. (2020). A scoping review of adoption of climate-resilient crops by small-scale producers in low- and middle-income countries. *Nature Plants*, 6: 1231–1241.

Ahmed, N., Lal, S., Das, B. and Mir, J.I. (2011). Impact of climate change on temperate fruit crops. pp. 141–150. *In*: Dhillon, W.S. and Aulakh, P.S. (eds.). *Impact of Climate Change on Fruit Crops*. Narendra Publishing House, New Delhi.

Akerstrom, A., Forsum, A., Rumpunen, K., Jaderlund, A. and Bang, U. (2009). Effects of sampling time and nitrogen fertilization on anthocyanidin levels in *Vaccinium myrtillus* fruits. *Journal of Agricultural and Food Chemistry*, 57: 3340–3345.

Alfonso, D.R. and Brent, M.S. (2014). Agricultural adaptation to climate change in the Sahel: expected impacts on pests and diseases afflicting selected crops. pp. 53–54. *In: African and Latin Americal Resilience to Climate Change Project*. USAID, Washington.

Ameglio, T., Alves, G., Bonhomme, M., Cochard, H., Ewres, F. et al. (2000). Winter functioning of walnut: Involvement in branching processes. pp. 230–238. *In: L'Arbre, Biologieet Development*. Montreal (CAN), Isabelle Quentin.

Anonymous, *Economic Times*. (2016). Climate Change Advances Wine Grape Harvest by Two Weeks. New Delhi.

Ashebir, D., Deckers, T., Nyssen, J., Bihon, W., Tsegay, A., Tekie, H., Poesen, J., Haile, M., Wondumagegneheu, F., Raes, D., Behailu, M. and Deckers, J. (2010). Growing apple (*Malus domestica*) under tropical mountain climate conditions in northern Ethiopia. *Experimental Agriculture*, 46: 53–65.

Balasimha, D. and Kumar, N.S. (2010). Net primary productivity, carbon sequestration and carbon stocks in areca-cocoa mixed cropping system. pp. 215–226. *In: Proceedings of the 16th International Cocoa Research Conference*. Bali.

Baumert, K.A., Herzog, T. and Pershing, J. (2009). *Navigating the Numbers, Greenhouse Gas Data and International Climate Policy*. World Resources Institute, Washington, DC, USA.

Bethere, L., Tija, S., Juris, S. and Bethers, U. (2016). Impact of climate change on the timing of strawberry phenological processes in the Baltic States. *Estonian Journal of Earth Sciences*, 65(1): 48–58.

Bhagat, R.M., Rana, R.S. and Kalia, V. (2009). *Global Climate Change and Indian Agriculture*. ICAR, pp. 48–53, Aggarwal, P.K. (ed.). New Delhi, India.

Bhatia, H.S. (2010). Evaluation of new apple cultivars under changing climate in Kullu Valley of Himachal Pradesh. pp. 1–6. *In: Proceeding of National Seminar on Impact of Climate Change on Fruit Crops* (ICCFC, 2010), 6–8 October, PAU, Ludhiana.

Bhut, J.B. and Jethva, D.M. (2015). Impact of weather factor on incidence of inflorescence pests of mango. pp. 291–296. *In: National Seminar on Water Management and Climate Smart Agriculture*, 13–14 February, JAU, Junagadh.

Bhutiyani, M.R. and Kale, V.S. (2002). *Climate Change in the Last Century are the Himalaya Warming*. Sapper, 13: 37–46.

Bindi, M., Fibbi, L. and Miglieta, F. (2001). Free air CO_2 enrichment (FACE) of grapevine (*Vitis vinifera* L.): II. Growth and quality of grape and wine in response to elevated CO_2 concentrations. *European Journal of Agronomy*, 14: 145–155.

Black, M.W. (1952). The problem of prolonged rest in deciduous trees. pp. 1122–1131. *In: Proceedings of 13th International Horticultural Congress*. London.

Blanke, M.M. and Kunz, A. (2011). Effects of climate change on pome fruit phenology and precipitation. *Acta Horticulturae*, 922: 381–386.

Borochov-Neori, H., Judeinstein, S., Harari, M., Bar-Yaaakov, I., Patil, B.S. et al. (2011). Climate effects on anthocyanin accumulation and composition in the pomegranate (*Punica granatum* L.) fruit arils. *Journal of Agricultural and Food Chemistry*, 59: 5325–5334.

CAIT, Climate Analysis Indicators Tool. (2011). CAIT version 8.0. Available from: http://cait.wri.org.

Cao, M.H. and Sun, Y.Z. (2007). Guangdong fruits and climate damages. *Modern Agricultural Sciences and Technology*, 4: 39–41.

CCPI, Climate Change Performance Index. (2019). *The New Climate Institute and the Climate Action Network*. website published by Germanwatch.

CDB, Coconut Development Board. (2019). *All India Final Estimates of Area, Production and Productivity of Coconut*.

Chabchoub, M.A., Aounallah, M.K. and Sahli, A. (2010). Effect of hydrogen cyanamide on bud break, flowering and fruit growth of two pear cultivars (*Pyrus communis*) under Tunisian conditions. *Acta Horticulturae*, 884: 427–432.

Chadda, K.L. and Kumar, S.N. (2011). Climate change impacts on production of horticultural crops. pp. 3–9. *In*: Dhillon, W.S. and Aulakh, P.S. (eds.). *Impact of Climate Change on Fruit Crops*.

Chadha, K.L. (2015). Global climate change and Indian horticulture, Chapter 1. *In*: Choudhary et al. (eds.). *Climate Dynamics in Horticultural Science: Impact, Adaptation, and Mitigation*, vol. 2, Apple Academic Press, Inc.

Chandio, A.A., Jiang, Y., Rehman, A. and Rauf, A. (2020). Short and long-run impacts of climate change on agriculture: An empirical evidence from China. *International Journal of Climate Change Strategies and Management*, 12(2): 201–221.

Chen, Q. (2012). Adaptation and mitigation of impact of climate change on tropical fruit industry in China. *Acta Horticulturae*, 101–104.

Chmielewski, F.M., Muller, A. and Bruns, E. (2004). Climate changes and trends in phenology of fruit trees and field crops in Germany 1961–2000. *Agricultural Forest Meteorology*, 121: 69–78.

Cleland, E.E., Chuine, I., Menzel, A., Mooney, H.A. and Schwartz, M.D. (2007). Shifting plant phenology in response to global change. *Ecological Evolution*, 22: 357–365.

Curry, E.A. (1997). Temperatures for optimal anthocyanin accumulation in apple skin. *Journal of Horticultural Science and Biotechnology*, 72: 723–729.

Darjazi, B.B. and Larijani, K. (2012). The effects of climatic conditions and geographical locations on the volatile flavor compounds of fig (*Ficus carica* L.) fruit from Iran. *African Journal of Biotechnology*, 11: 9196–9204.

Duan, H.L., Qian, H.S., Yu, F. and Song, Q.H. (2008). Temperature suitability of longan and its changes in south China area. *Acta Ecologica Sinica*, 28: 5303–5313.

Ehteshami, S., Sarikhani, H. and Ershadi, A. (2011). Effect of kaolin and gibberellic acid application on some qualitative characteristics and reducing the sunburn in pomegranate fruits (*Punica granatum* L.) cv. 'Rabab Neiriz'. *Plant Products Technology, Agricultural Research*, 11(1): 15–23.

Erez, A., Yablowitz, Z., Aronovitz, A. and Hadar, A. (2008). Dormancy breaking chemicals' efficiency with reduced phytotoxicity. *Acta Horticulturae*, 772: 105–112.

FAOSTAT, Food and Agriculture Organization Statistics. (2019). Data accessed: 28 August 2019.

Felicetti, D.A. and Schrader, L.E. (2008). Photo-oxidative sunburn of apples characterization of a third type of apple sun-burn. *International Journal of Fruit Science*, 8(3): 160–172.

Forster, E.J., Healey, J.R., Dymond, C. and Styles, D. (2021). Commercial afforestation can deliver effective climate change mitigation under multiple decarbonisation pathways. *Nature Communications*, 12: 3831.

Fraga, H. and Santos, J.A. (2021). Assessment of climate change impacts on chilling and forcing for the main fresh fruit regions in Portugal. *Frontier of Plant Science*, 12: 689121.

Frison, E.A., Cherfas, J. and Hodgkin, T. (2011). Agricultural biodiversity is essential for a sustainable improvement in food and nutrition security. *Sustainability*, 3: 238–253.

Gautam, H.R., Bhardwaj, M.L. and Kumar, R. (2013). Climate change and its impact on plant diseases. *Current Science*, 105: 1685–1691.

Gautam, H.R., Sharma, I.M. and Kumar, R. (2014). Climate change is affecting apple cultivation in Himachal Pradesh. *Current Science*, 106: 498–499.

George, K., Mavromatis, T., Stefanos, K., Nikolaos, M.F. and George, V.J. (2014). Viticulture-climate relationships in Greece: The impacts of recent climate trends on harvest date variation. *International Journal of Climatology*, 34(5): 1445–1459.

Gill, P.P.S. and Singh, N.P. (2012). Decline of mango diversity in sub-montane and Kandi zone of Punjab—An overview. *Indian Journal of Ecology*, 39(2): 313–315.

Glover, J.D., Cox, C.M. and Reganold, J.P. (2007). Future farming: A return to roots? pp. 82–89. *Scientific American*, August, 2007.

Gradziel, T.M., Lampinen, B., Connell, J.H. and Viveros, M. (2007). Winters' almond: An early-blooming, productive, and high-quality pollinizer for nonpareil. *Hort. Science*, 42: 1725–1727.

Griesbach, J. (2007). *Growing Temperate Fruit Trees in Kenya*. World Agroforestry Center (ICRAF), Nairobi, Kenya.

Haldankar, P.M., Parulekar, Y.R., Kireeti, A., Kad, M.S., Shinde, S.M. and Lawande, K.E. (2015). Studies on influence of bagging of fruits at marble stage on quality of mango cv. Alphonso. *Journal of Plant Studies*, 4(2): 12–20.

Hemanth, K.N., Ravishankar, K.V., Narayanaswamy, P. and Shivashankara, K.S. (2008). Influence of temperature on spongy tissue formation in Alphonso Mango. *International Journal of Fruit Science*, 8(3): 226–234.

Indiastat, India's most comprehensive e-resource of socio-economic data. (2019). Indiastat.com.

IPCC. (2007). *Climate Change: Mitigation. Contribution of Working Group III to the Fourth Assessment Report of the Intergovernmental Panel on Climate Change* [Metz, B., Davidson, O.R., Bosch, P.R., Dave, R. and Meyer, L.A. (eds.)], Cambridge University Press, Cambridge, United Kingdom and New York, NY, USA.

IPCC. (2009). *Climate Change: The Scientific Basis*. Cambridge University Press.

IPCC. (2014). *Climate Change: Mitigation of Climate Change, Fifth Assessment Synthesis Report of Intergovernmental Panel on Climate Change.*

IPCC. (2019). Intergovernmental panel on climate change, *Special Report on Climate Change, Desertification, Land Degradation, Sustainable Land Management, Food Security, and Greenhouse Gas Fluxes in Terrestrial Ecosystems*. SPM Approved Draft, Summary for Policy Makers.

IPPC, IPPC Secretariat. (2021). *Scientific Review of the Impact of Climate Change on Plant Pests—A Global Challenge to Prevent and Mitigate Plant Pest Risks in Agriculture, Forestry and Ecosystems*. Rome, FAO on Behalf of the IPPC Secretariat.

Issarakraisila, M., Considine, J.A. and Turner, D.W. (1993). Effects of temperature on pollen viability in mango cv. Kensington. *Acta Horticulturae*, 341: 112–124.

Issarakraisila, M. and Considine, J.A. (1994). Effects of temperature on pollen viability in mango cv. 'Kensington'. *Annals of Botany*, 73: 231–240.

Keukeleire, J., Janssens, I., Heyerick, A., Ghekiere, G., Cambie, J. et al. (2007). Relevance of organic farming and effect of climatological conditions on the formation of r-acids, a-acids, desmethylxanthohumol, and xanthohumol in hop (*Humulus lupulus* L.). *Journal of Agricultural and Food Chemistry*, 55: 61–66.

Khan, T., Qiu, J., Banjar, A., Alharbey, R., Alzahrani, A.O. and Mehmood, R. (2021). Effect of climate change on fruit by co-integration and machine learning. *International Journal of Climate Change Strategies and Management*, 13(2): 208–226.

Kliewer, M.W. and Lider, L.A. (1970). Effects of day temperature and light intensity on growth and composition of *Vitisvinifera* L. fruits. *Journal of the American Society for Horticultural Science*, 95: 766–769.

Kumar, N.S. (2009). Carbon sequestration in coconut plantations. *In*: Aggarwal, P.K. (ed.). *Global Climate Change and Indian Agriculture—Case Studies from ICAR Network Project*, ICAR Pub., New Delhi.

Kumar, S.N. (2013). Modelling climate change impacts, adaptation strategies and mitigation potential in horticultural crops, Chapter 3. *In*: Singh, H.P. et al. (eds.). *Climate-resilient Horticulture: Adaptation and Mitigation Strategies*. Springer Publications, India.

Li, Y.L., Su, Z. and Tu, F.X. (2002). The effects of climate factors on yields of lichee and longan in Guangxi. *Journal of Guangxi Academy of Sciences*, 18: 135–140.

Lobell, D.B., Field, C.B., Cahill, K.N. and Bonfils, C. (2006). Impacts of future climate change on California perennial crop yields: Model projections with climate and crop uncertainties. *Agricultural and Forest Meteorology*, 141: 208–218.

Lo Piero, A.R., Puglisi, I., Rapisarda, P. and Petrone, G. (2005). Anthocyanins accumulation and related gene expression in red orange fruit induced by low temperature storage. *Journal of Agricultural Food Chemistry*, 53: 9083–9088.

Mark, D.C.J. and Marin, R.A. (2016). Carbon sequestration potential of fruit tree plantation in southern Philippines. *Journal of Biodiversity and Environmental Sciences*, 8(5): 164–174.

Mohamed, A.W. (2016). Effect of bagging type on reducing pomegranate fruit disorders and quality improvement. *Egyptian Journal of Horticulture*, 41(2): 263–278.

Mori, K., Sugaya, S. and Gemma, H. (2005). Decreased anthocyanin biosynthesis in grape berries grown under elevated night temperature condition. *Scientia Horticulturae*, 105: 319–330.

Newman, J.A. (2004). Climate change and cereal aphids: the relative effects of increasing CO_2 and temperature on aphid population dynamics. *Global Change Biology*, 10: 5–15.

NICRA. (2015). *Climate Resilient Crop Varieties for Sustainable Food Production under Aberrant Weather Conditions*. ICAR, Central Research Institute for Dry Land Agriculture, Hyderabad, Bulletin No. 4, pp. 1–56.

Parashar, A. and Ansari, A. (2012). A therapy to protect pomegranate (*Punica granatum* L.) from sunburn. *Pharmacie Globale*, 3(5): 1–3.

Parmar, V.R., Shrivastava, P.K. and Patel, B.N. (2012). Study on weather parameters affecting the mango flowering in south Gujarat. *Journal of Agrometeorology*, 14: 351–353.

Parry, M.L., Rosenzweig, C., Iglesias Livermore, A.M. and Fischer, G. (2004). Effects of climate change on global food production under SRES emissions and socio-economic scenarios. *Global Environmental Change*, 14: 53–67.

Pathak, S., Pramanik, P., Khanna, M. and Kumar, A. (2014). Climate change and water availability in Indian agriculture: Impacts and adaptation. *Indian Journal of Agricultural Sciences*, 84(6): 671–679.

Patil, S.S., Kelkar, T.S. and Bhalerao, S.A. (2013). Mulching: A soil and water conservation practice. *Research Journal of Agriculture and Forestry Sciences*, 1(3): 26–29.

Patterson, D.T., Westbrook, J.K., Joyce, R.J.V., Lingren. P.D., Rogasik, J. et al. (1999). Weeds, insects and disease: Climate change: impact on agriculture. *Climate Change*, 43: 711–727.

Puri, M.G., Murai, A.M., Gholape, S.M., Shigwan, A.S. and Mesare, S.N. (2017). Climate smart agriculture: an approach to sustainably increasing agricultural productivity. pp. 39–42. *In*: *Proceedings of National Conference on Climate Change Adaption*. 24–25 February, Hyderabad, India.

Rai, R., Joshi, S., Roy, S., Singh, O., Samir, M. and Chandra, A. (2015). Implications of changing climate on productivity of temperate fruit crops with special reference to apple. *Journal of Horticulture*, 2(2): 1–6.

Rajan, R., Ahmad, M.F., Pandey, K., Aman, A. and Kumar, V. (2020). Climate change and resilience in fruit crops. pp. 337–3354. *In*: *Climate Change and its Effects on Agriculture*. BIOTEC BOOKS Publisher.

Rajatiya, J., Varu, D.K., Gohil, P., Solanki, M., Halepotara, F., Gohil, M., Mishra, P. and Solanki, R. (2018). Climate change: impact, mitigation and adaptation in fruit crops. *International Journal of Pure and Applied Bioscience*, 6(1): 1161–1169.

Ramos, C., Intrigliolo, D.S. and Thompson, R.B. (2011). Global change challenges for horticultural systems. *In:* Araus, J.L. and Slafer, G.A. (eds.). *Crop Stress Management and Global Climate Change*. CAB International.

Ravishanker, H. and Rajan, S. (2011). Possible impact of climate change on mango and guava productivity. pp. 151–156. *In*: Dhillon, W.S. and Aulakh, P.S. (eds.). *Impact of Climate Change on Fruit Crops*.

Reay, P.F. (1999). The role of low temperature in the development of the red blush on apple fruit (Granny Smith). *Scientia Horticulturae*, 79: 113–119.

Reddy, R.P.V., Verghese, A. and Rajan, V.V. (2012). Potential impact of climate change on honeybees (*Apis* spp.) and their pollination services. *Pest Management in Horticultural Ecosystems*, 18: 121–127.

Reddy, Y.N. and Singh, O. (2011). Role of heat shock proteins in adaptivity of plants to higher temperatures and alleviation of heat shock in mango. pp. 57–64. *In*: Dhillon, W.S. and Aulakh, P.S. (eds.). *Impact of Climate Change on Fruit Crops*.

Royle, D.J., Shaw, M.W. and Cook, R.J. (1986). Pattern of development of *Septorianodorume* and *S. tritici* in some winter wheat crops in Western Europe, 1983–84. *Plant Pathology*, 35: 466–476.

Ruck, H.C. (1975). Deciduous fruit tree cultivars for tropical and sub-tropical regions. *Hort. Rev 3*, Commonwealth Burr. Horticulture and Plantation Crops, East Malling, UK.

Samish, R.M. and Lavees, S. (1982). The chilling requirement of fruit trees. pp. 372–388. *In*: *Proc. of XVI International Horticultural Congress*. Brussels.

Samra, J.S., Singh, G. and Rama Krishna, Y.S. (2003). *Cold Wave of 2002–2003: Impact on Agriculture*. Natural Resource Management Division, ICAR, Krishi Bhavan, New Delhi.

Schrader, L.E., Zhang, J. and Duplaga, W.K. (2001). Two types of sunburn in apple caused by high fruit surface (peel) temperature. *Plant Health Progress*, 10: 1094.

Singh, H.P. (2010). Impact of climate change on horticultural crops. pp. 1–8. *In*: *Challenges of Climate Change – Indian Horticulture*. Westville Publishing House, New Delhi.

Singh, N., Sharma, D.P. and Hukam, C. (2016). Impact of climate change on apple production in India: A review. *Current World Environment*, 11(1): 251–259.

Staley, J.T. and Johnson, S.N. (2008). Climate change impacts on root herbivores. *In*: Johnson, S.N. and Murray, P.J. (eds.). *Root Feeders: An Ecosystem Perspective*. Wallingford, UK: CABI.

Sthapit, B.R., Ramanatha Rao, V. and Sthapit, S.R. (2012). *Tropical Fruit Tree Species and Climate Change*. Bioversity International, New Delhi, India.

Sthapit, S.R. and Scherr, S.J. (2013). Tropical fruit tree species and climate change. pp. 15–26. *In*: *Bioversity International*. New Delhi, India.

Sun, J., Chen, S.J. and Xin, J.W. (2009). Analysis on the effect of climate change on mango production in Changjiang City. *Journal of Anhui Agricultural Sciences*, 37: 4962–4963.

Ubi, B.W., Honda, C., Bessho, H., Kondo, S., Wada, M., Kobayashi. S. and Moriguchi, T. (2006). Expression analysis of anthocyanin biosynthetic genes in apple skin effect of UV–B and temperature. *Plant Science*, 170: 571–578.

UNFCCC, *United Nations Framework Convention on Climate Change Report*. (2019). Climate action and support trends, based on national reports submitted to the UNFCCC secretariat under the current reporting framework.

Vallat, A., Gu, H. and Dorn, S. (2005). How rainfall, relative humidity and temperature influence volatile emissions from apple trees *in situ*. *Phytochemistry*, 66: 1540–1550.

Varu, D.K., Viradia, R.R., Chovatia, R.S. and Barad, A.V. (2010). Response of different genotypes of custard apple to weather parameters. pp. 24–36. *In: AGRESCO Report-2010*, JAU, Junagadh.

Varu, D.K. and Viradia, R.R. (2015). Damage of mango flowering and fruits in Gujarat during the year 2015. *Survey Report of Department of Horticulture*, JAU, Junagadh.

Wand, S.J.E., Steyn, W.J., Mdluli, M.J., Marais, S.J.S. and Jacobs, G. (2002). Over tree evaporative cooling for fruit quality enhancement. *South Africa Fruit Journal*, 2: 18–21.

Wand, S.J.E., Steyn, W.J., Mdluli, M.J., Marais, S.J.S. and Jacobs, G. (2005). Use of evaporative cooling to improve 'Rosemarie' and 'Forelle' pear fruit blush color and quality. *Acta Horticulturae*, 671: 103–111.

Wang, S.Y. and Zheng, W. (2001). Effect of plant growth temperature on antioxidant capacity in strawberry. *Journal of Agricultural and Food Chemistry*, 49: 4977–4982.

Wei, J.H. and Mo, R. (2007). Agricultural meteorological disasters and prevention counter measures of affecting on the high-quality fruit project in Baise City. *Guangxi Agricultural Sciences and Technology*, 38: 212–214.

Xu, C., Zhang, Y., Zhu, L., Huang, Y. and Lu, J. (2011). Influence of growing season on phenolic compounds and antioxidant properties of grape berries from vines grown in subtropical climate. *Journal of Agricultural and Food Chemistry*, 59: 1078–1086.

Xue, J.J., Lu, H.C., Wang, H.S. et al. (2008). Investigation of chilling damage to Litchi and Longan in Nanning City and measures for remedy flowering. *China Fruits*, 5: 66–71.

Yadollahi, A. (2011). Evaluation of reduction approaches on frost damages of grapes grown in moderate cold climate. *African Journal of Agricultural Research*, 6(29): 6289–6295.

Yamane, T., Jeong, S.T., Goto-Yamamoto, N., Koshita, Y. and Kobayashi, S. (2006). Effects of temperature on anthocyanin biosynthesis in grape berry skins. *American J. Enology Viticulture*, 57: 54–59.

Zavalloni, C., Andresen, J.A., Black, J.R., Winkler, J.A., Guentchev, G. et al. (2008). A preliminary analysis of the impacts of past and projected future climate on sour cherry production in the Great Lakes Region of the USA. *Acta Horticulturae*, 803: 123–130.

Zekri, M. (2011). Factors affecting Citrus Production and quality. *Citrus Industry*. Online at crec.ifas.ufl.edu.

Zhong, S.Q. (2003). Causes of low flowering success of longan and litchi in Guangxi. *Chinese Journal of Agrometeorology*, 24: 55–57.

Part II
Rain Forest Tropical Fruits

3

Banana Cultivation for Resilience to Climate Change

*Ebtsam Labib Belatus** and *Adel Ahmed Abul-Soad*

1. Introduction

1.1 Climate Zones of Banana Cultivation

To evaluate the upshot of climate change on banana productivity, we first need to understand the climate in which banana can perform well, as also learn the climate sensitivity (Varma and Bibber, 2019). Bananas are mainly grown for being a profitable proposition ranging from the equator to 30° latitude or more, and perform well in warm, no-frost as well as costal climates (Turner and Lahav, 1983). The tropical kind of weather is perfect for banana to grow and produce throughout the year, except that some problems may occur at times. It yields marketable produce in the subtropics even when some dissimilarities in the weather are present.

Banana could be grown to produce a commercial yield in different environments although it is considered a tropical crop. In fact, it is grown and yields throughout the humid and sub-humid tropics, the tropical highlands and in the dry subtropics (Piet van and Charles, 2012). It is even grown and yields a satisfactory produce in the cool subtropics and Mediterranean countries. Moreover, it could be found even in small areas (Calberto et al., 2015) where the weather is unstable during the year, as in China and northern India (Van den Bergh et al., 2012) where a satisfactory

Tropical Fruit Department, Horticulture Research Institute, Agricultural Research Center, 9 Cairo University St., Orman, 12619, Giza, Egypt.
Email: adel.aboelsoaud@arc.sci.eg
* Corresponding author: ebtsamlabib@hotmail.com

yield for commercial production may be obtained. Regions and countries can vary widely in banana cultivation practices (Varma and Bibber, 2019). Growers often generate their own means and apply new and efficient techniques to avoid climate constraints during serious stages of plant development. These could be heat stress in the warm subtropical regions, severe cold winter months in the cool subtropics and even seasonal frost. This kind of ecological adaptability indicates the climatic flexibility of bananas (Robinson, 1996).

1.2 Common Cultivation Practices

1.2.1 Resources of Water Supply and Fertilization

In the humid tropical zone, water comes from plenty of rainfall due to the humid climate all-year round (Robinson and Galan Sauco, 2010). In such a nature of weather, banana articulates perfectly as climate is ideal at different phases of development. However, difficulties may be evident in the humid tropics owing to the heavy and continuous rainfall which makes application of fertilizers as also harvesting difficult. In the subtropical region, irrigation is essential; hence different systems and highly efficient water-use techniques, like drip irrigation and overhead irrigation are often utilized principally for raising young plantlets. Fertilizers are usually added at the time of irrigation. This method of irrigation, i.e., fertigation is appropriate for banana plantations (Robinson and Galan Sauco, 2010). Drip irrigation systems are applied in Egypt for the sandy-soil plantations in the new lands. In recent times, the traditional surface methods of irrigation in the River Nile delta and valley are being replaced by modified systems which are suitable for such claey and loamy soil (Abul-Soad and Belatus, 2020). Micro sprinklers are widely used, i.e., in Australia. In the Ord river irrigation area, micro sprinkler method produces a cooling effect and increases the relative humidity inside the canopy (Tara Slaven, 2019). However; shortage of water, salinity as well inefficiency of the drainage systems are apparent in the subtropical area.

1.2.2 Cultivars

Banana cultivar confusion is communal problem wide-ranging (El-Kheshin et al., 2009). From a single clone of Cavendish CV 'Grand Nain' six pesudostems with heights ranging from 1.5 to 4.8 m, were produced from *in vitro* progeny (Robinson et al., 1993b), which proved the cultivar uncertainty. Lately, off-type plants have been frequently viewed in plantations (Fig. 1). Other soma clonal variations also exhibit dwarfism, apart from leaf and pesudostem inconsistency (Kishk et al., 2016). Plantlets *in vitro* are habitually used for cultivation to a considerable extent worldwide. This practice, together with the conventional methods, is implemented in the subtropics and Mediterranean countries due to the

A B

Fig. 1. Off-type plants are viewed among bananas grown *in vitro* in Egypt: (A) compact leaves, (B) deformed bunch.

cheaper cost. There are more than five hundred varieties in the world, but the Cavendish subgroup AAA is the main exported banana cultivar (Piet van and Charles, 2012). Grand Nain cultivar is superior across-the-board and is considered the standard as it has higher harvest index; lower leaf area index; excellent bunch; and fingers marketing quality. Grand Nain is well known in Egypt in the new land and in the traditional areas too. Though Williams and, of late, Ziv Williams are common, but Grand Nain is still more favorable for breeding. Dwarf varieties like Hindi and Basraii are now less cultivated in Egypt since they are more prone to winter's cold stress and virus diseases; yet, dwarf varieties are resistance to wind harm and are mostly chosen under protected cultivation.

1.2.3 Banana Life Cycle and the Plantation Life

In appropriate circumstances, the life cycle of banana plant is of a year and could be lengthened owing to cold stress that is common during the winter months in subtropical countries, but the plantation life of banana differs extremely all over the world. Plantations with four to six crop cycles are found in Egypt and other subtropical countries where replacement in case of diseases; decrease productivity; soil problems; loss of spacing arrangement; climate extremes are common. Recently growers have put into practice the plantation life as two crop cycles with the mother plant and the sucker shosen for the next raton (follower). Single-cycle banana plantations at high densities are now a viable alternative in some areas of the Canary Island to channel production into the high price season (Galan and Cabrera, 2006). Other subtropical countries implement single cycle plantations, as does Israel, to keep away from weather problems by following the selection step as well for soil maintenance purposes, tillage and aeration.

1.3 *Phenology of Banana for Optimum Growth and Productivity*

1.3.1 *Tropical Humid and Semi-arid Tropical Regions*

Tropical areas have a unvarying average monthly temperature all through the year (Calberto et al., 2015) with mean temperatures of 22–31°C all along the year and which are in the optimal range of temperatures required for banana growth. The most favored temperatures for banana growth and flower initiation are 31°C and 22°C respectively (Turner and Lahav, 1983; Robinson and Anderson, 1991). Stability during the year is almost apparent and extremes may seldom occur in the tropics. Minor deviations in temperature are noticeable in winter and summer seasons. The winter cold problems and summer heat stress are virtually absent. A few years after establishing the plantation in the humid tropics, phenological phases straighten out into unvarying patterns and soon settle in one steady stream. Harvesting is not cyclic (Robinson and Galan, 2010); hence, bunches can be found throughout the year. In the tropical areas of Costa Rica and Honduras, time taken from planting to shooting and from planting to harvesting of Musa AAA subgroup cultivars is seven months and ten months, respectively (Robinson and Galan, 2010). These are considered brief time periods compared with the banana grown in the subtropics. Foremost it is the tropical crop that performs excellently in the tropical areas; yet, some constraints are present. Undo State in Nigeria has a tropical type of climate, yet faces restrictions due to heavy rainfall during the harvesting time (Opeyemi et al., 2016) as the method of handling bunches is not easy in such a situation and has an upshot on the yield. The period of continuous heavy rainfall creates a pessimistic brunt on the soil properties, particularly if the drainage system is not well-organized. It may cause waterlogging and consequently give rise to a shallow root system. In such a condition, leaching of nutrients and harsh fungal diseases are certainly evident.

Semi-arid tropical areas, like Carnarvon, Western Australia have a distinct advantage herein as the climate is devoid of the low temperature phenomenon. However, rainfall is not steady all over the year (Robinson and Galan, 2010). In such a climate, problems of winter stress, which often occur in the subtropics, mainly in the cool subtropics, are not present here. Problems of winter stress include a significant reduction in the leaf emergence rate (LER); malformed bunches; and cock throat. Yet, periods of shortage of water often occur; accordingly irrigation is needed during the period of low rainfall.

1.3.2 Subtropical Region

Temperature has a weighty outcome on banana development and becomes a significant issue for banana growers especially in the subtropics because climate is variable and differences in temperatures between winter and summer and between night and day occur. Phenological parameters of banana growth influence temperature, which has a clear effect on the length of each fruit. Banana is grown in dissimilar regions around the world, where variations in climate are present. In South Africa, the cool subtropical country located in the Southern Hemisphere, 1,000 heat units are needed for the bunch to be harvested, i.e., bunch development could be achieved within 100 days in summer while it requires up to 200 days during winter (Robinson, 1992). Growers in the subtropics often take these parameters into consideration, adjusting the appropriate time for the most important phenological phases; vegetative stage; flower initiation; bunch development stage and harvesting. Phenology of banana in the subtropics is more complicated than in the tropics (Robinson and Galan, 2010). In Egypt, it is agreed that the accurate time for planting usually is early spring (late February to mid-April); hence the proper time for selecting the follower is August in Upper Egypt (south of Egypt) and during July in the new reclaimed lands as also in the Nile delta (north of Egypt). So, bananas can reap the advantage of the warm weather in summer and autumn months, when there is enhanced rate of leaf emergence (LER). It is generally four to five leaves per month at that time and hastens inflorescence during July and August next year.

2. Recent Recorded Climate Change Impacts on Banana

It is clear that greenhouse gas (GHG) emissions from the agricultural sector will affect plants growing in most regions in the world, and bananas are no exception. Changes are highly likely to influence the Musa productivity (Van den Bergh et al., 2012).

2.1 Temperatures Brunt on the Banana

Wherever water is not limited, the rate of banana growth and development is determined by temperature (Stover and Simmonds, 1987). It is piece of information that temperature has an evident effect on banana. It has both direct and indirect outcome on growth (Sabiiti et al., 2016), affecting any process of development (Turner, 1995). For the reason that banana has a short life cycle. Also the influence of adverse conditions, which may not be so severe, is felt on the yield. Hence, in such well-defined type of weather changes, banana could be significantly affected. For example, Egypt geographically is a subtropical country as it is located in the north of Africa and has a climate characterized by scorching summer season, though within the normal borders. Lately, Egypt possibly will be considered a potential hot spot of climate change with warming and increased frequency of extreme temperatures, perhaps faster than the global level (Mostafa et al., 2019). A study of cultivars of Cavendish subgroup banana between the period 1996–1998 revealed the highest rate of all physiological activities in the summer season. There is a highly significant positive correlation between the rate of increase in the total leaf surface area and the maximum average air temperature at that time –33–30ºC in August and September (Ibrahim et al., 1999).

In the last few decades, the maximum temperatures have increased during the hot summer months—over 38ºC, along with clear sunshine, causing sunburn to the fruits, especially the top hands. Authors have observed that slowdown in LER was obvious and which can be an indicator to plant growth. Burned spots appeared on the leaves and did not recover. The root system is unable to lift water to compensate for the excess of water gone from the leaves; hence, these symptoms are the outcome of lack of cooling. Thus, in Pakistan, where the temperature in Upper Sindh can exceed 45ºC in mid-summer months, a sprinkler system was fixed to reduce the impact of elevated temperature and the cost was invisible. Lately, the peak of heat stress ranged between 40–47°C as was apparent in Egypt and other neighboring areas. What made the situation worse that the heat waves extended more than one day. Moreover, the year 2021 witnessed the repeat of the summer season (July–August). Such a situation put pressure on banana growing in the plantations and finally the output. It could be of short term, but damaging.

The impact of this unusual long-lasting heat wave during the summer months in Egypt could be noticed in the plantations during the plant development stage. In recently grown plants, the damage is observable because the root system is not yet established or well distributed compared to the suckers and also, the leaf's surface area is still limited. Heat waves cause burned spots; in addition, the extreme temperatures can also burn all the mate, the leaves, pesudostems and the suckers. It may be useful in the case of young and recently fixed plantations to slow down the rate of leaf emergence to avoid the impact of the harm heat tension. Also, the emergence of unfolded, thin and weak leaves is a sign for the growers to reduce the fertilizer doses, particularly nitrogen, to avoid such problems. Furthermore, irrigation needs to be more frequent and in little amounts. In well-developed plantations, sufficient microclimate is able to decrease evapotranspiration and eliminate the upshot of increasing temperature. The full-size leaves on the mother and big suckers have large surface areas which cover each other and the soil under the canopy, thus protecting against thermal stress. However, the influence of elevated temperature could also be observed as burned spots on the leaves, especially those on the borders. The extreme increase in temperatures affects the rate of leaves emerging, slows down RLE, causes delay in the time of emerging the bunch (shooting) and may shorten the period from shooting to harvest (E-H). The fruit quality is not good and deformed bunches are obtained.

Winter photo-oxidation in banana is a different cause for increase in temperature. Conversely this observable fact is viewed in sunny winter days. Due to the rise in winter temperature and withholding of irrigation periods in winter months, there is reduction in plant development, increased evaporation and greater loss of water from the leaves. The plant will wilt very rapidly and as the leaf temperature increases, the upper surface of the leaf becomes pale yellow in color, which is the photo-oxidation response (Robinson and Galan, 2010). In other words, in such cold winter months, the root system is less functional and incapable of absorbing and elevating water to cool the leaves. Conversely, in the usual summer days, temperature is often more than 30°C, but the situation is different with plants being well watered, the root system more functional for absorbing water since root elongation occurs as well as the appearance of new healthy roots. The corm works as a circulation pump which lifts the water needed to the areal parts for cooling the leaves; its phloem vessels form the link between the leaves (Lahav, 1995).

The phenomenon of cock throat is one of the wider problems during the cold winter weather in the subtropics. It is common in Egypt where the bunch may come out but fails to take the downward direction as is usual (Fig. 2) in severe cases the whole bunch fails to emerge completely and normally. The phenomenon of cock throat commonly occurs with short varieties, like cultivar Hindi and this is not often seen with Giant Cavendish cultivars. It seems that slowdown of LER leads to the presence of old leaves which are less flexible. This sees a stiff peduncle which is not

Fig. 2. The problem of cock throat, where the bunch fails to take the downward direction during the duration of cold winter months in Egypt (HRI farm).

limber; hence the bunch finds difficulty in emerging normally (Turner and Trevorow, 2003). In addition, during the sunny winter days, the fingers may possibly get exposed to direct sunburn. Slows down the pushing out of leaves. It may also be due to abiotic factors, like cold winter months, stress of hot weather in summer, prolonged drought.

Hot weather problems are not only confined to the subtropics but at times also arise in the dry tropics or the humid tropics, where extremely high temperatures take place (Robinson and Galan, 2010). The problems of climate change and weather extremes are evident on banana output worldwide. In southwest Uganda, the decline in banana yield was attributed to climate variations which affected all the ecosystems, leading to extreme rainfall (floods and droughts), hailstorms and high surface air temperature (Van Asten et al., 2010; Nyombi, 2010, 2013). If this severe weather occurs during the entire duration of the significant phases of plant development, then the impact poses a threat, depending on the repetition and severity of the adverse weather (Turner, 1995). In such a case, banana is under controversial stress as growth slows down and the yield is greatly affected. In Nigeria, low productivity is expected under extremely high temperatures, particularly in Undo State which is very vulnerable (Rosenzweig and Liverman, 1992). Despite the usefulness of banana in such tropical regions as a daily item of food and income, banana production faces a crash due to the noticeable increase in temperatures. Hot weather causes dry soil, while heavy floods occur in important banana-producing and exporting regions.

2.2 *The Atmospheric Gases and Banana Productivity*

Extensive information is recorded considering the climate change issue, particularly what would happen to our planet in the coming decades and in the long term too. The three most harmful greenhouse gases are carbon dioxide (CO_2), methane (NH_4) and nitrous oxide (NO_2) which are responsible for stratospheric ozone-wearying (Thompson et al., 2019).

Carbon footprint evaluation in the banana sector from the total greenhouse gases generated on the farm worked out to 24–49% emissions from fertilizers based on nitrogen (Vallejo Chavarri et al., 2017). It seems that the high percentage of NO_2 emission is a result of the extensive use of nitrogen fertilizers as nitrogen is one of the three major elements needed for banana-plant nutrition. The other two are phosphorus and potassium and the three combined are known as NPK. This increase in applying synthetic fertilizers in agriculture since the last few decades has improved the plant growth and accelerated marketable production. However, synthetic fertilizers have adverse effects too. The most obvious challenge is the emission of wearying substances compared to the traditional organic manure. Considerable differences were found between the ecological performances of the farms where synthetic fertilizers were applied in comparison to organic manure (Roibas et al., 2015). The use of surplus nitrogen fertilizers in growing banana is common as it leads to a promising vegetative growth but not to much of a yield. Furthermore, negative effects of the soil begin to appear. Use of excess of N-fertilizers within a few seasons causes complexities in the soil (Robinson, 1996), influencing the soil's chemical properties and changing the balance between the nutrients. As a result, deficiencies of certain elements appear. The increase in concentration of certain minor nutrients, which the plant needs in low percentages, may become toxic to the plant. Hence, methods of cultivation should be such that the negative effects are avoided.

No information is available on the effect of CO_2 concentration on banana productivity (Piet van Asten and Charles, 2012). Despite banana being an important product, merely few carbons footprint analyses have been published. Also, the results have varied, depending on the methodology and data—from 324 g to 1124 kg CO_2e/kg of banana (CO_2 Living, 2019), yet banana being C3 plant may perform very well under high CO_2 concentration. It was observed that CO_2 concentration increased the photosynthetic rate and biomass accumulation of banana plant (Ravi and Mustafa, 2013) and subsequently, the increase in output could be observed. Bananas are environment responsive; it is a common practice for banana to be grown in the open field under the natural solar energy. It is easy to handle as a bunch or hands; the consumer can keep it at home at room temperature; it needs less protection during the postharvest processes as well as during transportation.

2.3 *Growing Zones*

With the global warming of 1.5°C in 2018 and according to Climate Change Agriculture and Food Security (CCAFS) database portal (Ramirez and Jarvis, 2008), the impact of this unexpected increase in temperature will affect Musa productivity in different ways according to the prevailing situation.

Tropical regions by 2020 will not be appropriate for growing banana, especially in the lowland areas but, some subtropical and highland regions will be apposite (Julian Ramirez et al., 2011). Lowland tropical regions would face heavy rains, recurrent floods causing certain soil problems, increase in the danger of fungus disease called Black Sigatoka and frequent periods of water scarcity. An additional problem is the lack of resources that are required to mitigate and overcome the crash of climate in countries like Nigeria and Uganda, as well as in the Philippines and Costa Rica which the two are of the largest countries producing and exporting bananas. Climate change may endanger progress of banana industry in the most important productive regions.

Temperatures are expected to increase in northern and southern areas and may probably encourage growth in both yield and areas suitable for Musa (Van Den Burgh et al., 2012). Cold winter pressure in such areas may perhaps not be so serious at that time, since the influence of obvious increase in temperatures could be more favorable to growing bananas. The cool subtropical countries may not be the threshold since overall the monthly mean temperature will lie below the 22°C in areas like Canary Island, Spain, Turkey and Morocco (Robinson and Galan, 2010). Conversely, there are places like Salta in Argentina, Sindh in Pakistan and Uttar Pradesh in India which by now may have extreme summer temperatures that are marginal for banana production (Calberto et al., 2015). Climate change is expected to have a downbeat effect on Indian agriculture; yet, subtropical areas located in north India will be appropriate for banana production and increased area is projected to be covered with this fruit (Ravi and Mustaffa, 2013). Not long ago in Egypt, continuing waves of excessive heat stress came to pass during the summer season, particularly in Upper Egypt may growing bananas then will be insignificant. However, if (IPCC, 2021) the world reaches or exceeds 1.5°C of warming within just the next two decades (Kelly Levin et al., 2021), then measures would have to be taken to eliminate this inconvenient situation.

2.4 *Water Availability and Plant Water Status*

Warming of 1.5°C is associated with two extremes: serious rainfall and conversely, scarcity of water (IPCC, 2018). There are regions where rainfall is expected to increase, as in East Africa highlands, or alternatively regions where rainfall is supposed to decline by 150–200 mm per year, as in some parts of the Caribbean and Central America (Piet van and Charles, 2012). However, owing to the uncertainty of rainfall in the subtropical region, irrigation is employed skillfully for achieving a profitable production and so the limitation of rainfall in the subtropics is not significant. On the other hand, excessive and continuous rainfall in the tropical regions causes lessening of banana output. Heavy rainfall together with poor drainage and waterlogging may lead to soil problems with shallow root system becoming evident. More than 1500 mm/year of rainfall is suitable for banana growth and productivity; growth limitation occurs if it is below this amount (Calberto et al., 2015).

In the tropics, like in Nigeria, the production is limited when both rainfall and temperature are very low with poor humidity (Opeyemi et al., 2016). Above all, if shortage of rainfall occurs together with increase in temperature, then a low yield is obtained. Threat of flooding was seen, in 2021, in different areas around the world—west of Africa as in Ethiopia and Sudan; the Far East, and other places in South America as well as many countries in Europe and China. The damage to agricultural lands and human resources was enormous due to the flow of rivers like Nile, which comes mainly from Ethiopia and through Sudan in the autumn months, particularly in 2021. Egypt can at present be able to deal with such a high flood as such water can be stored. Conversely, shortage of rainfall causes desertification which covers large areas in Africa.

The instability and limitation of water supply has its adverse effects on banana's phenological phases of growth, like delay in shooting; lengthening of the duration of E-H period; expansion of the life cycle. The plant wilts temporarily during the sweltering day hours and the two halves of the leaf blade fold downward. The folding of the leaf blade reduces the amount of water required for cooling by more than half, reducing the leaf temperature by 6–8°C (Turner and Trevorrow, 2003). Protracted period of drought causes slowdown in LER and leaves become compact, acquiring the rosette shape.

Bananas show a rapid physiological response to soil water deficit (Robinson, 1996) through squashy makeup of gigantic leaves and a pesudostem. In the case of shortage of water, leaves soon wilt and the plant slowdown pushes the leaves. The most sensitive indicator of soil water deficit in banana is the LER (Kallarackal et al., 1990; Hoffmann and Turner, 1993; Turner and Thomas, 1998) and the bunch also may be affected. The yield declines due to the increase in duration of the vegetative phase caused by low soil moisture (Piet van and Charles, 2012). In the case of prolonged drought, temperature increases on the surface of the leaf to a danger point and the leaves get sunburnt (Turner and Lahve, 1983).

The stomata open out in the epidermis bound by two specialised epidermal cells called the guard cells, which, by changing shape, bring about the opening and closing of the aperture (Esao, 1960). In the Williams variety, a lesser stomata number was found on the upper leaf surface (adaxial with 30.60 stomata/mm^2), compared with the lower leaf surface stomata number (abaxial with 100 stomata/mm^2) (Belatus, 1998). Such stomata differences may lead to less transpiration and more tolerance to water deficiency. Hence, the Williams variety as also the Grand Nain variety is growing and producing a valuable production in the newly reclaimed lands, northwest of Egypt. Plants with low stomata count, like soybean, are more resistant to water stress (Buttery et al., 1993). As the soil starts to dry, the stomata close and leaves remain hydrated through root pressure (Thomas and Turner, 2001). Closure of stomata during such a period of drought is a response to the lack of water, indicating the resilience of banana to excessive evaporation. When the evaporation stops, the root system may be obligated to lift up the water to keep the leaf cool. This could occur through the efficient roots' osmotic pressure (O.P.) Banana and certain palms have pressure of a kind that could force water up to the aerial parts of the plant. Therefore, as in large monocotyledons,

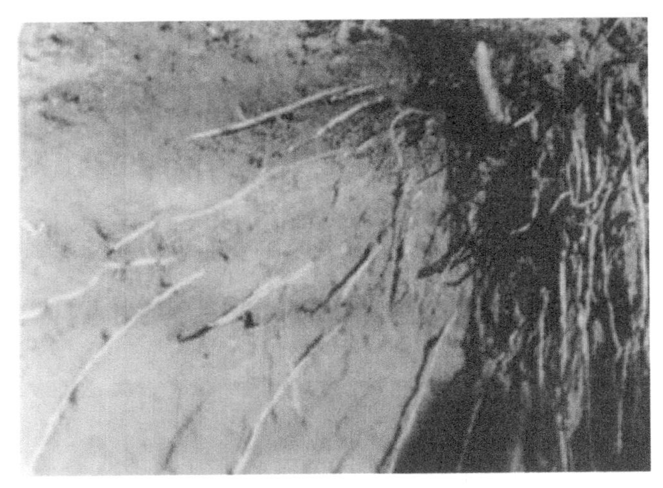

Fig. 3. In the left side, the roots are shown imbedded in the soil; new white roots are distributed in all directions—vertical, oblique and horizontal and are of uniform diameter along their length. In the right side, soil has been excavated out (HRI-El-Kanater farm).

root pressure may play a major role in the supply of water to the aerial parts (Davis, 1961). Banana roots (Fig. 3) are fleshy, supple, cylindrical, rope-like and 5–10 mm in diameter. They are yellowish white when young and become brownish corky when old.

Furthermore, the three types of root segments—the apical new white root segment; the intermediate yellowish root segment; and the old brownish basal root segment (Fig. 4) play an equal role in the supply of water to the aerial parts called as 'the hall root' (Belatus, 2018). What is more imperative is that the root system of banana and plantain is extensively distributed with richly branching superficial roots giving rise to an intense carpet, archetypical of monocotyledons (Wardlow, 1961) and indicative of the potential of banana roots during the time of prolonged drought.

2.5 *Pests and Diseases Outbreak*

The projected pest-and-disease episodes may be prevalent as a warning of the climate change. Climate change has raised the risk of fungal diseases that devastate the banana crop (Piet van and Charles, 2012). The remarkable increase in temperatures has its impact on pathogenic organisms. Pests and harmful microorganisms, which cause diseases, are affected by the recent change in climate. Increasing temperatures may possibly influence these pathogenic organisms to disappear in certain regions, but the worse is the increase in their population, causing severe problems in other areas. The leaf spot disease as well cucumber mosaic virus disease were not found in India earlier, but recently have begun to appear due to the fluctuations in climate (Ravi and Mustafa, 2013). The appearance of diseases which was not seen

Fig. 4. Banana main roots are shown growing directly from the corm and possessing abundance of hairy roots along their length. The main root tips are soft and easily broken. Behind this region, branch roots, smaller in diameter, are found in a cluster.

before in certain areas makes the situation more serious. Black Sigatoka is caused by fungus, *M. fijiensis*, whose life cycle is strongly determined by the weather and microclimate (Bebber, 2019). Black Sigatoka is the most dreadful fungus disease to invade the banana plantations in humid tropical regions, where the climate is suitable for the pathogen to grow and multiply in population. The discrepancy in the climate is thought to be the reason for the invasion by Sigatoka in new lands, as in Latin America and the Caribbean. The disease now occurs not very far from Florida, drier conditions in some parts of Mexico and Central America have reduced infection risk (Bebber, 2019). But in such condition shortage of the water required for optimum growth and productivity may cause reduction in the banana yield.

Banana aphid *Petalonia Nagronervosa*, the vector of banana bunchy top virus disease (BBTV), during the hot summer months increases in number and invades the plantations in Egypt. The insect could be found inside the rolled unfolded leaves during folding. In this case, the entire mate should be removed as the bunch top virus will reach the mother, followers and the growing suckers. Growers add special pesticides regularly during the hot summer months, starting from May to September. In the recent few years, there has been an atypical rise in the temperature with the aphid increasing in population and increase in the phenomenon of cock throat caused by punchy top virus disease can be expected (Abul-Soad and Belatus, 2020). In order to tell between the reason (biotic or abiotic) for cock throat in banana, will via the symptoms lest BBTV disease are observed, like the new leaves appearing erected and thinner; the new leaves being smaller than the older; the leaves having dark blue strings perpendicular to the main leaf vines as well the pesudostem; the plant emitting the cucumber smell; and the presence of the vector aphid, *Pentalonia Nagronervosa*, on the leaves. One more problem related to banana growth in the subtropics is the occurrence of the fungus disease, fusarium wilt caused by *Fuzarium oxysporum*, which enter the plant through the roots. The infected soil becomes biologically diseased soil, which is opposed to the growth of bananas. While the increase in temperature may reduce the incidence of milder strains of fusarium, the tropical Race 4 fusarium may become even more violent than those currently existing in the subtropics (Bebber, 2019).

2.6 *Socio-economic Impacts*

Climate change is recognised as a critical threat to land productivity in developing countries (World Bank, 2020a). Recent changes in climate occur radically in certain banana-growing regions of the world as bananas constitute the staple food in the poorest parts of the world. These days, extreme climate-generated incidents, such as droughts and insect-pest infestations harm small farms of millions of families in the developing countries (Nthambi et al., 2021). An uncommon increase in temperature can be expected to create more frequency and severity of extreme climate which influences the banana-growing areas. The prospect of banana plant to yield fruits continuously throughout the year in the humid tropics would be a reason to consider banana an essential food crop and source of income for small farmers in tropical countries, such as Uganda (Wairegi et al., 2010; Nyombi, 2013), Nigeria (Rosenzweig and Liverman, 1992) and the Philippines (Andria, 2012). The highland banana areas of Uganda and the great lakes of central Africa are highly vulnerable as the people here depend on bananas for food security (Piet van and Sharles, 2012).

Growers of inadequate banana plantations will have to endure the approaching warm climate (Allan Brown et al., 2020). Above all, these are the regions where rainfall is the essential source of water (Piet van and Charles, 2012) because lack of adequate income prevents availability of suitable irrigation systems; cultivation practices which counter the current situation; and unusual techniques which are needed to control the prevailing pests and diseases. They may possibly be forced to shift to other locations which are appropriate for growing bananas. In sub-Saharan Africa (SSA), approximately 80% of poor people continue to depend on the agriculture sector for their livelihood (FAO, 2014). In the subtropics, like Egypt, small owners of banana plantations in the rural areas often resort to extra jobs beside cultivation to increase their income. Besides the agri-food industry, private sectors with aid from the government are seen to provide jobs. In the Caribbean, a major threat is posed by the increased hurricane activity (Valerie Nelson et al., 2010). Philippines in Asia is at risk too since it is one of the most exposed countries to climate change distress, like storms and heavy rains which adversely affect banana production. About 5 million Filipino farmers depend on banana as a source of food, livelihood, income and export (Andrea, 2012).

Migration of banana production areas could highly be projected in some significant cases (Julian Ramirez et al., 2011) owing to the pressure of climate change which causes lasting changes in agriculture lands, like desertification, floods that change geographic locations, insects and diseases which may lead to endemics. Inadequate technology and exceptionally low yield due to agro-ecological features are seen (FAO, 2014). It is observed that the impact of weather change on banana is evident in countries having low economic constitution as they depend on a few and uncertain natural resources (Vallejo Chavarri et al., 2017).

3. Cultivation Management Strategies of Banana under Climate Change

3.1 Natural Adoption

Adoption approaches can be specified as actions needed to be carried out to enhance the grower's dynamic outcomes to climate change and (Li et al., 2021) and selecting proper sites for cultivation in order to bounce back to profitable production. The growers avoid establishing plantations in spots subject to banana pathogens, or where sufficient supply of water is not found, or where recent temperature data are not available and strong winds do not prevail. Above all, the other points to consider are the soil properties, physical, chemical and biological assets (Robinson and Galan, 2010); soil nature and drainage system which are significant for well distribution of the root system essential for banana growth and yield; soil salinity and acidity (pH); and soil which is free from soil-borne diseases.

Adaptation to weather change is through suitable adoption of common resources; proper use of water; and promotion of environmentally suitable practices (Vallejo Chavarri et al., 2017). Drip irrigation systems are widely used in banana plantations in the subtropics, where supply of water is essential. This technique of irrigation promotes sharing of water and fertilizers. Use of drip irrigation and sprinklers instead of surface irrigation methods improve water use efficiency (Valerie Nelson et al., 2010). Genetic improvement promotes (Van den Bergh et al., 2012; Julian Ramirez, 2011) new varieties which are tolerant or resist drought; make efficient usage of water; counter salinity, avoid cold stress; confront elevated temperatures and prevailing pests and diseases. *In vitro* propagation and genetic upgrading are favored. Each variety is grown for commercial production in certain areas which are acclimatized to a certain environment. Adjusting the cultivar in each area can help maintain the current production levels (Calberto et al., 2015; Bernard Kilian et al.,

2012). Williams and Grand Nain cultivars are grown successfully in Egypt in the newly reclaimed areas as they appear tolerant to such conditions. However, of late, growers have begun to switch to cultivar Ziv Williams which is considered more appropriate to such weather and more stable against wind. It was observed that high doses of nitrogen and potassium fertilizers increase the yield; improve the quality; and shorten the life cycle of Ziv cultivar (Hosny et al., 2020) and can be substituted for other cultivars if needed. Furthermore, it is more productive, more forbearing to climate constraints and more tolerant to common pests and diseases.

Growers have relocated to another area if the original region has not been found appropriate for cultivating banana as a result of the climate change. This has helped in obtaining a positive yield (Van den Burgh et al., 2012; Bernard Kilian et al., 2012). However, the warm climate in cool subtropical countries, like Turkey, Spain, Canary Island, South Africa, encourages cultivation of banana in open-air plantations. Hence the main production here is under protected cultivation. The situation may be different in semi-dry subtropical countries for relocation from such cultivated areas. Egypt, which geographically extends from the Mediterranean in the north to Tropic of Cancer in the south, experiences extremely hot but less humid summer months, may be probable to migrate growing banana locations to the north. Favors growth of banana crop in the higher altitudinal border in the highland tropics will expand because of increasing temperature. Conversely in the lowland tropics, the long periods of temperatures above 30°C lead to decline in productivity (Piet van and Charles, 2012) and so could necessitate transition of cultivation areas from the lowland tropics to the highland tropics where constructive temperatures prevail. Dole Banana Producing Company is considered one of the largest banana-producing enterprises in the world. As temperatures increase, the Dole Company is obligated to relocate cultivation from certain banana-producing areas, which become too hot for commercial production, to new regions that provide suitable climate (www.dole.com).

3.2 Physical Strategies

Spreading the practice of banana cultivation under net shades to reduce the temperature; diffuse the light intensity; cold wind blows protection in winter and preventing frost to accumulate on the leaves are some of the physical strategies that are adopted. In the cool subtropics, like Turkey, frost damage can be seen from time to time in open-field plantations; however, it is rarely seen under protected cultivation (Gubbuk et al., 2018). Growing bananas under protected covers is used in dissimilar countries in the subtropics, with dissimilar materials, colors and sizes. It is applicable to countries with cold winter, as in some Mediterranean regions (Robinson, 1996) as well as in Canary Island where there is severe cold climate (Galan

Fig. 5. Cold stress impacts banana during mid-winter in the north of Egypt when the plants are at adult pre-floral stage. The mates get burnt entirely as the temperature falls below 4°C for a few days.

et al., 1992) and in hot summer countries, like in Egypt (Refaie et al., 2012; El Shahed, 2017). Protected cultivation shows encouraging results—bunches are extra-large; rate of emergence of leaves is faster; there is reduce crop round; less evapotranspiration and high water use efficiency (WUE). Banana covering can help in protecting bananas from severe cold in subtropical regions as well as in countries with high temperatures. Cold stress impacts banana in mid-winter in the north of Egypt when banana leaves get burned including the pesudostem and suckers (Fig. 5).

In Egypt, temperatures under the net house of 6m height decrease by about 2–3°C less than the outside ambient conditions (El Shahed, 2017). It is believed that this temperature decrease which is associated with low sun radiation and protection from pests, has a positive impact on the overall banana fruit productivity and quality. Protected cultivation help to acclimatize the *ex vitro* banana for two months (Fig. 6). These acclimatized plantlets could be fixed to the plantation directly since they are juvenile and can grow perfectly (Abul-Soad and Belatus, 2020).

Fig. 6. Using protected cultivation to acclimatize the *ex-vitro* banana plantlets.

However, in the scorching summer regions, growing banana under greenhouses with polyethylene covers may have an outcome on the plants. Special aeration control is needed to protect the plants from sunburn effect (Gubbuk et al., 2018). Growers usually adopt their personal techniques to avoid the adverse upshot of polyethylene covers during the hot months. They often use two distinct kinds of material for covering; one of them is polyethylene which always is lifting during the hot months and the other cover commonly used is transparent shade screens which allow aeration and lessen light, i.e., protect the plants additionally when the hot hours decrease. Other methods are used for cooling during the period of hot months, like overhead sprinkler systems. Painting a solution of water and calcium carbonate on the covers during summer (Gubbuk et al., 2018) helps. Shading powder is applied to the covers (Gubbuk et al., 2018). Conditioned greenhouses, where temperature, light and humidity are monitored to overcome the hot temperature could be used effectively for growing bananas although at high initial costs.

Under greenhouse cultivation, leaves are sheltered from the effect of winds, dust, etc.; hence, too much healthy leaves are produced. In order to avoid excessive shade, it is necessary to remove the old leaves after the bunch has emerged, keeping hold of the youngest eight leaves (Robinson and Galan, 2010). Undue shade for the period of bunch development may perhaps lengthen the period to collect the accumulated heat units needed for bunch development and consequently, lengthen the E-H period. Tunnels, almost two-meter in height are used for raising the acclimatized plantlets for six months during the stage of *ex vitro* acclimatization. Low tunnels are also practiced for growing young plantlets *in vitro* (Abul-Soad and Belatus, 2020).

Wind breaks are often fixed before establishing the plantation, to be all set for protecting the recently growing plantlets. They are often planted against the prevailing wind, to eliminate wind devastation. Wind breaks are essential, especially in the spots that not sheltered by any natural protection. In Turkey, the northern borders of plantations are mountainous, which provides protection against wind and harm by frost (Gubbuk et al., 2018). Another point to be considered is that if there are fruit orchards near the plantains, the former could perhaps provide the kind of protection needed for the plantation. Wind breaks are extensively used as shelters to banana plantations in the newly reclaimed regions of Egypt. When the prevailing winds are serious, certain shortcomings may need to be tackled like the shade or wind breaks provided, that enough nutrient and water are available and that the plants have adequate plantation space (Eckstein, 1994). What in more, these shelter plants often have woody, deep-rooted and widely spreading root systems, which may intermingle with banana's flashy and lithe roots. Hence growers in Egypt in order to avoid these shortcomings often excavate a trench between the wind-break trees and the plantation to prevent the interaction between the roots.

Bunch covers with polyethylene material are extensively used particularly in the subtropical regions that have cold winter months (Robinson, 1996) like Egypt and other neighbors in the area. In Egypt, bunch covers with blue polyethylene carrying 1–2 mm pores are used as they are considered one of the essentials in the plantation to increase the yield by 1–2%. Bunch covers have physiological benefits as they improve the microclimate as well as have physical effects as they reduce rasping against dust and leaves and help in the production of a larger fruit (Robinson and Galan, 2010). They fasten the fruit development by 10–15 days and improve the fingers' quality while protecting the bunch during harvest and transportation. In hot weather, it may be more practical to tie one leaf or two over the bunch save for, using Kraft paper (Abul-Soad and Belatus, 2020). Care is needed to avoid sunburn and it may be helpful to provide protective covering by placing paper between the fruit and the cover (Turner and Trevorrow, 2003) and also between the hands, Growers often adopt diverse methods based on their skill and available resources.

3.3 *Cropping System and Cultivation Efficiency*

Agriculture is more concerned with encouraging ecological effects on cropping patterns than other cropping methods as the prime aim is to obtain the maximum yield (Humberto, 2010). To obtain a high yield, often growers add an overdose of nutrients, especially nitrogen which pollutes the groundwater and its release. Nitrous gases like NO_2 as well as the use of a lot of pesticides and fungicides often lead to decrease in soil productivity. For optimum cropping systems, fertilizers in banana plantations should be added in low doses and more frequently (Vallejo Chavarri et al., 2017). Adding surplus fertilizers causes toxicity to banana crop; antagonism with other nutrients; pollutes the soil and the environment; and is expensive (Abul-Soad and Belatus, 2020). In Costa Rica, shifting to other cultivation methods can lead to a high marketable yield together with reduced reliance on agrochemicals, which are thought to be necessary for banana exportation (Bellamy, 2012).

Influence of the climate variations on banana is apparent in some regions as well as countries, though in other areas, this might not be so noticeable. This could be attributed to the dissimilarity in the cultivation efficiency. Use of constructive technology in banana production may prevail over less significant and unhelpful climate changes (Varma and Bebber, 2019; Ravi and Mustaffa, 2013). In Egypt, farmers adopt practical methods as well as unusual utensils; proper time for planting; ideal time for harvesting and using shielding covers. Drip irrigation with fertilization are excessively resorted to in almost in all the newly reclaimed regions where 43% of total banana production is witnessed. Use of apposite post-harvest techniques improves the fruit's marketable quality (Abul-Soad and Belatus, 2020).

Harvesting and transportation methods under controlled temperature and humidity are essential to avoid spoiling of the bunch quality (Ogbonnaya et al., 2011). Since the loss of the yield during post-harvest and handling of the fruit needs to be eliminated, techniques are adopted to keep the fruit cool during all the processing stages, starting from pre-cooling in the farm which is one of the many. The pressure of climate change has made the tropical region increasingly prone to plant infection and damaging disease of this major tropical food plant—the Black Sigatoka disease of banana (Bebber, 2019). Natural adoption of cropping practices like reduced use of synthetics; cultivar selection to ensure that it is able to resist disease or is tolerant; crop managing; migration to adjoining regions with conditions not suitable for the pathogen may possibly help to overcome this dangerous pathogen.

3.4 *Agronomic Strategies*

Currently techniques of cultivation need to be more concerned with changes in climate which influence banana. Tillage is one of these techniques which has been widely used to plough the soil and which has many advantages, though the essential fact is that tillage might cause aeration of the topsoil by wind, particularly in locations which are exposed to strong prevailing winds. The disadvantage however is that tillage may injure the banana roots, especially if done during critical periods of development. Extensive information about the advantages has led to a shift towards reduced tillage in recent cropping systems. Hence, for weed control, it is better to eliminate the weeds by hand, if banana plant is still less than 1 m in height, but subsequently herbicides could be used if needed. Now the banana canopy will reduce the growth of weeds (Robinson, 1996; Abul Soad and Belatus, 2020).

Maintaining the soil moisture is advantageous and is needed but will encourage weeds to grow and spread. Mulching with cover crops can maintain the moisture as well as prevent the growth of weeds, only if they are not competing with banana (Vallejo Chavarri et al., 2017). Managing crop residues is important in most systems. Some of the nutrients contained in these dead tissues are made available to crops during decomposition, thus reducing the need for fertilizer inputs (Brennan and Smith, 2005). It is of value to allow the banana residues to remain on the soil surface of the plantation during different cultivation practices, like the removal of old leaves and surplus suckers, etc. which comprise large quantities of organic matter. For each ton of fresh banana fruits that are harvested, one ton of dry matter is added to the soil as trash (Robinson and Galan, 2010).

Carbon in organic matter is the main source of energy for the activities of soil microorganisms; nitrifying bacteria; cellulose-dissolving fungi; soil warming, etc. The most probable advantage of organic farming lies in carbon removal from soil (El-Hage and Muller, 2010), leading to elimination of banana carbon footprint. Organic fertilizers are an additional source of nutrients besides accelerating useful microbes; improving physical properties of the soil; holding nutrients; and providing buffering effect.

Shared challenges could be found in the desert-land reclamation in Egypt in the form of water scarcity; high water salinity; excessive cost of chemical fertilizers; low level of microbial biodiversity and organic residue. Recent decades have seen increase in use of effective microorganisms (EM) in agriculture. EM technology is one of the many used for increased cultivation worldwide. A combination of useful natural microorganisms existing everywhere (Higa, 1994) could be fermented with organic matter, and this fermented product is called *bokashi*. It is a Japanese word widely used as an organic alteration on banana in Latin America (Robinson and Galan, 2010) as well in several countries of Africa and Asia. The Egyptian farmer with a well-established tradition in farming is readily accepting this recent technology without hazard, complexity and large investment. This is being done also through research work (Habashy, 2016; El-Desoky et al., 1999) and technology transfer training; regular contacts with agriculture engineers and the other technical staff who collaborate directly with the farmers.

3.5 *Reduce Banana Nitrogen, Water and Carbon Footprint*

Methods of cultivation are managed to avoid the negative effects by giving attention to maintenance of resources and soil conservation; efficient use of available water; well-organized cultivation practices as suggested by experts. In this concern, improved nitrogen use efficiency (NUE) is needed (Davidson et al., 2015). A balance should be maintained between obtaining profitable production of such daily foodstuff crop for millions of the poorest in the world and the downbeat effects on the environment because of NO_2 release. In other words, balance between N input and N removal through harvest should be the concern (Thompson et al., 2019). This will require improved and developed methods of cultivating banana to reduce the enhanced release of NO_2 and take steps toward decrease of such emissions which depend on the ability and efforts of concerned countries. Europe and North America have succeeded in controlling the increase in nitrous oxide emission. Regions in East Asia and South America have made the largest contribution to the global increase (Thompson et al., 2019). Transfer of knowledge and experiences from one location to another with less knowhow is needed.

Ecuador is a major banana producer and exporter; it is concerned with reducing the banana carbon and water footprint in the country. What is more imperative is that they have adjusted developed procedures in calculated carbon footprint of Ecuadorian bananas and this can be useful in potential situations with similar characteristics (Roibás et al., 2015). This will be of great significance for the growers exposed to similar conditions requiring reduction in carbon, adoption of NUE in their plantations and putting this into practice through a banana cost chain by means of adjusted and confirmed methods and tools (Vallejo Chavarri et al., 2017). Elimination of banana water footprint can be brought about through resourceful water use and prevention of pollution of water as also groundwater. The main reasons for water pollution are agricultural chemicals including pesticides, synthetic fertilizers; waste including effluent from livestock operations; industrial and household effluents and acid rain (Higa, 1994).

3.6 *Awareness*

Adjustment to climate change may be through consciousness to promote the sustainable use of natural resources, protection and upkeep of soil, adoption of efficient agronomic practices, water conservation and well-organized and effective water management, creating methods and tools for saving water during heavy rainfall for use in subsequent periods of shortage of water. Additionally, sponsor efficient irrigation techniques and reduce use of synthetic fertilizers which pollute the soil and the groundwater. Yet, to mitigate the adverse effects, rely on the capability and facilities of the country, the methods to reap advantage from the situation, increase in areas appropriate for producing banana and the ways to counter difficult situations. The ability to derive benefit depends on how countries put efforts at preserving and developing their cultivation effectiveness (Varma and Bibber, 2019). Growers in the subtropical countries like Egypt, which produces commercial banana, are adopting certain techniques to avoid the effects of summer heat and cold winter stress by maintaining long records on how to keep away from the unhelpful seasonal climate conditions.

Growers' alertness and training in becoming more involved in facing the severe and unexpected clime are the most important dynamics in banana production system. Their previous long cultivation practices and historical experiences in growing banana makes them well versed, educated and conscious of the intense pressures imposed by changes in the weather. It is essential for the farmers to be up to date through knowledge transfer and adoption of best practices. It is of major value to exchange experiences and tools used by growers located at sites with a similar environment. Commercial plantations with large-scale production, as in the new lands northwest of Egypt, are considered as ideal plantations to transfer their experiences on to the small and limited ones in the area.

In the subcontinent of India, owing to the influence of weather changes on banana production, farmers are being informed and trained to use new technologies as well as adopt shifting farming and cropping practices (Ravi and Mustafa, 2013). In the Davao region of the Philippines, the farmers are being prepared and educated to be responsive and ready for the outcome of warm climate difficulties and threats to their banana production (Andrea, 2012). Moreover, the Dol Food Company has resorted to reforestation and sponsoring climate change awareness in its banana-growing regions (www.dole.com).

It seems that alterations in the environment have a clear effect on banana production worldwide, posing a serious challenge to banana producers. The adverse changes in climate have become true and are expected to increase militantly because of man's irresponsible use of the natural resources of the earth. The earth and the environment is suffering because of the methods and tools applied since the past many decades which have adversely affected the earth (Higa, 1994).

4. Conclusion and Prospects

Banana is cultivated in dissimilar regions around the world, where variations in climate are seen. The tropical crop performs excellently in the tropical area with some constraints. Tropical areas have unvarying average monthly temperature all through the year. Banana is grown to yield a marketable production in the subtropical climate. A few years after establishing the plantation in the humid tropics, phenological phases straighten out into a fixed pattern. But due to the climate change, phenological parameters become a significant issue for banana growers, especially in the subtropics. Hence, adjusting to the appropriate time for the most important phenological phases, vegetative stage, flower initiation, bunch development stage and harvesting becomes important.

The outcome of the atmospheric variation on banana is clear in countries which do not preserve their natural resources and thereby face difficulties. Owners of limited banana plantations get affected by the incoming warm weather, particularly those depending on rainfall and who lack technology, suffer from shortage of adequate funds needed to acquire suitable implements to tackle the current situation and are unable to use the techniques needed to control the prevailing pests and diseases which may become aggressive.

Adjusting the cultivar growing in each area can help maintain the current production levels. Substitution of other cultivars, if needed, can be done if they are more productive, more forbearing to climate constraints and tolerant to common pests and diseases. If certain regions are not found conducive for cultivating banana because of the pressures of the changes in the environment, then it may be helpful to translocate to another area where a positive situation is found.

The implementations of technologies which assist in soil and water preservation are aspects needed to confront climate variations. Growing bananas under protected covers is practiced in different countries of the subtropics by using dissimilar materials, colors and sizes. Wind breaks are often initiated to eliminate wind devastation. Bunch covers are extensively used, particularly in the subtropical regions. It is important to manage the banana residuals on the soil surface of the plantation. Carbon in organic matter is the main source of energy for the soil where microorganisms are active like the nitrifying bacteria, cellulose-dissolving fungi, earthworms, etc. Drip irrigation is widely practiced and plantlets from *in vitro* fertilization are habitually used for cultivation to a considerable extent, together with the conventional methods which are cheaper.

Using excess of nitrogen in growing banana is at times a widespread practice, regardless of the promising vegetative growth but not so much of a yield. Negative effects on the soil's chemical properties are not taken into consideration. It is also associated with a number of environmental problems, among which is the rise in NO_2 emissions. A balance should be maintained in the production of such a daily foodstuff crop meant for millions of the poorest in the world and the downbeat effects to the environment. Improved and efficient use of nitrogen (NUE) is needed.

Low carbon strategies in the banana value chain should be kept in mind and so should the elimination of banana water footprint by means of a standardized and proven methodology and tools. Adaptation to climate change through promotion of sustainable use of natural resources; soil protection and upkeep; water conservation in a well-organized manner; less use of synthetic fertilizers which pollute the soil as well the groundwater; sharing knowledge and know-how between growers in locations with different circumstances.

Trying to surmount the adverse effects of the warm weather changes through limited use of synthetic fertilizers in banana plantations to conserve soil and water; cultivation of banana under protected greenhouses; efficient use of new tools to counter the outburst of pests and diseases; loss of the yield during post-harvest and handling of the fruit; using techniques to keep the fruit cool during all the processing stages, starting from pre-cooling on the farm, are some points to keep in mind. In the subtropics, the date of planting as well as the time of planting the following crop may need to be modified to cope with the recent changes in weather.

Innovative research together with the history and present experiences may be of significance to face the current extraordinary increase in temperatures; the phenology of banana mainly in the subtropics; organic fertilizers and use of biotic manure; decrease in NO_2 emission; reduced release of CO_2 from the banana chain production; protected cultivation; efficient water use; improved post-harvest methods to decrease the loss of the yield during different processes; transfer of experiences and adoption of new technologies.

Genetic improvement can be anticipated to make an input to adoption of new varieties which are not less than, tolerant to drought; water efficient usage; salinity; cold stress; crash of high temperatures; and prevailing pests and diseases. These might be the central target for *in vitro* propagation and genetic upgrading. Restricted roles should be followed to eliminate the regular appearance of off-type plants from *in vitro* propagation as it may lead to more cultivar confusion.

References

Abul-Soad, A.A. and Belatus, E.L. (2020). *Cultivation and Production of Banana in New Lands.* www.dkhbooks.com. 94 Abbas Akkad Street – Nasser City – Cairo – Egypt, pp. 180.

Allan Brown, Sebastien C. Charpentier and Rony Swennen. (31 March 2020). Chapter first online: *Genomic Designing of Climate-smart Fruit Crops.* https://link.springer.com/book/10.1007/978-3-319-97946.

Andrea, M. Ortiz. (2012). *Climate Change Impacts on Banana Production in the Davao Region of the Philippines.* Thesis for MSc, MAg, Advisor: John R. Porter, Willibald Lois Kandi.

Bebber, P.D. (2019). Banana disease posted by climate change. *American Association for Advanced Sciences (AAAS).* https:\\www.eurekalert.org\login.

Bebber, P.D. (2019). Climate change effects on black sigatoka disease of banana. *Philosophical Transactions of the Royal Society B.* Doi: 10.1098/rstb.2018.0269.

Belatus, E.L. (1998). *Investigating the Root Dynamics of Banana and Their Relation to Shoot Growth.* Ph.D. thesis, Faculty of Agriculture, Cairo University.

Belatus, E.L. (2018). Morphological features, cation exchange capacity and osmotic pressure for different banana root segments. Review Article, *Middle East Journal of Agriculture Research,* 7(1): Jan.–Mar. 2018, pp. 21–26, CC-CC.

Bellamy Sanderson, A. (2012). Banana production systems: identification of alternative systems for more sustainable production. *AMBIO, A Journal of the Human Environment,* 42(3). Doi: 10.1007/s13280-012-0341-y.

Brennan, Eric and Smith, Richard. (2005). Winter cover crop growth and weed suppression on the central coast of California. *Weed Technology,* 19(4): 1017–1024. Doi:10.1614/WT-04-246R1.1–via NALDC.

Buttery, B.R., Tan, C.S., Bussell R.I., Gaynor, J.D. and MacTavish, V.C. (1993). Stomatal numbers of soybean and response to water stress. *Plant and Soil,* 149: 283–288.

Calberto, G., Staver, C. and Siles, P. (2015). An assessment of global baanana production and suitability under climate change scenarion. pp. 264–291. *In:* Asis Elbehri (ed.). *Climate Change and Food Systems: Global Assessments and Implications for Food Security and Trade. Food Agriculture Organization of the United Nations* (FAO), Rome.

CCAFS: *Climate Change Agriculture and Food Security.* CCAFS Research Program on Climate Change, Agriculture and Food Security. CCAFS is a collaboration among CGIAR Centers and Reserch Programs. https://ccafs.cgiar.org/.

Davidson, E.A., Suddick, E.C., Rice, C.W. and Prokop, L.S. (2015). More food, low pollution (Mo Fo Lo Po): A grand challenge for the 21st century. *Journal of Environment Quality,* 44: 305–7.

Davis, T. (1961). High root pressure in palms. *Nature,* 192: 277–278.

Dole. (2016). *Focus on Dole.* www.dole.com, posted 4 Nov. 2016.

Eckstein, K. (1994). *Physiological Responses of Banana (Musa AAA; Cavendish Subgroup) in the Subtropics*. PhD Thesis, *Institute fur Obstbau und Gemusebau, Universitas* Bonn, Germany, 203 pp.

El-Desoky, I.M., Riad, T.M. and Belatus, E.L. (1999). Influence of Effective Microorganisms on Papaya in Egypt, *Sixth International Conference on Kyusei Natural Farming*, Pretoria, South Africa, pp. 213–217.

El-Hage, Scialabba, N. and Muller-Lindelöf, M. (2010). *Organic Agriculture and Climate Change*, United Nations (FAO), *Viale delle Terme di Caracalla*, 00153 Rome, Italie.

El-Kheshin, D.A., Belatus, E.L., Hamid, A.A. and Radwan, K.H. (2009). Molecular characterization of banana cultivars (Musa Spp.) from Egypt using AFLP. *Research Journal of Agriculture and Biological Sciences*, 5(3): 272–279.

El Shahed. (2017). *Microclimatic and Productivity of Screen Banana Crop in Egypt*, Ph.D. Thesis. Genetic Engineering and Biotechnology Research Institute: Sadat City University.

Esao, K. (1960). *Anatomy of Seed Plants*. New York, John Wiley & Sons.

FAO. (2014). *Adapting to Climate Change through Land and Water Management in Eastern Africa: Results of Pilot Project in Ethiopia, Kenya, and Tanzania*. Food and Agriculture Organization of the United Nations, Rome, Fingerprint.

Galan Sauco, V., Cabrea Cabrea, J. and Hernandez Delgado, P.M. (1992). Phenological and production differences between greenhouses and open-air bananas (Musa AAA, cv. Dwarf Cavendish) in Canary Islands. *Acta Horticulture*, 296: 97–112.

Galan Sauco, V. and Cabrea Cabrea, J. (2006). *El cultivo del platano (banano, Musa acumitata Colla AAA, subgrupo Cavendish) en las Islas Canarias*. pp. 289–301. *In*: Sporano, E., Tcacenco, F.A., Lechtemberg, L.A. and Silva, M.C. (eds.). *Banana: A Sustainable Business, Proceedings XVII ACROBAT International Meeting*. Joinville, Brazil, 15–20 October 2006, ACROBAT/ACAFRUTA, Joinville Brazil.

Gubbuk, H., Altınkaya, L. and Balkıç, R. (2018). Banana: A very Profitable Tropical Crop for Turkey, XXX International Horticultural Congress. www.ihc2018.org.

Habashy, S.E. (2016). Effect of reducing mineral nitrogen fertilizer in williams banana orchards by using organic and bio-fertilization on growth and yield. *J. Biol. Chem. Environ. Sci.*, 11(1): 447–462.

Higa, T. (1994). *Earth Saving Revolution*, Sun Mark Publication Inc., Japan; English translation Anja Kanal, (1996). ISBN4 -7631-8 C0030.

Hoffmann, H.P. and Turner, D.W. (1993). Soil water deficits reduce the elongation rate of emerging banana leaves, but the night/day elongation ratio remains unchanged. *Sci. Hort.*, 54: 1–12.

Hosny, S. Samia, Mahdy, H.A. and El-Kholy, M.F. (2020). Nitrogen and potassium requirements (fertigation) of ZIV cv. Banana on growth, yield and fruit quality in sandy soil. *Future J. Agric.*, 2: 32–42 DOI: 10.37229/fsa.fja.2020.04.19.

Humberto, B. (2010). Principles of Soil Conservation and Management, *Springer Science*, pp. 167–193. ISBN 978-9048185290.

Ibrahim, F.A., Elezaby, A.A. and Belatus, E.L. (1999). An endoscopy examination of banana root Dynamics and their relation to shoot growth. Recent Technologies in Agriculture, *Proceedings of the 1st Congress*, Cairo University, Faculty of Agriculture, vol. 1.

IPCC. (2018). *The Special Report on Global Warming of 1.5°C*. https:\\www.ipcc.ch\\2018\ sumurry-for-policymaker-of IPCC-special-report-on-global-warming of 1.5°C-approved-by-governments.

IPCC. (2021). The Intergovernmental Panel on Climate Change. The Intergovernmental Panel on Climate Change (IPCC) is the United Nations body for assessing the science related to climate change. https://www.ipcc.ch/.

Julian Ramirez, Andy Jarvis, Inge Van den Bergh, Charles Staver and David Turner. (2011). *Changing Climates: Effects on Growing Conditions for Banana and Plantain (Musa* spp.) and Possible Responses. https://doi.org/10.1002/9780470960929.ch29 Citations.

Kallarackal, J. Milburn and Baker, D.A. (1990). Water relations of banana, 111. Effect of controlled water stress on water potential, transpiration, photosynthesis, and leaf growth. *Australian Journal of Plant Physiology*, 17: 79–90.

Kelly Levin, David Waskow and Rhys Gerholdt. (2021). *5 Big Findings from the IPCC's 2021 Climate Report*. https://www.wri.org/insights/ipcc-climate-report.

Kilian, B., Hettinga, J., Jiménez, G.A., Molina, S. and White, A. (2012). Case study on Dole's carbon-neutral fruits. *Journal of Business Research*, 65(12): December 2012, pp. 1800–1810. ISSN 0148-2963, http://dx.doi.org/10.1016/j.jbusres.2011.10.040.

Kishk, A.D., Abul-Soaud, A.A. Adel, Abbas S. Mohamed, Hattem M., El-Shabrawi, El-Sayed I. Gaber, Noor El-Dene and Tarek, M. (2016). Effect of diethyls sulphate and sodiumazide on tolerance of *ex vitro* banana to salt stress. *International Journal of Chemtech. Research*, 9(12): 81–99.

Lahav, E. (1995). Banana nutrition. pp. 258–316. *In*: Gwen, S.R. (ed.). *Banana and Plantains*. Chapman & Hall, London.

Li, W., Ruiz-Menjivara, J. et al. (2021). Climate change perceptions and the adoption of low-carbon agricultural technologies: Evidence from rice production systems in the Yangtze River Basin. *Total Environ.*, 759: 143354.

Mostafa, N. Amira, Wheida, A., EI Nazer, M., Adel, M., EL Leithy, L., Siour, G., Coman, A., Borbon, A., Adel Wahab, M., A. Omar, M. Saad-Hussein, A. and Stephane C. Alfaro (2019). Past (1950–2017) and future (–2100) temperature and precipitation trends in Egypt. *Weather and Climate Extremes*, vol. 26, December 2019, 100225.

Nthambi, M., Markova-Nenova, N. et al. (2021). Quantifying loss of benefits from poor governance of climate change adoption projects: A discrete choice experiment with farmers in Kenya. *Ecol. Econ.*, 179: 106831.

Nyombi, K. (2010). *Understanding Growth of East Africa Highland Banana: Experiments and Simulation*. PhD Thesis, Wageningen University, the Netherlands.

Nyombi, K. (2013). Towards sustainable highland banana production in Uganda: Opportunities and challenges. *African Journal for Food, Agriculture, Nutrition and Development*, 13: 7544–61, official URL: http//www nri. Org/docs/d4679-10_ftf_climate_agric.

Ogbonnaya, C., Sunmonu, M.O. and Egbujor, E.C. (2011). Effect of storage condition on nutritional compositions of banana. *Quality Assurance and Safety of Crops and Food*, 3: 135–139.

Opeyemi, R. Salaul, Marvelous Momohl, Oluwatosin A. Olaleye and Rufus S. Owoeye. (2016). *Effects of Changes in Temperature, Rainfall and Relative Humidity on Banana Production in Undo State*. Nigeria. WSN 44 143-154 EISSN 2392-2192.

Piet van Asten and Charles Staver. (2012). *The Importance of Banana (Musa* sp.*) for Food and Nutrition Security*. https:\\cgspace.cgiar.org\handle\10568\80801.

Ramirez, J. and Javis, A. (2008). *High Resolution Statistically Downs called Future Climate Surface*. International Center for Tropical Agriculture (CIAT), CGIAR Research Program on Climate Change, Agriculture and Food Security (CCAFS), Cali, Colombia.

Ravi, I. and Mustafa, M. (2013). Impact, adaptation and mitigation strategies for climate resilient banana production. Singh, H.P. et al. (eds.). *In*: *Climate-Resilient Horticulture*. 45 Doi: 10.1007/978-81-322-0974-4_5, © Springer India 2013.

Refaie, K.M., Esmail, A.A.M. and Medany, M.A. (2012). The response of banana production and fruit quality to shading nets. *Journal of Applied Sciences Research*, 8(12): 5758–5764.

Robinson, J.C. and Anderson. (1991). The influence of temperature on dry matter assimilation and distribution in young banana plants. *Newsletter of International Group on Horticultural Physiology of Banana*, 14: 37.

Robinson, J.C. (1992). Phenology of banana in the subtropics. *Subtropics*, 13(11): 26–32.

Robinson, J.C., Nel, D.J. and Eckstein, K. (1993b). A field comparison of ten Cavendish subgroup cultivars and selections (Musa AAA) over four crop cycles in the subtropics. *Journal of Horticultural Science*, 68: 511–521.

Robinson, J.C. (1996). *Banana and Plantains, Crop Production Science in Horticulture*. Wallingford, UK: CAB International.

Robinson, J.C. and Galán Saúco, V. (2010). *Bananas and Plantains*, 2nd ed., Crop Production Science Horticulture Series No. 19. https://www.cabi.org/bookshop/book/9781845936587/.

Roibás, L., Elbehri, A. and Hospido, A. (2015). Carbon footprint along the Equadorian banana supply chain: Methodological improvements and calculation tool. *Journal of Cleaner Production*. Doi: 10.1016/j.jclepro.2015.09.074.

Rosenzweig, C. and Leverman, D. (1992). Predicted effects of climate change on agriculture: A comparison of temperature and tropical regions. pp. 342–361. *In*: Majumdar, S.K. (ed.). *Global Climate Change: Implications, Challenges and Mitigation Measures*. PA, USA: The Pennsylvania Academy of Science.

Sabiiti, G., Ininda, J.M., Ogallo, L., Opijah, F., Nimusiima, A., Otieno, G., Ddumba, S.D., Nanteza, J. and Basalirwa, C. (2016). Empirical relationship between banana yields and climate variability over Uganda. *Journal of Environmental & Agricultural Sciences*, 7: 3–13.

Stover, R.H. and Simmonds, N.W. (1987). *Bananas*. 3rd ed., Longman, London, 468 pp.

Tara Slaven. (2019). *Banana Irrigation and Soil Water Management in the ORIA*, Government of Western Australia, Department of Primary Industries and Regional Development. https://agric.wa.gov.au/n/2767.

Thomas, D.S. and Turner, D.W. (2001). Banana, Musa sp. Leaf gas exchange and chlorophyll fluorescence in response to soil drought, shading and lamina folding. *Sci. Hort.*, 90: 93–108.

Thompson, R.L., Lassaletta, L., Patra, P.K. et al. (2019). Acceleration of global N_2O emissions seen from two decades of atmospheric inversion. *Nat. Clime Change*, 9: 993–998. https://doi.org/10.1038/s41558-019-0613-7.

Turner, D.W. and Lahav, E. (1983). The growth of banana plants in relation to temperature. *Australian Journal of Plant Physiology*, 10: 43–53.

Turner, D.W. (1995). The response of the plant to the environment. pp. 206–229. *In*: Grown, S.R. (ed.). *Bananas and Plantains*. Chapman & Hall, London.

Turner, D.W. and Thomas, D.S. (1998). Measurements of plant and soil water status and their association with leaf gas exchange in banana (*Musa* spp.): A laciferous plant. *Sci. Hort.*, 77: 177–193.

Turner, D.W. and Trevorrow, N. (2003). Bananas – Response to temperature. *Ag fact H6.2.6*. www.agric.nsw.gov.au.

Valerie Nelson, John Morton, Tim Chancellor, Peter Burt and Barry Pound. (2010). Climate Change, Agricultural Adaptation and Fair-trade Identifying the Challenges and Opportunities, *Working Paper, Natural Resources*, Natural Resources Institute [website]/ Fairtrade Foundation [website] NrI_FTF_Climate_Agri_WEB_D4679-10_92) pdf-Published Version. Official URL: http//www nri. Org/docs/d4679-10_ftf_climate_agri_.

Vallejo Chavarri, A.L., Vallejo Solis, M.A., Najera Fernandez, J. and Gamier Zamora, L.A. (2017). *Methodological Guide to Reduce Carbon and Water Footprint in Banana Plantations*. Food and Agriculture Organization of the United Nations (FAO).

Van Asten, P.J., Fermont, A.M. and Taulya, G. (2010). Drought is a major yield loss factor for rain-fed East African highland banana. *Agricultural Water Management*, 98: 541–552.

Van den Burgh, I., Ramirez, J., Staver, C., Turner, D., Javis, A. and Brown, D. (2012). Climate change in the subtropics: The impact of projected averages and variability on banana productivity. *Acta Hort. (ISHS)*, 928: 89–99.

Varma, V. and Bibber, D.P. (2019). Climate change impacts on banana yields around the world. *Nat. Clime Change*, 9: 752–757. Doi:10.1038 /s41558-019-0559-9.

Wairegi, W.I., Piet, J.A. van Asten, Moses M.T. Enya and Mateete A. Bekunda. (2010). Biotic constraints override biotic constraints in East African highland banana systems. *Field Crops Research*, 117(1): 8 May 2010, pp. 146–153.

Wardlow, C.W. (1961). Banana Deseases Including Plantains and Abaca. Longmans, London. https:\\www.scirp.org>.

World Bank. (2020a). *Annual Report 2020: Supporting Countries in Unprecedented Times*. World Bank, 2020a. http://hdl.handle.net/10986/34406.

4

Potential of Coconut Palm (*Cocos nucifera* L.) in Resilient Development Solutions

Bioethanol

Sarah L Hemstock,[1,*] *Vineet V Chandra*[1] and *Teuleala Manuella-Morris*[2]

1. Introduction

There are many varieties, but there is only one species of coconut palm (*Cocos nucifera* L.) and its importance to tropical countries cannot be overstated. It is appropriately referred to as the 'Tree of Life'. The scale of reliance on the coconut palm in many coconut-producing countries is vast, with an estimated 80–90 million people reliant on its cultivation or value-added products (e.g., nuts, copra, oil, fiber, toddy, timber, thatch) for their livelihoods. In addition, as a conveniently packaged and portable provider of drinking water, food and fuel, the coconut has been instrumental in the exploration and settlement of the tropics and continues to play a fundamental role in their development. Globally, around 12.13 million hectares of land is currently cultivated with coconut across 92 countries. However, just three countries accounted for 73% of this cultivation area—the Philippines, 29%; Indonesia, 26%; and India,

[1] Resilient Development Solutions, Longdales Road, Lincoln, UK.
[2] Live and Learn, Funafuti, Tuvalu.
Emails: vineetvishalchandra@gmail.com; teuleala.manuella@live&learn.org
* Corresponding author: sarah_hemstock@hotmail.com

17% (based on data from FAOSTAT, 2020—a five-year annual average, 2014–2018). Globally, almost all of the crop is produced by resource-poor farmers on smallholdings of under 4 ha. For example, in Tuvalu, a Pacific small island, developing state, with a population of around 11,000 people, plots are small and fragmented which impact the harvest and collection of coconuts. In Funafuti, Tuvalu, there are an estimated 1,800 plots with 100 plot holders; each holding is around 3 ha, but each individual plot is only 0.15–0.25 ha (Rosillo-Calle et al., 2003). As land is inherited, plots tend to become smaller and more fragmented with each successive generation.

Coconut is grown throughout the tropics; its history of domestication and its population genetic structure relates to global human dispersal patterns (Gunn et al., 2011). Gunn et al. (2011) propose two geographical origins of coconut cultivation: the islands of southeast Asia and the southern margins of the Indian subcontinent. Their genetic analyses revealed that, although coconut is a single species, it has two subpopulations that correspond to the Pacific and Indo-Atlantic oceanic basins, suggesting that coconut cultivation occurred twice. This distinct population structure reflects ancient Austronesian trading routes and later Arab trading routes. This remains evident today despite long-term human cultivation and dispersal (Gunn et al., 2009). Human cultivation of coconuts in the Pacific region may have led to an additional genetic substructure in the Pacific, which corresponds to phenotypic and geographical subgroups—dwarf varieties, self-pollination, and a 'round' as opposed to 'triangular oblong' fruit morphology. It is thought that these varieties evidence a genetic admixture between the Pacific and Indo-Atlantic groups in the south-west Indian Ocean. This is consistent with human introduction of Pacific coconuts along the ancient Austronesian trade route, connecting Madagascar to Southeast Asia, with coconuts of Philippine origin being introduced to the Pacific basin and New World coast around 2,250 years ago (Gunn et al., 2011).

Based on the scale of cultivation and the geographical reach of the coconut, it is evident that this palm will have an enormous impact on resilient development in the tropics. However, since this chapter is examining the cultivation of coconut palms for climate change resilience, it will focus on examples from the Pacific Islands Region (PIR, covering 26 countries and territories) as this region has often been described as 'the canary in the coal mine' in terms of facing the adverse impacts of climate change. Resilient development in the Pacific context has the following definition:

'Development processes and actions that address the risks and impacts of disasters and climate change while progressing to stronger and resilient communities' (FRDP, 2016).

In this definition 'resilient development' parallels development processes and interventions that address the risks and impacts of disasters and climate change (which can be viewed as a slow-acting and progressive disaster), while building stronger and more resilient communities (FRDP, 2016; Hemstock et al., 2018).

Around 5 million people in the PIR depend on coconut palms for food, livelihoods and raw materials for many uses (e.g., pharmaceutical products, building and construction materials, handicrafts, domestic uses—such as string, mats, blinds, bags, trays, plates and fuelwood and charcoal for cooking). In Pacific Small Island Developing States (PSIDS), such as Vanuatu and Kiribati, coconuts are a high-value commercial crop with copra, the dried white flesh of the mature nut, being a premium product. For PSIDS, like Tuvalu, coconut is the most important nationally produced crop for food security, not only in terms of being a source of food and drink for humans, but also as the main source of fodder for livestock. Just a single product—toddy—provides around one-quarter of the total household income for produces in Tuvalu's outer islands (Hemstock, 2013).

1.1 Toddy Production and Impact on Livelihoods in Tuvalu

The 'Tree of Life' endures environments and conditions considered to be marginal for many other crops. For example, the Fiji case study below noted that coconut can be cultivated throughout Fiji, whereas sugarcane and cassava are dependent on suitable soil and weather regimes and can only be grown in favorable areas. In Tuvalu, where 57% of the total landmass of 26,007 ha is under coconut (Hemstock, 2013), soils are generally of poor quality and only support limited flora. The majority of Tuvalu's soil is derived from corals, calcareous algae, foraminifera and Phylum Mollusca—all composed mainly of calcium carbonate (McLean et al., 1986). This means that soils are alkaline, gravelly, have large grain size and low organic matter content. Soils are, therefore, not very fertile as fertility depends on the proportion of organic matter in the substrate and the grain size of constituent materials. Around half of Tuvalu's soil can be classed as gravelly with substrates composed of coral rubble. Such stony soils are of poor quality and difficult to cultivate. However, the substantial hardiness of the coconut palm contributes to its reliability as a stabilizing influence for livelihoods dependent on the farming systems of marginal and fragile environments, such as those found in the majority of Pacific Small Island Developing States, like Tuvalu.

In Tuvalu, since the collapse of the copra industry in 2002 (*see* Woods et al., 2006 for details), coconut products, such as coconut sap (fresh toddy, boiled toddy, red toddy or toddy syrup, and sour toddy—an alcoholic beverage), have been a staple of livelihoods for many producers, providing on average of around 22% of household income (Table 1). Current Gross Domestic Product (GDP) per person in Tuvalu is around US$3937, so the sale of coconut products makes a significant economic contribution for

Table 1. Household income for toddy producers from all toddy products.

Tuvalu Island group	Total household income from toddy products (US$)
Funafuti	2897
Nanumea	2654
Nanumaga	2654
Niutao	0
Nui	2878
Vaitupu	1929
Nukufetau	909
Nukulaelae	2897
Niulakita	2119

Source: After Hemstock, 2013

producers. Around 10% of all households in Tuvalu produce toddy and its domestic sale is equivalent to over 7% of the country's GDP (World Bank, 2020).

2. An Estimate of Global Standing Stock and Carbon Sequestration Potential

Considering the total area under coconut globally is just over 12 million hectares, coconut palm is already a considerable carbon store and has great potential for further development for carbon sequestration. Tables 2 and 3 below take a broad approach to estimating the carbon sequestration potential of the two subpopulations of coconut palm that correspond to the Pacific and Indo-Atlantic oceanic basins.

The general assumptions for Table 2 are based on Bhagya et al. (2017), who undertook a field experiment on the Indo-Atlantic subpopulation in a coconut garden with red sandy loam soil at Kasaragod, Kerala, India (12°30′ N latitude and 75°00′ E longitude, at an elevation of 10.7 m above mean sea level). The general assumptions for Table 2 include the following:

a) A tree height of 20.8 m and a diameter at breast height (DBH) of 0.85 m.

b) Above ground biomass (dry weight of stem wood) of 575 kg/palm.

c) Above ground carbon stored 287 kg/palm.

d) 12.1 million ha under cultivation globally (FAOSTAT 2020—five-year annual average 2014–2018).

e) The average annual rainfall is 3,500 mm, of which 86% is received during the four monsoon months (June–September) (Bhagya et al., 2017).

Table 2. An estimate of global above ground carbon sequestration potential of coconut cultivation based on Bhagya et al. (2017) (Indo-Atlantic oceanic basin subpopulation of coconut palm).

High planting density 400 palms per hectare			
Number of palms globally	Above Ground Biomass – AGB (Gt)	Total above ground carbon storage (Gt)	Above ground carbon storage (t/ha)
4,852,987,120	2.8	1.4	115
Number of palms (medium planting density 178 palms/ha)			
Number of palms globally	Above Ground Biomass – AGB (Gt)	Total above ground carbon storage (Gt)	Above ground carbon storage (t/ha)
2,156,883,164	1.2	0.6	51
Number of palms (low planting density 100 palms/ha)			
Number of palms globally	Above Ground Biomass – AGB (Gt)	Total above ground carbon storage (Gt)	Above ground carbon storage (t/ha)
1,213,246,780	0.7	0.3	29

Gt = Gigatonnes

Table 3. An estimate of global above ground carbon sequestration potential of coconut cultivation, based on Hemstock (2013) (Pacific oceanic basin subpopulation of coconut palm).

High planting density 638 palms per hectare			
Number of palms globally	Above Ground Biomass – AGB (Gt)	Total above ground carbon storage (Gt)	Above ground carbon storage (t/ha)
7,740,514,456	2.8	1.4	117
Number of palms (medium planting density 338 palms/ha)			
Number of palms globally	Above Ground Biomass – AGB (Gt)	Total above ground carbon storage (Gt)	Above ground carbon storage (t/ha)
4,100,774,116	1.5	0.8	62
Number of palms (low planting density 173 palms/ha)			
Number of palms globally	Above Ground Biomass – AGB (Gt)	Total above ground carbon storage (Gt)	Above ground carbon storage (t/ha)
2,098,916,929	0.8	0.4	32

Gt = Gigatonnes

f) Standing stock was measured using standard non-destructive methods as outlined in Rosillo-Calle et al. (2015).

g) Wood density values were sourced from the Global Wood Density database (Zanne et al., 2009).

The general assumptions for Table 3 are based on Hemstock (2013) who carried out field studies in Tuvalu, a Pacific Small Island

Developing State located approximately 1000 km north of Fiji (8° 310000 S, 179° 130000 E; 0.8 m above mean sea level), where the majority of palms constituted the Pacific Ocean basin subpopulation. Tuvalu's total land mass is just 26 km^2, spread across its 900,000 km^2 and its ocean exclusive economic zone. Ground surveys were undertaken in all of Tuvalu's nine island groups (Funafuti and the 'outer islands' of Vaitupu, Nanumea, Nanumaga, Niutao, Nui, Nukufetau, Nukulaelae, and Niulakita). The general assumptions for Table 3 include the following:

a) An average tree height of 16.4 m and a diameter a DBH of 0.73 m.
b) 12.1 million ha under cultivation globally (FAOSTAT 2020—5 year annual average 2014–2018).
c) Above ground biomass (kiln dry weight of stemwood) of 367 kg/palm.
d) Above ground carbon stored 184 kg/palm.
e) Tuvalu's climate is sub-tropical, with temperatures ranging between 28–36°C, uniformly throughout the year. The mean rainfall ranges between 2,600–3,550 mm per year across Tuvalu, but there are significant variations from island to island. Funafuti has the highest annual rainfall with over 3500 mm and a marked dry season from June to September (winter months), with an increased chance of drought during this period. All three of the northern islands are 'dry islands' with Nanumea, Nanumaga and Niutao having an annual rainfall of around 2700 mm. Northern islands have higher rainfall from December to February and lower, from April to October.
f) Standing stock was measured using standard non-destructive methods as outlined in Rosillo-Calle et al. (2015).
g) Wood density values were obtained by destructive harvesting and oven drying samples (Hemstock, 2013; Hemstock and Singh, 2015).

In Tables 2 and 3, there does not appear to be a large difference in the sequestration potential of the two subpopulations. An above-ground carbon (AGC) storage of 50–55 tons of carbon per hectare for coconut plantation appears realistic and demonstrates what a hugely productive species coconut is. The AGC for coconut compares favorably with mixed species forest in Africa and the sub-tropics (Liu et al., 2018; McNicol et al., 2018).

The sequestration potential of coconut is impressive. To put this in perspective, according to the FAO Global Forest Resources Assessment (FAO, 2005), the carbon content of the current global standing stock of coconut is equivalent to 4–14% of the carbon that all the forests of the EU27 countries contain (9.8 Gigatonnes of carbon). If this does not impress, look at it this way—the maximum theoretical potential carbon content of current global coconut standing stock is equivalent to 120% of the total carbon emissions of EU27 countries in 2018.

Tables 2 and 3 do not account for soil carbon stock of area under coconut, which Bhagya et al. (2017) estimated to be 48 tons of carbon per hectare considering a soil depth of 0–60 cm.

3. Main Impacts of Climate Change on Coconut Cultivation

The situation with regard to the coconut palm and its role in resilient development across the tropics, like all organisms constituting our Gaia, is both tangential and direct, complex and obvious. For example, simply put, climate change is caused by elevated levels of CO_2, which will increase temperatures, which will, in turn, cause changes in precipitation. Changes in temperature, CO_2 concentration and water availability are the climate change effects with the most impact on crop productivity. These impacts will affect the biosphere and have diverse and opposing effects on photosynthesis. For instance, elevated temperatures escalate both photorespiration and dark respiration which cause an overall reduction in the primary productivity of the coconut; elevated levels of CO_2 on the other hand increase the rates of photosynthesis and hence, increase plant productivity. In that regard, Naresh et al. (2013), who did their experiments in India, predicted that climate change will increase coconut productivity by 7–32% by 2080, depending on the cultivar. However, coconut palm crown initiation and growth are very sensitive to CO_2 concentration, temperature and fresh water availability; so climate change impacts are difficult to predict.

According to Hebbar and Chaturvedi (2015), around 45% of India's coconut is grown in Kerala in coastal areas and hilly terrain, which is highly vulnerable to climate change. In Kerala, the higher temperatures associated with climate change influence both growth and productivity of coconut palms, and these effects can be either positive or negative. For example, Hebbar and Chaturvedi (2015) have observed that negative climate change impacts, such as added heat stress, are already adversely affecting coconut production in Kerala's low to mid-latitudes, but predict that higher temperatures could increase productivity in currently cold-limited high latitudes. In open top chamber experiments Hebbar and Chaturvedi (2015) observed that at 550 and 700 ppm CO_2 coconut palm biomass increased by 8 and 25% respectively against an ambient CO_2 concentration of 380 ppm. Increased levels of CO_2 also went some way to reducing the impacts of increased temperature and drought.

Predicted changes in climate for the Pacific Islands Region include increased incidence of severe drought, such as the one experienced in Tuvalu, when a state of emergency was declared in September 2011 (IFRCRCO, 2011). As the coconut there is rain-fed, it was severely adversely affected by the three-month dry period and the effects were further exacerbated by the fact that most of the coconut in Tuvalu is coming to the end of its productive lifespan as the majority was planted in the late 1940s and early 1950s. Coconut requires between 50–100 liters of water per day to produce and maintain its crown and canopy. Water stress reduces root growth, which limits water uptake further. The canopy of the coconut palm reduces and the yield may reduce by as much as 50% (Hebbar and Chaturvedi, 2015).

4. Case Study: The Potential of Coconut Toddy for Use as a Feedstock for Bioethanol Production in Fiji

Sugar-yielding palms, such as coconut palm (*Cocos nucifera*), sugar palm (*Arenga pinnata*), palmyra palm (*Borassus flabellifer*) and nipa palm (*Nypa fructicans*) are mainly believed to produce sap with a higher concentration of sugars than sugarcane (Tamunaidu et al., 2013). Whilst existing literature on palm toddy (sap) is sparse, it indicates that coconut toddy has a good potential for bioethanol production (Tamunaidu et al., 2013; Hemstock, 2013) and thus its use could be beneficial for Fiji. However, of the species listed above, sugar palm has the maximum production yield of 20,160 liters per hectare per year of bioethanol. In comparison cassava, which is also grown in Fiji, has a production yield of around 4500 l/ha/year; and sugarcane, which is also grown in Fiji, has 5025 liters per hectare per year (Ishak et al., 2013).

This case study investigates the toddy (sap) production potential of coconut palms in Fiji and its potential conversion into ethanol and further compares the ethanol production potential of coconut with that of sugarcane and cassava.

4.1 Background Information

The increasing demand for and use of fossil fuels and a growing urgency to mitigate anthropogenic climate change demands the development of clean energy that can help reduce greenhouse gas emissions (Cerqueira Leite et al., 2009). Ethanol produced from biomass feedstock is one such biofuel considered to be clean and sustainable. The current global bioethanol supply is predominantly produced from sugar (sugarcane) and starch (corn) feed stocks. Sugar crops produce around 61% of the total bioethanol production while 31% is produced from starch feed stocks (Tamunaidu et al., 2013). Brazil and the US are the leading producers (Elobeid and Tokgoz, 2008). In Fiji, the two potential sugar and starch feed stocks are sugarcane and cassava. However, producing these feedstocks relies heavily on the use of fossil fuels with associated negative environmental impacts as these feed stocks require land preparation after each harvest but bioethanol produced from coconut toddy (sap) does not have such negative environmental impacts in comparison (Ishak et al., 2013).

Fiji is a small island developing state in the Pacific with a land mass of 18,272 km^2 spread over more than 300 islands. The area occupied by coconut is approximately 70,000 hectares (Republic of the Fiji Islands, 2013).

Fiji has a population of 890,000 (World Bank, 2020). Around 100,000 people from farming communities and rural smallholdings are economically dependent on coconut (Republic of the Fiji Islands, 2013). Coconut palm products are used in a myriad of ways (e.g., food and drink, mats are woven from leaves, fish traps, handicrafts, husks are used for string and cooking fuel, shells are used for handicrafts and cooking fuel, flesh is used for copra and oil, trunk as timber, etc.). However, coconut toddy, which is a source of sugar, is not presently extracted for any use in Fiji (Dalibard, 1999; Tamunaidu et al., 2013).

Fiji's main income is generated from tourism, agriculture, forestry and fisheries and is heavily dependent on imports, such as fossil fuels. In 2014, 876 million liters of fossil fuels were imported at a cost of US$667,544,108 (Fiji Bureau of Statistics, 2014) of which 115 million liters was motor spirit. The price of all imported goods increased due to high transport costs as Fiji is geographically isolated. The lack of access to safe, sufficient, affordable, reliable and environmentally friendly energy is a development constraint. For electricity generation in Fiji, renewable energy plays an important role. In 2014, a total of 44.96% (400.94 GWh – Gigawatt hours) was produced from hydro; 3.65% (32.54 GWh) was produced from biomass (sugarcane bagasse and wood chips); 0.48% (4.28 GWh) was produced from wind and 50.91% (454 GWh) was produced from diesel and heavy fuel oil (Dean, 2014).

However, although the use of existing traditional biomass energy still provides for basic needs, such as cooking, it will not resolve the problem of accessing wide-scale modern energy services required for steady economic growth and higher standards of living. The transformation of biomass to modern energy carriers is possible via bioethanol (examined here), biogas, gasification and biodiesel production. The move away from traditional biomass energy in rural areas will involve social and cultural changes. Additionally, policy changes are required for successful execution of modern bioenergy initiatives as the Fiji Electricity Authority currently has the monopoly on electricity generation, transmission and distribution (Prasad and Raturi, 2017).

4.2 Toddy (Sap) Collection

In many Asian, African and Pacific island locations, palm sap is collected and consumed as such or as fermented beverage (Doner, 2003). In many locations it is culturally and/or economically important. The technique of collecting toddy is very ancient (Lal et al., 2003) and usually differs in different locations. However, in general, toddy can be collected when a palm is ready for bearing coconuts as the stem that will eventually, once

Fig. 1. Cutting toddy (photo by Sarah L. Hemstock, Niutao).

Fig. 2. Toddy (sap) collection (photo by Sarah L. Hemstock, Niutao).

fertilized, develop into coconuts, known as the spadix (which contains the immature polygamomonoecious inflorescences) is 'tapped' for the sap. Tapping usually takes place when the inflorescence is one-month old. As represented in Fig. 2, in order to collect the sap (toddy), the 'spathe' (leaf that is wrapped around the inflorescences) is sliced through or removed and the inflorescence is bound tightly with string to prevent it from opening. The cut and bound spadix is known as 'tap'. The inflorescence is gradually bent over, so that the coconut palm juice can flow from the vascular bundles into either a plastic or glass bottle (known as 'cup'). The juice flow increases slowly and the 'cup' is changed twice daily. During this time, the thin slice is shaved from the end of the inflorescence to continue the flow of sap (Lal et al., 2003; Hemstock, 2013; Tamunaidu et al., 2013).

4.3 *Case Study Methodology*

The following method was used to estimate the total theoretical bioethanol potential from coconut, sugarcane and cassava. Using the data source for coconut, the total harvest area was obtained and using the general assumptions listed below, the number of trees producing coconuts was calculated. The number of trees and the sap yield factors were used to calculate the potential annual sap yield (liters per year) and subsequently the bioethanol content was estimated. Similarly, for cassava and sugarcane, the bioethanol factors and production values were used to calculate the potential bioethanol yield (liters per year). The potential production of bioethanol (liters) from coconut, cassava and sugarcane was used to calculate the reduction in the use of fossil fuels by comparing energy content of bioethanol and gasoline. The total reduction in fossil fuel use was calculated by converting bioethanol theoretically available to equivalent gasoline (1 liter bioethanol = 0.65 liters of gasoline) and using the energy content factor of gasoline, the energy value of non-renewable energy was calculated and is indicated by a negative sign. Similarly, considering the reduction in fossil fuels and CO_2 emission factors, the greenhouse gas mitigation potential of bioethanol from all three sources was calculated.

Data on coconut and cassava production were obtained from FAOSTAT. Sugarcane production data were obtained from Fiji Sugar Corporation (Khan, 2014) and substantiated with other published sources to allow for more inclusive assessment (Fiji Bureau of Statistics, 2015; FAOSTAT, 2020). Ten year annual average data were collected and analyzed. The number of coconut palms was obtained using the area planted under coconut and general assumptions on plantation spacing.

There are five main groups of vegetation cover in Fiji (LRPD, 2012), namely: cropland; grassland or pasture; weedland; scrubland and forest:

a) Cropland is covered by sugarcane, rice, cereals, market vegetables, coconuts, bananas, cocoa, citrus, pineapples and ginger.
b) Grassland is covered by para-grass, blue grass, grass-legume, mission grass and native grasses.
c) Weedland is covered by navua sedge, lantana, solanum, noogoora burr, mint weed, hibiscus burr and swamp vegetation.
d) Scrubland is covered by braeken fern, reeds, guava, vaivai and mangrove.
e) Forestland is covered by native forest, hardwood, softwood, exotic forest, pine and mahogany.

This case study is only considering cropland area under coconut, cassava and sugarcane as follows:

a) Area occupied by coconut farming in 2014 was 62,000 hectares (FAOSTAT: 10-year annual average 2006–2016).
b) Area occupied by cassava farming in 2014 was 6,820 hectares (FAOSTAT: 10-year annual average 2006–2016).
c) Area occupied by sugarcane farming in 2014 was 37,560 hectares (FAOSTAT: 10-year annual average 2006–2016).
d) Other crops occupy an area of 76,320 hectares—details of other crops can be found in Chandra and Hemstock (2015).

Since there is no history of toddy production or use in Fiji, the production potential of toddy in Fiji is theoretical and based on findings from Hemstock's (2013) field study in Tuvalu. In Tuvalu, toddy is collected at dawn and dusk and consumed before it begins to ferment. In order to delay the fermentation of toddy for human consumption, toddy is filtered, boiled and then cooled. For the production of alcoholic 'sour toddy', the collected toddy is left outside in the sun to heat and allowed to ferment by the action of naturally occurring yeast. Fermentation usually takes two to four days. The alcohol/ethanol content is noted to be around 6–8% (Hemstock, 2013). General assumptions are as follows:

a) Average sap yield per palm/tree = 1.3–2.0 liters per day (Doner, 2003; Hemstock, 2013).
b) Trees per hectare = 197 (Doner, 2003; Lal et al., 2003).
c) Average sucrose content of sap = 13%.
d) Net calorific value of ethanol = 26.8 megajoules per kg (Yüksel and Yüksel, 2004; Hemstock and Radanne, 2006; Lin et al., 2010; García et al., 2011; Hemstock, 2013).
e) Cassava (fresh root) – 1 ton can produce 180 liters of bioethanol (Kishore and Srinivas, 2003; Silalertruksa and Gheewala, 2010; Chandra and Hemstock, 2016).
f) Sugarcane – 1 ton of sugarcane can produce 70 liters of bioethanol and 1 ton of molasses can produce 250 liters of bioethanol (Kishore and Srinivas, 2003; Silalertruksa and Gheewala, 2010; Chandra and Hemstock, 2016).
g) Alcohol content in toddy by volume fraction is considered to be 8% (Hemstock, 2013).
h) 1 US$ = 2 FJ$.
i) The energy content of 1 liter of bioethanol (21.8 megajoules per liter) is equal to 0.65 liter of gasoline (32.19 megajoules per liter) and gasoline's fuel-cycle GHG emissions is assumed as 2.918 kg of CO_2

equivalent per liter (Hemstock and Radanne, 2006; Nguyen et al., 2007; Silalertruksa and Gheewala, 2010).

j) Molasses % of cane: 3.5–7.0% depending on impurity loading and quality of cane (Charan, 2015).

k) The bioethanol potential is only calculated from areas where coconuts are harvested. According to the article published (Republic of the Fiji Islands, 2013), 30% of the coconuts produced are not harvested and are not considered in any analysis.

4.4 Case Study Results and Discussion

Coconut is grown widely across Fiji in all Divisions and is noted to be an important source of income for many who rely on various coconut products for their main source of income. Within the different areas (Divisions) of Fiji, the potential of coconut bioethanol production can be seen as represented in Table 4. It is noted that Northern Division has the highest production potential and Central Division has the lowest. However, in Fiji, the economy of the Central Division is boosted by other commercial activities. For example, Suva City, the capital city and seat of government is in the Central Division. This brings in foreign investment, infrastructural development and employment opportunities. The Western Division's economy is boosted by the sugarcane industry, extensive tourism development and other commercial activities. The Eastern Division has some economic activity from tourism, but its economy is mainly reliant on agricultural activities.

The Northern Division, on the other hand, currently relies almost entirely on the sugar industry and the population is almost totally dependent on this industry for their economic security. There are always issues of rural to urban drift for better employment opportunities. People move from the Northern and Eastern Divisions to the Central and Western Divisions to find employment in industrial, commercial and/or tourism sectors. This brings added difficulties for the population of urban area as it increases the cost of housing which, in turn, leads to informal settlement development on land at risk of flooding, and with an increase

Table 4. Coconut production potential by division (administrative area).

Division	Area Occupied (%) (Republic of the Fiji Islands, 2013)	Sap (Liters)	Ethanol (Liters)	Trees/ person	Ethanol (Liters/ person)
Central	3.43	195,579,028	15,646,322	1	46
Western	3.76	214,899,812	17,191,985	1	54
Northern	69.60	3,973,055,773	317,844,462	61	2338
Eastern	23.21	1,324,878,855	105,990,308	71	2696

in population the jobs become hard to find and wages remain low. The last population census in Fiji, in 2017, indicated that the population of Fiji was 884,887—an increase of just under 10% from 2007 when the population was 837,271 (Fiji Bureau of Statistics, 2012, 2019). 38.16% of the population reside in the Western Division, 4.69% in the Eastern Division, 16.23 in the Northern Division, and 40.9% in the Central division. It is noted that majority of the people are located in the Central Division and if we look at the share of benefit from coconut (as shown in Table 4), it can be deduced that someone from either the Central or Western Division will not benefit a great deal from the production of bioethanol from coconut. However, Northern and Eastern Divisions will have 61 and 71 trees per person, respectively. Therefore, the population from the Northern and Eastern Divisions will derive the maximum benefit from coconut bioethanol production. Additionally, this would diversify the agricultural sector in the Northern and Eastern Divisions and bring the much needed employment opportunities to alleviate the rural to urban population drift. Development of a coconut bioethanol industry would also uplift the living standards of people in the Northern and Eastern Divisions, enabling them to enjoy the same benefits as those in the developed Central and Western Divisions. In Table 4, the potential of bioethanol production in Northern and Eastern Divisions shows that bioethanol from coconut could bring high returns for the farmers where economic activity is currently the lowest in Fiji, for example, the sugarcane industry has been in decline for the past two decades and now contributes less than 5% of Fiji's current GDP. However, it provides the main source of income for more than 25% of the population (Chandra and Hemstock, 2015). Additionally, incomes from sugarcane are below the poverty line for Fiji. Sugarcane currently generates a profit of around FJ$16.95 per mt, cane yields are around 47–55 mt per hectare in Fiji and smallholding sizes are 4ha on average. Therefore, the profit for a farmer selling 200 tons of cane from their 4-ha farm would be around FJ$3,390 per year or FJ$65 per week—significantly lower than Fiji's poverty line income of FJ$10,880 per year or FJ$209 per week for a family of four (Singh, 2020). There is therefore a strong economic case for diversification of the agricultural sector in Fiji (Singh, 2020).

For cane farmers in Fiji, exiting from sugarcane farming in the short term may not be economically feasible as they are under contractual agreements with the Fiji Sugarcane Corporation to produce sugarcane. However, according to research reported by Singh (2020), approximately 30% of arable land currently remains fallow. There is a clear opportunity for the Ministry of Agriculture to encourage farmers to use land for crop diversification and to encourage farmers to plant coconuts on land that are currently not used (around 30% of palms). From studies carried out in Fiji (Woods et al., 2006), coconut resources were noted to be underutilized and could provide biofuel sustainably over the short term. In order to ensure long-term sustainability of any future coconut bioethanol or biodiesel initiative, appropriate re-planting schemes are required.

It is assumed from the analysis that there are around 12,030,376 coconut palms which are harvested for coconut under an area of 60,857 hectare as shown in Table 4. These trees produce around 200,063 mt of coconut annually (average). In comparison, sugarcane and cassava production areas are smaller as not all areas are suitable for planting these crops. However, coconut is found everywhere in Fiji, as shown in Table 4. This indicates that coconut bioethanol could benefit all.

Coconuts are harvested annually from 12,030,376 palms in Fiji. From this it is possible to obtain around 456 million liters of ethanol—around 544 liters per person annually. Existing coconuts do not require cultivation and preparation of land, such as would be required for planting sugarcane and cassava; and neither would coconuts require the use of fossil fuels to harvest in the field. Therefore, coconut bioethanol can be considered 'cleaner' than that from sugarcane or cassava. Once the coconut tree is planted, it is assumed that no fertilizers and cultivation are required which leads to pollution and the plant can be tapped for the next 25–35 years (Doner, 2003). However, cassava needs yearly cultivation of seeds, and sugarcane and cassava need annual land preparation and/or the use of pesticides and fertilizers.

The annual bioethanol yield (liters per hectare) is attractive for coconut in Fiji having 7,504 liters per hectare. However, the bioethanol yields of cassava and sugarcane are reported higher than that of coconut in Fiji in some research (Tamunaidu et al., 2013; Ishak et al., 2013; Quintero et al., 2008), but this was not the case in this analysis. The difference could be due to climate, soil type and varieties. Additionally, sugarcane and cassava are currently used for food in Fiji, so bioethanol conversion from these feed stocks needs careful analysis to ensure that food supplies are not affected by bioethanol production.

Coconut is currently used for the production of copra, normal oil and virgin oil. From Table 5 it can be seen that the use of coconut bioethanol could result in decrease of the use of non-renewable energy (NRE) by 9,555 terajoules which will lead to reduction in emissions by 863,797 tons CO_2 equivalent. However, Fiji's demand for gasoline is only 115 million liters which represents just 25% of coconuts' bioethanol production potential. This would mean that, theoretically, Fiji could be an exporter of bioethanol. However, since Fiji is currently not ready to eliminate the use of gasoline completely, an immediate option would be 10% coconut bioethanol blending. This would provide a ready market for the production of coconut bioethanol as well as existing products.

Considering that the current possibility of E10 blending with gasoline will require 11.5 million liters of ethanol, cassava is not considered competitive as the current theoretical bioethanol potential from cassava

Table 5. Annual bioethanol production potential.

Annual Average Production (2005–2014)			
Parameter	**Cassava**	**Sugarcane**	**Coconut**
Area harvested (hectares)	5,588[A]	47,856[B]	60,857[A]
Production (mt)	63,860[A]	2,154,189[B]	200,063[A]
Yearly sap (liters)	-	-	5,708,413,466[C]
Bioethanol (liters)	11,494,800[D]	150,793,230[D]	456,673,077[D]
Yield (liters per hectare)	2,057	3,151	7,504
NRE (terajoules)	–240.5	–3,155	–9,555
GWP (mt of CO_2 equivalent)	–21,742	–285,225	–863,797

Note: A: The harvested area data was obtained from FAOSTAT. B: The data was obtained from FSC (Khan, 2014). C&D: The values were calculated using assumptions listed in Case Study Methodology.
NRE = Non-Renewable Energy; GWP = Global Warming Potential

in Fiji is only 11.45 million liters. In addition, cassava is currently only used as food so its use as a fuel would cause difficulties for food supply. If cassava is to be used for bioethanol production for E10 blending, then the cassava production has to be doubled which indicates that the land requirement will be doubled as well. This would have a knock-on effect of increasing emissions due to land preparation and use of fertilizers and other agrichemicals to sustain production.

Sugarcane (juice) currently has the capacity to produce the bioethanol required for E10 blending, but if used for this purpose, the current sugar production would be affected. However, sugarcane has another advantage over cassava—in that, sugar has a by-product known as molasses. The current molasses production from existing sugarcane is approximately 84,000 tonnes which has the potential to produce 21 million liters of bioethanol. This is sufficient for the E10 blend without affecting the sugar production and so the surplus could be exported. However, it would be interesting to see if the conversion of current exported molasses to ethanol and the capital costs associated with it are profitable. This economic analysis would require further investigation.

It should be noted that the production of copra, the traditional product of coconut cultivation in Fiji, is declining and new products, such as virgin oil, are still emerging. Therefore, since coconut has a significant bioethanol production potential, a bioethanol industry for coconut could be a good option for producers over the short term. In order for any coconut-based industry to be successful and sustainable over the long term, appropriate coconut replanting schemes would need to be put into place. In a Tuvalu case study (Hemstock, 2013), bioethanol from toddy was successfully demonstrated at a very small scale. However, the cost of production was not competitive with available fossil fuels.

5. Conclusion and Prospects

Scenarios outlining future anthropogenic climate change impacts predict temperature increases for the end of this century of up to 4°C (Collins et al., 2013), accompanied by changing rainfall patterns and an increase in the incidence and severity of extreme weather events. This will result in a climate that is considerably different from the one experienced at present. While changes in coconut productivity will occur as a result of these changes, the full impacts remain unclear. From the research reviewed in this chapter, increased drought conditions, and higher temperatures will reduce coconut productivity, while elevated CO_2 concentrations will increase productivity. One thing is for sure—climate change impacts will certainly influence subsistence and commercial coconut production on a vast scale across the tropics. More research needs to be done on these impacts.

The coconut is a phenomenally productive plant and well deserving of the name, the 'Tree of Life'. The productivity of coconut may be a key to its development as a crop for carbon storage. The analysis presented in this chapter demonstrates the current and significant potential value of coconut forest as a carbon sink. By factoring in a soil carbon of 48 tC/ha (Bhagya et al., 2017), the potential of coconut forest to store carbon is almost doubled. Additionally, the 7–32% increase in coconut productivity by 2080, predicted by Hebbar and Chaturvedi (2015) and Naresh Kumara and Aggarwal (2013), will also impact the coconut's value as a tool for carbon storage.

From the Fiji case study, it is clear that the decline of the copra industry can make space for the development of coconut biofuels for Fiji's future agricultural development and mitigate climate change impacts. Coconut resources in Fiji are not fully exploited and the resource has the capability to provide biofuel sustainably over the short term. However, long-term sustainability can only be realized by a suitable large-scale replanting scheme which Fiji is believed to be working on. The annual replanting target is currently 123,000 palms, which is inadequate to replace currently the senile palms. The production of coconut bioethanol will help mainly the Northern and Eastern Divisions, which are currently the least developed regions in Fiji. Since toddy bioethanol is not commercially produced anywhere, the Fiji case study only analyzed the theoretical bioethanol production potential from toddy. It is anticipated that production of toddy bioethanol will face many challenges if undertaken on a large scale, and which will require detailed further investigation. However, the case study demonstrates that coconut bioethanol is a feasible option to provide a significant source of clean renewable energy which could be sustainably produced. The introduction of a coconut bioethanol industry in Fiji and similar small island states in the PIR would greatly improve energy security by reducing dependency on imported oil, reduce trade balances and increase incomes for rural farmers. It would also be a sustainable mitigation and adaptation strategy, help achieve Paris Agreement Nationally Determined Contributions and reduce the burden of greenhouse gas emissions associated with the use of fossil fuels. The Green Climate Fund could be explored as a possible source of investment to establish a coconut bioethanol industry in Fiji and countries facing similar issues. The UNFCCC COP21 Paris Agreement committed a minimum US$100 billion per year in climate finance for all developing countries by 2020. Resilient development, which includes adapting to climate change, reducing disaster risk and sustainable energy provision (FRDP, 2016), is central to development efforts in the PIR. Finances for adaptation are already flowing into the PIR and totaled US$1,914 million in 2015. Over 70% of these aid flows are funding current climate change adaptation, disaster risk reduction and sustainable energy initiatives (OECD, 2017; Hemstock et al., 2018).

Large aid flows into the PIR mean that the energy sectors of many Pacific Small Island Developing States (P-SIDS) are very much in the hands of external donors. Donors have to be made aware of energy service needs and available renewable resources, such as coconut bioethanol, in order to make appropriate technology choices that would support energy needs and agricultural sector diversification. The current lack of modern biomass renewable energy applications in P-SIDS, such as Tuvalu and Fiji, are due to lack of appropriate technology selection—such as a focus on solar and wind, or introducing 'new' energy crops where there are already abundant biomass resources. These technology choices are made for a variety of reasons—from donor agenda through to lack of technical expertise, inappropriate technical assistance and weak recipient institutional structure to plan, manage and maintain renewable energy programs. There are several successful examples of coconut oil biodiesel projects operating throughout the PIR (Woods et al., 2006). However, coconut bioethanol has only been demonstrated on a small scale as part of the NGO Alofa Tuvalu's 'small is beautiful' project in Tuvalu and which concluded that the concept is technically feasible (Hemstock et al., 2013). The majority of the barriers listed above can be broken down with effective recipient and donor planning and appropriate technology selection and implementation—with the first step being training and capacity development of recipient communities and local governance structures (Hemstock et al., 2016, 2018).

For Tuvalu, in terms of the existing coconut resource, it would easily be possible to substitute 50% (271 toe) of Tuvalu's current petroleum use and still have enough coconuts to feed humans and pigs, and produce 1,200 tons of copra which would provide 400 toe coconut oil biodiesel, which would substitute around 40% of current diesel used to fuel the government's inter-island boats (Hemstock, 2013). For Fiji, the current coconut resource could provide the 11.5 million liters of bioethanol required for a 10% blend for all gasoline. However, in terms of the human resource required, it is difficult to predict how much effort people are prepared to put in to collecting toddy, since although it is less arduous than planting and cutting sugar cane, collecting toddy has to be done on a daily basis.

There are many options for the coconut palm to provide resilient development solutions and the improvement of rural livelihoods should be at the forefront of any implementation. For example, if large-scale coconut bioethanol production were to be implemented in the PIR as a means of providing novel and appropriate mitigation and adaptation options, it should be done on a more formal basis than past efforts for copra production. The introduction of toddy bioethanol production should involve long-term contracts for toddy production; realistic 'living wage' minimum payments for production; simplification of land lease renewals; and accompanying coconut woodland rehabilitation and replanting should be negotiated with producers and landowners. Investment in infrastructure and equipment would be required (possibly kick-started through the Global Climate Fund) and mid- to long-term contracts between coconut growers (farmers) and petroleum suppliers would ensure a consistent supply.

Acknowledgments

Alofa Tuvalu: Gilliane Le Gallic; Eti Esela; Risassi Filikaso; John Hensford; Fanny Heros; Christopher Horner; Nala Ielemia; Pierre Radanne; Gilles Vaitilingom. The French Ministry for Foreign Affairs (Pacific Fund); ADEME - Agence de l'Environnement et de la Maıˆtrise de l'Energie; UNDP; The University of the South Pacific Intra-ACP European Union Global Climate Change Alliance Project. Antoine N'Yeurt; David Manuella; Helene Jacot des Combes; Isaia Taape; Joeli Veitayaki; Kaio Taula; Kapuafi Lifuka; Kausea Natano; Maatia Toafa; Mataio Tekinene; Panapasi Nelesone; Pasivao Maani; Saufatu Sopoaga; Tataua & Tuvalu Red Cross; Willy and Senati Telavi. Alpha Pacific Navigation; The Government of Tuvalu; Mama's Petrol Station; Kaupule; TANGO; TMTI; Tuvalu National Council of Women; Tuvalu Electricity Corporation.

References

Bhagya, H.P., Maheswarappa, H.P., Surekha and Bhat, R. (2017). Carbon sequestration potential in coconut-based cropping systems. *Indian Journal of Horticulture*, 74(1): 1–5. Doi: 10.5958/0974-0112.2017.00004.4.

Cerqueira Leite, R.C.D., Verde Leal, M.R.L., Barbosa Cortez, L.A., Griffin, W.M. and Gaya Scandiffio, M.I. (2009). Can Brazil replace 5% of the 2025 gasoline world demand with ethanol? *Energy*, 34: 655–661.

Chandra, V.V. and Hemstock, S.L. (2015). A biomass energy flow chart for Fiji. *Biomass and Bioenergy*, 72: 117–122.

Chandra, V.V. and Hemstock, S.L. (2016). The potential of sugarcane bioenergy in Fiji. *Sugar Tech.*, 18: 229–235.

Charan, N. (2015). RE: Commodified from Sugarcane; personal communication to Chandra, V.

Collins, M., Knutti, R., Arblaster, J., Dufresne, J.-L., Fichefet, T., Friedlingstein, P., Gao, X., Gutowski, W.J., Johns, T., Krinner, G., Shongwe, M., Tebaldi, C., Weaver, A.J. and Wehner, M. (2013). Long-term climate change: projections, commitments and irreversibility. *In*: Stocker, T.F., Qin, D., Plattner, G.-K., Tignor, M., Allen, S.K., Boschung, J., Nauels, A., Xia, Y., Bex, V. and Midgley, P.M. (eds.). *Climate Change 2013: The Physical Science Basis*, Contribution of Working Group I to the *Fifth Assessment Report of the Intergovernmental Panel on Climate Change*. Cambridge University Press, Cambridge, United Kingdom and New York, NY, USA.

Dalibard, C. (1999). Overall view on the tradition of tapping palm trees and prospects for animal production. *Livestock Research for Rural Development*, 11: 5. http://www.lrrd.org/lrrd11/1/dali111.htm.

Dean, N. (2014). *Annual Report 2014*, Fiji Electricity Authority. www.fea.com.fj.

Doner, L.W. (2003). Sugar Palms and Maples A2 – Caballero, Benjamin. *Encyclopedia of Food Sciences and Nutrition* (2nd ed.), Oxford: Academic Press.

Elobeid, A. and Tokgoz, S. (2008). Removing distortions in the U.S. ethanol market: what does it imply for the United States and Brazil? *American Journal of Agricultural Economics*, 90: 918–932.

FAO, Food and Agriculture Organization. (2005). *Global Forest Resources Assessment 2005: Progress towards Sustainable Forest Management*, Food and Agriculture Organization of the United Nations, Rome, Italy. http://www.fao.org/3/a0400e/a0400e00.htm.

FAOSTAT. (2020). http://faostat.fao.org: Food and Agriculture Organization of the United Nations.

Fiji Bureau of Statistics. (2012). Population by age, race and province of enumeration, *Fiji: 2007 Census* [Online]. http://www.statsfiji.gov.fj/indephp/2007-census-of-population: Fiji Bureau of Statistics. http://www.statsfiji.gov.fj/indephp/2007-census-of-population.

Fiji Bureau of Statistics. (2014). *HS Import and Export*. http://www.statsfiji.gov.fj/indephp/search?searchword=import&ordering=newest&searchphrase=all.

Fiji Bureau of Statistics. (2015). *Key Statistics 2014*. www.statsfiji.gov.fj.

Fiji Bureau of Statistics. (2019). 2017 Fiji population and housing census: administration report/Fiji Bureau of Statistics. Suva, Fiji: ISBN 978-982-510-055-3.

FRDP, Framework for Resilient Development in the Pacific. (2016). *Framework for Resilient Development in the Pacific: An Integrated Approach to Address Climate Change and Disaster Risk Management (FRDP), 2017–2030*, Pacific Community (SPC), Secretariat of the Pacific Regional Environment Programme (SPREP), University of the South Pacific (USP), Pacific Islands Forum Secretariat (PIFS), United Nations Office for Disaster Risk Reduction (UNISDR) and United Nations Development Programme (UNDP), The Pacific Community, Geoscience Division, Suva, Fiji.

García, C.A., Fuentes, A., Hennecke, A., Riegelhaupt, E., Manzini, F. and Masera, O. (2011). Life-cycle greenhouse gas emissions and energy balances of sugarcane ethanol production in Mexico. *Applied Energy*, 88: 2088–2097.

Gunn Bee, F., Baudouin, L. and Olsen, K.M. (2009). Independent origins of cultivated coconuts (*Cocos nucifera* L.) and human explorations in the old world tropics. *In: Proceedings of the 9th International Plant Molecular Biology Congress*, October 25–30, 2009, St Louis, USA, by Pierre Gustafon (ed.). Lancaster: DEStech Publications, *Résumé*, p. 333. ISBN 978-1-60595-018-1 IPMB Congress. 9, St Louis, États-Unis, 25 October 2009/30 October 2009.

Gunn, B.F., Baudouin, L. and Olsen, K.M. (2011). Independent origins of cultivated coconut (*Cocos nucifera* L.) in the Old World tropics. *PLoS ONE* 6(6): e21143. DOI:10.1371/journal.pone.0021143.

Hebbar, K.B. and Chaturvedi, V.K. (2015). Impact and adaptation strategies of coconut to climate change. *In:* Vinod, T.R., Sabu, T. and Ambat, B. (eds.). *Proceedings of the Kerala Environment Congress – 2015,* Centre for Environment and Development, Thozhuvancode, Vattiyoorkavu, Thiruvananthapuram, Kerala, India-695013.

Hemstock, S.L. and Radanne, P. (2006). Tuvalu renewable energy study: Current energy use and potential for renewable energies. *An Alofa Tuvalu Report for the Government of Tuvalu,* The French Ministry for Foreign Affairs (Pacific Fund) ADEME – *Agence de l'Environnement et de la Maîtrise de l'Energie,* Funafuti: Government of Tuvalu.

Hemstock, S.L. (2013). The potential of coconut toddy for use as a feedstock for bioethanol production in Tuvalu. *Biomass and Bioenergy,* 49: 323–332.

Hemstock, S.L. and Singh, R.D. (2015). The assessment of biomass consumption. Chapter 5. *In*: Rosillo-Calle, F., de Groot, P., Hemstock, S.L. and Woods, J. (eds.). *The Biomass Assessment Handbook.* 2nd Edition. Routledge. ISBN9781315723273.

Hemstock, S.L., Buliruarua, L.-A., Chan, E.Y.Y., Chan, G., Jacot Des Combes, H., Davey, P., Farrell, P., Griffiths, S., Hansen, H., Hatch, T., Holloway, A., Manuella-Morris, T., Martin, T., Renaud, F.G., Ronan, K., Ryan, B., Szarzynski, J., Shaw, D., Yasukawa, S., Yeung, T. and Murray, V. (2016). Accredited qualifications for capacity development in disaster risk reduction and climate change adaptation. *Australasian Journal of Disaster and Trauma Studies,* 20(1): 15–34. ISSN 1174-4707.

Hemstock, S.L., Jacot Des Combes, H., Buliruarua, L.-A., Maitava, K., Senikula, R., Smith, R. and Martin, T. (2018). Professionalizing the 'resilience' sector in the Pacific Islands Region: Formal education for capacity building. *In:* Klepp, S. and Chavez-Rodriguez, L. (eds.). *A Critical Approach to Climate Change Adaptation: Discourses, Policies and Practices.* Routledge Advances in Climate Change Research. Routledge, UK. ISBN 978-1-13-805629-9.

IFRCRCO, International Federation of Red Cross and Red Crescent Organizations. (2011). Information Bulletin - Tuvalu Drought, *Information Bulletin, No. 2 GLIDE Number,* DR-2011-000146-TUV-14 October 2011.

Ishak, M.R., Sapuan, S.M., Leman, Z., Rahman, M.Z.A., Anwar, U.M.K. and Siregar, J.P. (2013). Sugar palm (*Arenga pinnata*): Its fibers, polymers and composites. *Carbohydrate Polymers,* 91: 699–710.

Khan, A. (2014). *2014 Annual Report,* Fiji Sugar Corporation, http://www.fsc.com.fj/AnnualReport.html.

Kishore, V.V.N. and Srinivas, S.N. (2003). Biofuels of India. *Journal of Scientific and Industrial Research,* 62: 106–123.

Lal, J.J., Kumar, C.V.S. and Indira, M. (2003). Coconut palm. *Encyclopedia of Food Sciences and Nutrition* (2nd ed.), Amsterdam Elsevier Science Ltd.

Lin, W.-Y., Chang, Y.-Y. and Hsieh, Y.-R. (2010). Effect of ethanol-gasoline blends on small engine generator energy efficiency and exhaust emission. *Journal of the Air & Waste Management Association,* 60: 142–148.

Liu, Trogisch, S., He, J.S., Niklaus, P.A., Bruelheide, H., Tang, Z., Erfmeier, A., Scherer-Lorenzen, M., Pietsch, K.A., Yang, B., Ku¨hn, P., Scholten, T., Huang, Y., Wang, C., Staab, M., Leppert, K.N., Wirth, K., Schmid, B. and Ma, K. (2018). Tree species richness increases ecosystem carbon storage in subtropical forests. *Proclamations of the Royal Society B,* 285: 20181240. http://ddoi.org/10.1098/rspb.2018.1240.

LRPD, Land Resource Planning & Development. (2012). *Land Use Capability Classification System: A Fiji Guideline for the Classification of Land for Agriculture,* Land Use Planning Section, Land Resource Planning & Development, Department of Agriculture,

https://pafpnet.spc.int/pafpnet/attachments/article/183/Land%20Use%20Capability%20Guideline_web.pdf, 1–41.
McLean, R.F., Holthus, P.F., Hosking, P.L., Woodroffe, C.D. and Kelly, J. (1986). *Tuvalu Land Resources Survey*, vols. 1–9, UNDP/FAO Contract: DP/TU/80/001-1/AGOF, University of Auckland, New Zealand.
McNicol, I.M., Ryan, C.M., Dexter, K.G., Ball, S.M.J. and Williams, M. (2018). Aboveground carbon storage and its links to stand structure, tree diversity and floristic composition in South-eastern Tanzania. *Ecosystems*, 21: 740–754. https://doi.org/10.1007/s10021-017-0180-6.
Naresh Kumara, S. and Aggarwal, P.K. (2013). Climate change and coconut plantations in India: Impacts and potential adaptation gains. *Agricultural Systems*, 117: 45–54. https://doi.org/10.1016/j.agsy.2013.01.001.
Nguyen, T.L.T., Gheewala, S.H. and Garivait, S. (2007). Energy balance and GHG-abatement cost of cassava utilization for fuel ethanol in Thailand. *Energy Policy*, 35: 4585–4596.
OECD, Organization for Economic Co-operation and Development. (2017). *Development Aid at a Glance: Statistics by Region – Oceania.* http://www.oecd.org/dac/stats/documentupload/Oceania-Development-Aid-at-a-Glance.pdf.
Prasad, R.D. and Raturi, A. (2017). Grid electricity for Fiji islands: Future supply options and assessment of demand trends. *Energy*, 119: 860–871.
Quintero, J.A., Montoya, M.I., Sánchez, O.J., Giraldo, O.H. and Cardona, C.A. (2008). Fuel ethanol production from sugarcane and corn: Comparative analysis for a Colombian case. *Energy*, 33: 385–399.
Republic of the Fiji Islands. (2013). *Fiji Coconut Industry, High-level Consultation on Coconut Development in Asia and the Pacific Region*, Bangkok, Thailand: Fiji Government. http://www.fao.org/fileadmin/templates/rap/files/meetings/2013/131030-fiji.pdf.
Rosillo-Calle, F., Woods, J. and Hemstock, S.L. (2003). *Consultancy Report for SOPAC – South Pacific Applied Geoscience Commission on Biomass Resource Assessment, Utilization and Management for Six Pacific Island Countries: Fiji, Kiribati, Samoa, Tonga, Tuvalu and Vanuatu.* Imperial Centre for Energy Policy and Technology/EPMG, Dept. of Environmental Science and Technology, Imperial College London.
Rosillo-Calle, F., De Groot, P., Chandra, V.V. and Hemstock, S.L. (2015). General introduction to the basis of biomass assessment methodology. *In*: *The Biomass Assessment Handbook*, Routledge, 2 Park Square, Milton Park, Abingdon, Oxon OX14 4RN. ISBN 978-1-138-01965-2 (pbk) ISBN: 978-1-315-72327-3 (ebk).
Silalertruksa, T. and Gheewala, S.H. (2010). Security of feedstocks supply for future bio-ethanol production in Thailand. *Energy Policy*, 38: 7476–7486.
Singh, A. (2020). Benefits of crop diversification in Fiji's sugarcane farming. *Asia-Pacific Policy Studies*, 7: 65–80.
Tamunaidu, P., Matsui, N., Okimori, Y. and Saka, S. (2013). Nipa (*Nypa fruticans*) sap as a potential feedstock for ethanol production. *Biomass and Bioenergy*, 52: 96–102.
Woods, J., Hemstock, S.L. and Bunyeat, J. (2006). Bio-energy systems at the community level in the South Pacific: Impacts & monitoring, Greenhouse gas emissions and abrupt climate change: Positive options and robust policy. *Journal of Mitigation and Adaptation Strategies for Global Change*, 4: 473–499.
World Bank. (2020). *Data Catalogue*. https://data.worldbank.org/country/TV.
Yüksel, F. and Yüksel, B. (2004). The use of ethanol-gasoline blend as a fuel in an SI engine. *Renewable Energy*, 29: 1181–1191.
Zanne, A.E., Lopez Gonzalez, Gcomes, D.A., Ilic, J., Janson, S. and Lewis, S.L. (2009). *Global Wood Density Database*. https://dryad.figshare.com/articles/Global_Wood_Density_Database/4172847.

5

Climate Change is Challenging Oil Palm (*Elaeis guineensis* Jacq.) Production Systems

Alain Rival[1,*] and *Cécile Bessou*[2]

1. Introduction

The agricultural sector has a direct impact on climate as it generates significant amounts of greenhouse gas emissions. Indeed, the recorded increase of GHG concentration in the atmosphere, together with a steady rise in temperatures together with disturbance of precipitation regimes, are currently influencing not only the volume, quality and stability of agricultural productions, but also the natural environment surrounding agricultural activities.

The major agriculture-related determinants of climate change (captured by CO_2 emission) are linked to crop production, burning biomass residues, enteric fermentation, manure management, or synthetic fertilizers. This situation necessitates the immediate implementation of adaptation strategies at both the farm and landscape levels. It is important to decrease vulnerabilities of both individual farmers and agricultural systems in order to mitigate the adverse effects of climate change.

[1] Cirad Senior Project Manager. Graha Kapital 1, Jalan Kemang Raya 4, Jakarta 12730, Indonesia.

[2] Cirad. UMR AbSYS, TA B-34/02, Avenue Agropolis 34398, Montpellier, Cedex5, France. Email: cecile.bessou@cirad.fr

* Corresponding author: alain.rival@cirad.fr

Sustained mitigation efforts are also required in a global effort to reduce climate change, ensure sustainable food production and improve rural livelihoods.

2. The Oil Palm: A Champion or a Clay-footed Colossus?

In terms of sustainability and resilience, the rapid and large-scale development of oil palm cultivation reminds the careful observer about the image of a clay-footed colossus; as a sector showing a very strong growth and commercial success, which is nevertheless masking several worrying weaknesses. The vegetable oil market is very competitive and researchers and decision-makers should be aware of the fragilities carried by both the crop itself and its cultivation systems.

Indeed, besides the intrinsic risks caused by the large-scale monoculture of one single species on extended areas, it must be noted that the global palm oil production is limited—in its vast majority—to a rather narrow geographical region that is limited to both sides of the Malacca Strait. Indeed, Indonesia and Malaysia combined accounted for 87% of global oil palm production in 2019. This situation may pose serious problems in case of either an extreme climatic event or a pest/disease invasion, which could be connected to each other (Paterson, 2019). It is paramount to remember that the African palm oil was almost wiped out from Brazil because of the Bud Rot syndrome that occurred a few decades ago (De Franqueville, 2003).

The poor climatic resilience of present palm oil cultivation systems is observable in Fig. 1, which clearly shows a sizeable drop in global palm oil production as the consequence of the severe the El Niño–Southern Oscillation (ENSO) episode, which occurred in Southeast Asia in 2015. Also, a similar extreme drought did occur in 2019, with similar consequences on yields at the global level. Plantations had to face the double effect of extreme drought—directly through water stress-induced responses, such as changes in sex ratio and drop in fruit development and oil synthesis, and indirectly through the impact of bushfires surrounding estates as they hampered photosynthetic activity for months when thick haze covered the entire region.

Indeed, Stiegler et al. (2019) showed that haze conditions during ENSO 2015 resulted in a complete pause of oil palm net carbon accumulation, which lasted for almost 1.5 months and this situation was found to cause a decline in oil palm yield by 35%. The model developed by the author also demonstrated that an increase in drought was able to stimulate the net CO_2 uptake. Besides, a more severe haze, in combination with drought, can induce some pronounced losses in productivity and net CO_2 uptake by oil palm stands.

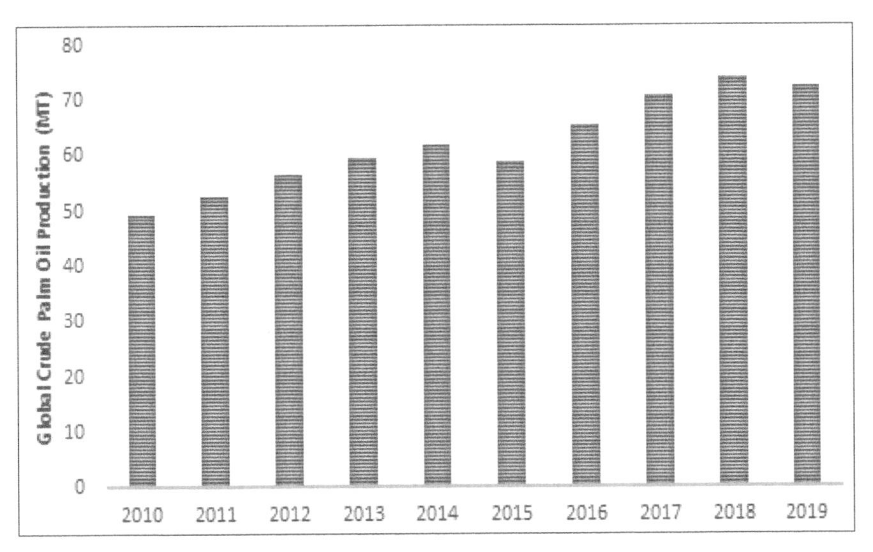

Fig. 1. Changes in global crude palm oil production between 2010 and 2019 (United States Department of Agriculture, Foreign Agricultural Service).

Furthermore, another source of vulnerability lies in the genetic basis of present palm oil hybrid material that is selected, produced and cultivated worldwide. The exploited and cultivated agro-biodiversity remains quite narrow when compared to other industrial crops of major importance. Would this restricted genetic diversity be wide enough to enable the oil palm to survive a major pathology/agronomy/climate-related disaster (Paterson et al., 2013; Paterson and Lima, 2018)?

Today, the competitive advantages of palm oil compared to competing vegetable oil still rely on cheap production costs that are structurally based on the abundance of arable land, the natural high productivity of the crop and cheap labor costs (Corley and Tinker, 2015).

Now it is about time for research to address the long-term resilience of such a fragile situation, the 'so far so good' current behavior might not be an option any more.

2.1 *Keys for Improved Resilience*

Agriculture is often described as one of the sectors most vulnerable to future climate change; building resilience in both the crops itself and in its cultivation systems (including its natural environment) is the key.

Eycott et al. (2019) postulated that biodiversity losses related to the expansion of oil-palm plantation at the expenses of rain-forests have a series of measurable impacts on ecosystem functions inside the oil-palm plantation and on the resilience of functions to changing rainfall patterns, with impact on the yield.

Approaches based on 'ecological intensification' and 'climate-smart agriculture' are highly complementary. Indeed, ecological intensification is an essential means of adapting to climate change, which results in lower emission rates. Climate-smart agriculture provides the foundations for incentivizing and enabling intensification, as it improves risk management, information flows and local institutions to support adaptive capacity. Both approaches are also crucial for global food and nutritional security. They are only part of a multi-pronged approach, that includes reducing consumption and waste, building social safety nets, facilitating trade and enhancing diets (Lipper et al., 2014).

Growing evidence suggests that ecological intensification of mainstream farming can safeguard food production, with accompanying environmental benefits even if the approach needs to be fully adopted by farmers. Kleijn et al. (2019) reviewed the evidence for replacing external inputs with ecosystem services. Their research showed that scientists tend to focus on processes (e.g., pollination) rather than outcomes (e.g., profits). Scholars usually express benefits at spatio-temporal scales that are not always relevant to farmers.

Nelson et al. (2018) modeled the ways in which oil palms can influence the biophysical state of their environment, not only within the field itself but also in the wider environment, in particular via exchanges with the atmosphere and hydrosphere (Fig. 2). This vision puts the agricultural parcel at the center of multiple interactions between the plant, soil, water, the atmosphere and the living organisms living in these areas.

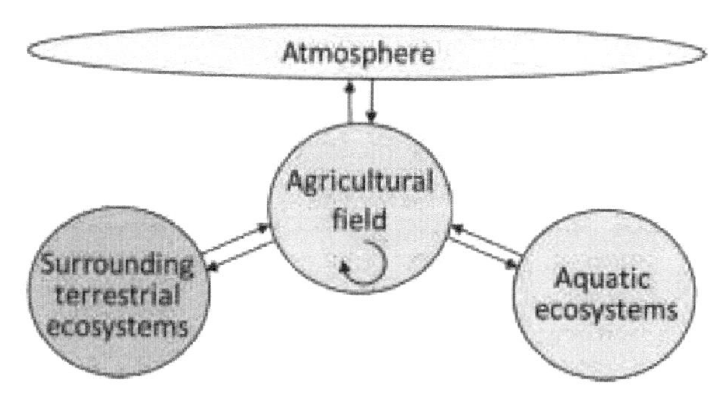

Fig. 2. Simplified scheme of movement of energy, materials and organisms (arrows) between the cultivated plot and the atmosphere, aquatic ecosystems and surrounding terrestrial ecosystems, and within the field itself (Nelson et al., 2018).

2.1.1 Building on Ecosystem Functions

Oil-palm plantations generally show a reduced ecosystem functioning when compared to rain-forests (Dislich et al., 2017)—an ecosystem which replaces palm oil in frontier zones. These authors found, from exhaustive literature review, that some functions show decreases with potentially irreversible global impacts (e.g., reductions in gas and climate regulation, habitat and nursery functions, genetic resources, medicinal resources and information functions). Such impacts become more serious when the forest is cleared to establish new plantations and immediately afterwards, especially on peat soils. Several specific changes in plantation management can prevent or reduce losses of some ecosystem functions and synergistic mitigation measures can improve multiple ecosystem functions simultaneously.

2.1.2 Innovative Planting Designs

In the context of climate change, especially in the inter-tropical regions, agricultural management strategy must innovate in order to improve biodiversity and ecosystem functions within oil-palm production landscapes. Production systems based on oil-palm-intensive monoculture certainly need to be explored with a new perspective embracing climate change (Rival, 2017; Paterson and Lima, 2018). Reforestation experiments and rehabilitation of riparian areas have proven as key in retaining wildlife and improving local ecosystems in plantations-dominated landscapes (Horton et al., 2018; Lee et al., 2019). In addition, tree crop plantations are increasingly implementing agro-ecological approaches, relying on the incorporation of environmental services (Bessou et al., 2017). Research from Azhar et al. (2017) suggests that retailers and consumers who are interested in promoting sustainable palm-oil production should consider small-scale farmlands in which palm oil is mixed with other crops. Diversifying livelihood options through intercropping or other means is important for smallholders. Potential benefits rely on the diversification of crop and rural activities, bringing improvements in food safety, nutrition and financial resilience. Characterizing and assessing intercropping and agroforestry-based systems need solid research (Nchanji et al., 2016; Brandão et al., 2018) as more evidence is needed to identify appropriate crops, systems and markets. Changes in agricultural practices must involve innovative planting designs for a precise and long-term estimation of the benefits of simultaneous cultivation of forest species and perennial plantation crops. From demonstration plots of various sizes, compositions and shapes, a series of precise measurements of bioclimatic parameters can provide data on both the agronomic performance and the climatic resilience of such mixed agroforestry systems.

Changes in the biodiversity and abundance of wildlife need to be monitored together with the impact of agro-forest designs, not only on the connectivity between plantations, corridors and riparian forests, but also on crop yields and resistance to both biotic and climatic stress. New collaborative research is needed in order to document such changing paradigms in tree-crops plantation management that must lead to a sustainable combined management of wildlife and agricultural development.

In the aim of improving and fine-tuning the many possible conservation strategies, there are opportunities for partnerships between oil-palm producers, conservation practitioners and rural communities. Such partnerships are expected to enable the channeling of financial resources from oil-palm plantations into forest conservation efforts, such as local capacity building in legal aspects of forest law and enforcement (Fitzherbert et al., 2008). There might be also opportunities for palm-oil producers to preserve the forest remnants located within their own plantations. Strategic alliances between multiple stakeholders, such as oil-palm producers, environmental organizations, rural communities, government agencies and carbon offsetters have reasonable chances of success (Ancrenaz et al., 2021).

Among various systems designed for improving agro-ecological practices, the alley-cropping system is probably the most easy to install from traditional mono-cropping systems. These adaptations often show a good potential to enhance faunal biodiversity, ecosystem services and food security in agricultural landscapes. Alley cropping relies on the association of a main crop intercropped with a secondary species. The secondary crops (for food, timber, spices, medicinal plants, or horticultural species) are cultivated in the alleys in between the main crop. Ashraf et al. (2018) compared arthropod taxonomic richness, arthropod predators and decomposers between five alley-cropping treatments (pineapple, bamboo, black pepper, cacao, Bactris palms). In all tested cultivation systems, the oil palm was intercropped with another species. The authors described an increase in beta-diversity of oil-palm production landscape: they found that the number of arthropod orders, families and abundance were significantly greater in alley-cropping parcels. In addition, alley-cropping treatments harbored a larger numbers of predators and decomposer families of arthropods.

The adoption of new models of palm-oil production based on tree intercropping/agroforestry systems is still at its infancy, despite the paramount interest of such system in terms of climatic and socio-economic resilience, when compared to monoculture-based plantations that have hardly changed since colonial times. Comparisons between conventional and agroforestry-based production are made difficult by information asymmetry. Miccolis et al. (2019) shed light on oil palm farmers' interest in diversification in Brazil—major motivations included potentially greater resilience to market risks and fluctuations, climatic resilience and enhanced food security.

Potential barriers were also identified, which included high costs and the knowledge-intensive nature of agroforestry in comparison to standard monoculture. Scaling up agroforestry experiments from oil-palm demonstration sites to large plantation area will require a large body of research, which should also focus on buffer zones between forests and plantations, such as riparian areas. Indeed, Bhagwat and Willis (2008) suggested that oil-palm plantation managed as agroforestry systems could foster conservation efforts through the provision of habitat for forest-dwelling species, the establishment of connections between biodiversity-rich areas and the securing of livelihood for local people.

The brutal conversion of complex native rain-forests into intensive oil-palm monocultures leads to dramatic losses of biodiversity and ecosystem functions. In order to alleviate negative ecological impacts, enrichment with native tree species may rapidly provide solutions. Indeed, experiments are under way (Messier et al., 2022; Zemp et al., 2019) aimed at restoring some degree of structural complexity in existing oil-palm plantations. The authors measured structural complexity from terrestrial laser scanning in a biodiversity-enrichment experiment with multiple tree species planted in an oil-palm monoculture, forming agroforestry plots of varying tree species diversity and plot size. It was found that three years after tree planting, structural complexity in plots increased by one-third, thus representing 25% of the increase needed to restore the structural complexity of tropical forests. Changes in structural complexity were associated with denser and more complex filling of three-dimensional space, whereas vertical stratification was mainly influenced by oil palm.

Korol et al. (2021) identified the landscape-wide density and distribution patterns of scattered trees and estimated their size in an oil-palm dominated landscape in Indonesia. The authors found a considerable number of scattered, mostly small-statured trees. They concluded that, in order to ensure the survival of trees and further provision of related ecosystem services, scattered trees in the oil-palm landscape need to be preserved and/or restored.

Furthermore, without profoundly changing the planting designs, it is also of paramount importance to avoid the total destruction of the whole plantation biotope at replanting time. Such current practices jeopardize all efforts engaged during the whole plantation life (20 years) in integrating agro-environmental services into best agricultural practices.

2.1.3 Breeding New Palms for a Changing Climate

Breeding the oil palm for climate change requires projects that are established and secured in the very long term, as the needed multidisciplinary and collaborative research must be based on strong partnerships (Rival, 2017). Indeed, besides genetics and structural and functional genomics, this approach involves almost all disciplines related to life sciences, ranging from physiology to plant architecture and anatomy. Such an integrative research work relies on the identification of genetic variation governing the various strategies of response to stress developed by the plant. A wide diversity of resources needs to be explored, covering not only the diverse genotypes generated by natural variation, but also germplasm collections, selected genitors from breeding programs as well as plant material of interest collected from farmers.

The phenotyping of selected plant material under biotic/abiotic stress involves new methods for high-throughput phenotyping. The efficient selection of targets relies on genomic approaches aimed at identifying genes and genes families that govern the variation of traits.

Oil-palm breeding programs are increasingly depending on the understanding on how climate change interferes in the various chemical and physical processes occurring in soils, how this affects nutrient composition and availability and how the plant responds to such changes.

Various patterns of plant breeding, especially for perennial species, are optimized through molecular approaches and tools. Polymorphisms controlling various traits of interest are now identified, thanks to an exponential increase in molecular resources and methods. Research efforts now focus on the exploration of the mechanisms linking identified genetic polymorphisms to stable phenotypes. With genomic resources becoming increasingly affordable for the oil-palm research community (sequencing, resequencing and chips development), it is now increasingly feasible to explore the genetic basis of complex traits, such as oil yield, resistance to diseases or growth speed (Singh et al., 2013; Rival, 2017). The frontier between crop plants and model species, such as Arabidopsis, is constantly fading away. Consequently, the securing and sharing of big data generated by high throughput approaches is currently reshaping the modern strategies for oil-palm breeding.

2.1.4 Adapting the Sex Ratio: Water, Sex and Sun

The oil-palm reproductive biology is governed by a monoecious (also referred to as 'temporally dioecious') distribution which generates successive cycles of male and female flowers on the same plant. Increasing the sex ratio of oil-palm flowers (the number of female inflorescences in relation to the total number of inflorescences) is a key breeding target. An ideal oil palm in the planter's eye would produce only female inflorescences whenever possible...with just a little help from a small number of adequately dispersed pollinating palms.

In-depth research on oil-palm flowering is complicated but vital to understand the biology of this important crop plant, despite its large size and long life cycle. Adam et al. (2011) reviewed the current understanding of the process of sex determination in oil palm (Fig. 3). The sex ratio of an oil-palm population is known to be influenced by both genetic and environmental factors. In particular, several authors have documented the role of water stress on the higher production of male inflorescence (Corley et al., 1976; Legros et al., 2009). A mature palm alternates between male and female flowering phases during its lifetime, but the proportion of time spent in each phase will vary considerably, depending on both environmental and genetic factors. In regions of the world with high and regular rainfall (e.g., Malaysia and Indonesia), oil-palm sex ratios tend to be very stable and to vary little throughout the year. In regions with a marked dry season, such as in West Africa, the sex ratio undergoes extensive fluctuations. Indeed, Corley (1976) noted that in such places, the period of lowest sex ratio (high male inflorescence production) occurs during the rainy season. This author speculated that this such changes are an adaptation to the reduction in airborne pollen density caused by a higher atmospheric humidity.

When combined with appropriate and targeted field studies, innovative approaches based on genomics, molecular and biochemical cytology can provide a new understanding of the complex processes that

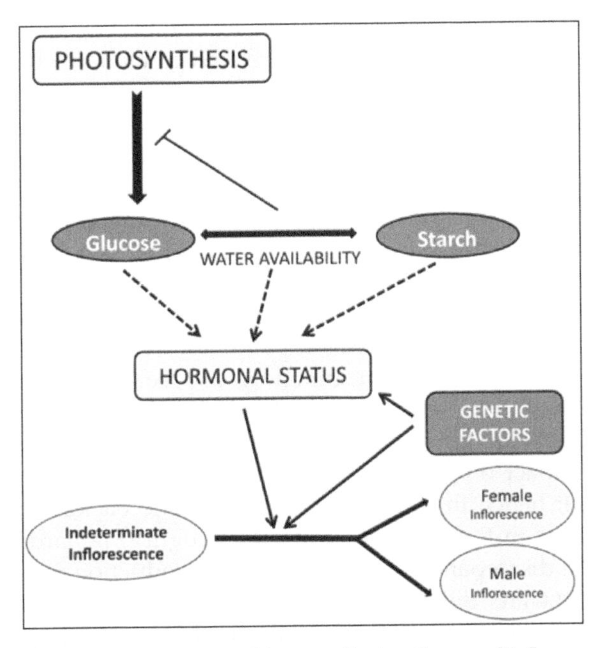

Fig. 3. A putative mode of interaction of factors affecting the sex of inflorescences in the oil palm, showing various targets for molecular studies (Rival, 2017; adapted from Adam et al., 2011). The suggested mechanism relies mostly on phenological and physiological data; it suggests various points of interaction with genetic factors.

govern the determination of sex in the oil palm. Singh et al. published the genome sequence of the oil palm in 2013, thus providing a key milestone for genetics and genomics studies in this species. This boosted the research efforts developed by various groups on the epigenetic determinism of flower structure in the *mantled* oil-palm variants (Jaligot et al., 2014; Ong-Abdullah, 2015). Such research works have paved the way for a deeper molecular analysis of sex ratio and the further elucidation of its determinism in relation with changes in the plant environment.

Given that the sex ratio in oil palm is dependent on the genotype makes it possible to identify the genes in charge of determining the sex of inflorescences. Indeed, the EgACCO1 (encoding aminocyclopropane carboxylate (ACC) oxidase) on chromosome 10 and EgmiR159a (microRNA 159a) at chromosome 6 were identified by Somyong et al. (2016). These are the most linked QTL genes or determinants for FFB yield and/or female inflorescence number in the oil palm. Markers-assisted breeding for a higher sex ratio will require the identification of both genetic and epigenetic determinisms of sex ratio and of their respective regulatory mechanisms. Adam et al. (2011) have discussed the hypothesis of a control of sex determination genes by plant growth regulators, thus suggesting a possible crosstalk between sex determination genes, plant growth regulators and various environment factors (Fig. 3).

In a context of high uncertainties related to climate change and its impact on stress and sex determination for oil-palm crops, short- and long-term impacts of combined genetic and environmental determinants must be anticipated as precisely as possible. It is of uttermost importance to decipher the physiological and molecular mechanisms at stake, as previous efforts in genetic selection have led to the generation of elite varieties and clones bearing a high proportion of female inflorescences in order to reach high yields. The process of genetic selection creates issues linked to the reduced number of male inflorescences and pollen available in the field, calling for the planting of pollinator trees.

Several authors have contributed to a precise description of the impact of drought on oil palm (Cochard et al., 2005). Their research shows that the behavior of oil palm is strongly affected by water deficit through its impact on growth, sex ratio and the rate of aborted female inflorescences; hence the yields. Different genetic backgrounds display clear differences in their susceptibility to drought (Maillard et al., 1974) although such discrepancies are not linked to production potential. In the absence of any water deficit, the production of oil palm is spread over the year, however, when water deficit occurs, a relative peak in production is recorded during the wet season. It is strongly recommended to develop planting material displaying a wide distribution of yield over the year or showing a staggered production peak. Some substantial genetic diversity linked to production sequences was evidenced, although production cycles were found to be governed by the same seasonal variations when water deficits were pronounced (Cochard et al., 2005). More, when genetic crosses showing a high root density and mechanisms of resistance to cell dehydration were studied, authors revealed that drought-related mortality was low (Cornaire et al., 1994).

At the agro-ecosystem level, Eycott et al. (2019) found that all ecosystem functions within mature oil-palm plantation were not very sensitive to changes in rainfall. Even if functions, such as seed removal and herbivory, were not altered by rainfall, decomposition and predation were found to show more complex effects, as levels of both processes increased with current rainfall levels, when rainfall in preceding time periods was low. Such findings suggests that the oil palm may be robust to future changes in the precipitation regime.

2.1.5 Climate-smart Pest and Diseases Management

In a vast range of crops and across all land uses and landscapes, climate change is affecting the biology, distribution and outbreak potential of pests. Indeed, up to 40% of the world's food supply is already lost to pests. In this context, reducing their impact is gaining more attention than ever in order to ensure global food security, a limited application of inputs and decreased GHG emissions. Climate-smart pest management (CSPM) has emerged as a cross-sectoral approach aimed at reducing pest-induced crop losses, enhance ecosystem services, reduce the greenhouse gas emissions intensity per unit of food produced and strengthen the resilience of agricultural systems in the context of climate change (Heeb et al., 2019). Through the implementation of CSPM, crop production, extension, research and policy act in coordination towards more efficient and resilient food production systems (Fig. 4).

Several authors highlighted the role and importance of maintaining functional biodiversity within the oil-palm landscape (Foster et al., 2011; Bessou et al., 2017) through integrated pest management strategies. Besides pathogenic species, climate change may also affect various populations of insect (Paterson et al., 2019) that play key roles in oil palm productivity, such as pollinators.

Mariau and Gentil (1988) reviewed issues on pollination of the oil palm in Africa, South America and Indonesia. In Africa, the curculionids *Elaeidobius kamerunicus, E. subvittatus, E. plagiatus* and *E. singularis* ensure the major part of pollination with the first species being the most important. In South America, palm-oil pollination is mainly ensured by *E. subvittatus* (Colombia, Brazil and Peru) and the nitidulid *Mystrops costaricensis* (Central America, Colombia and Ecuador). Pollination is generally insufficient, due to the poor role played by *E. subvittatus*. In Indonesia,

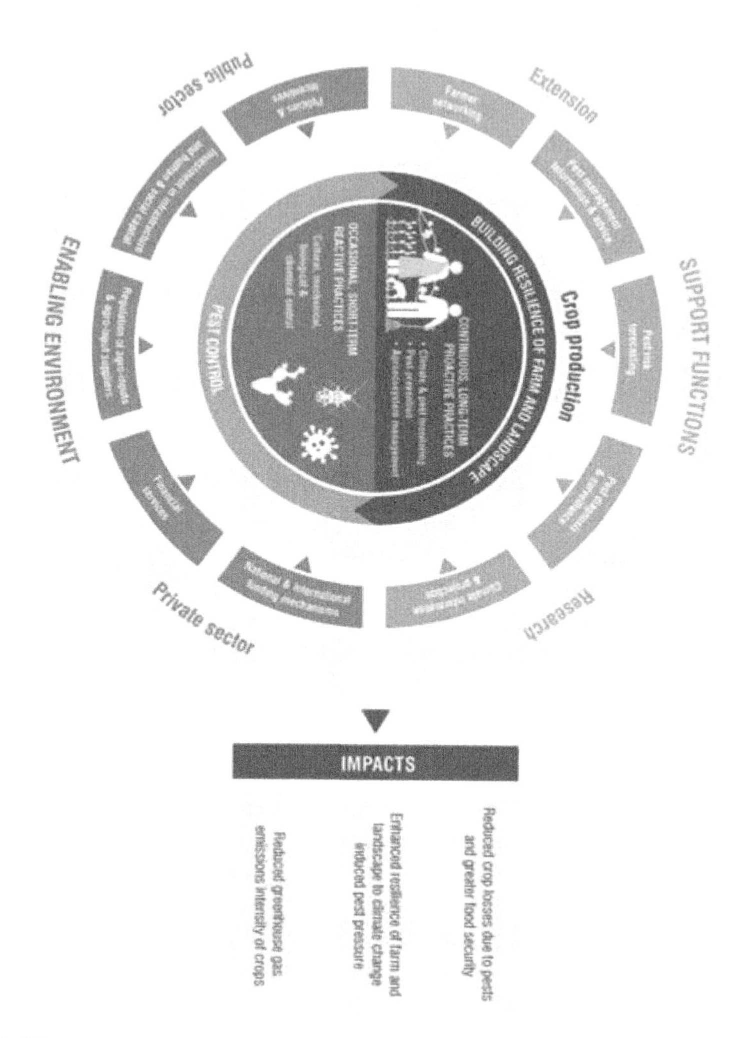

Fig. 4. Climate-smart pest management (CSPM): An interdisciplinary approach to increase the climatic resilience of farms and landscapes. Figure from Heeb et al. (2019), distributed under the terms of the Creative Commons Attribution 4.0 International License (http://creativecommons.org/licenses/by/4.0/), no changes were applied to the original figure.

the low pollinating capacity of *Thrips hawaiiensis* led to the introduction of *E. kamerunicus* in 1983. This introduction provided a huge boon to the oil-palm industry, through the elimination of the need for hand pollination. This resulted in an increasing yield by around 20% (Foster et al., 2011).

Despite an overall success, a decline in pollination efficiency was documented in several regions and there may be high levels of inbreeding depression within the present weevil populations now living in oil palm-producing countries. In order to explore the genetic diversity of *Elaeidobius kamerunicus* in both its native and introduced ranges, Haran et al. (2020) recently reconstructed a multilocus phylogeography of the world populations. These authors found that the African populations of *E. kamerunicus* are shared between two differentiated mitochondrial clusters in west and central Africa, following a contact zone along the Cameroon Volcanic Line. A sex-biased dispersal in this species is suggested, with males dispersing more than females. This is coherent with a differential genetic structure between mitochondrial and nuclear genes, and the strong level of genetic structure of the mitochondrial gene. The genetic structure inferred from Asian and South American populations of *E. kamerunicus* suggests that they originate from populations of both western and central regions of tropical Africa.

However, relying on a single species for such a pivotal service remains a risky strategy. Indeed, the pollinating capacity of *Elaeidobius* is known to be sensitive to climatic conditions. The fruit set is generally hampered by reduced flying capacity during the rainy season and the pollinator performance is also reduced under dry conditions. In oil-palm plantations, a high abundance of native pollinator species can support adequate pollination and fruit set, thus suggesting a higher diversity of oil-palm pollinators than is generally realized.

2.1.6 Drivers of Oil Palm Environmental Impact

The main approach to quantifying all environmental impacts of the production of palm oil is life cycle assessment (LCA) (Fig. 5), which proposes an integrated model of environmental impacts (Nelson et al., 2018).

LCA of oil-palm products enables the assessment of various facets of the crop cultivation, together with several parts of the commodity chain. This approach also facilitates comparisons between the production of palm oil and other alternatives (Schmidt, 2015). When its impact on climate change is assessed through LCA, palm oil generally shows poorer level of performance than other vegetable oil crops, mainly because of the induced land use change from systems with high carbon stock. Oil palm was found to perform better than rapeseed oil in terms of eutrophication, acidification, ozone depletion and photochemical ozone impacts (Schmidt, 2010).

Over the last decade (Nelson et al., 2018) about 70 full or partial LCAs of palm-oil products have been published. Most of these studies focused on palm oil as biofuel, and they principally considered GHG emissions (or climate change impact) and energy balance (or fossil resource depletion) (Bessou et al., 2013; Manik and Halog, 2013). The focus was on the life

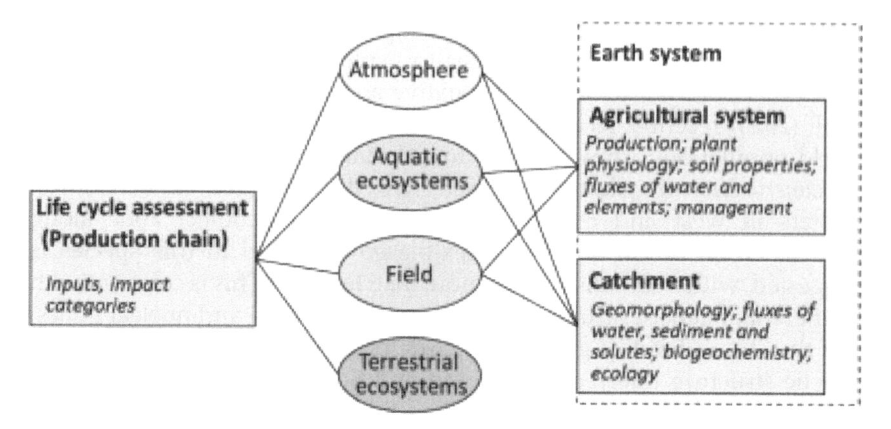

Fig. 5. Integrated biophysical modeling of environmental impacts of agriculture follow two different approaches: life cycle assessment on the one hand, or dynamic process-based modeling of the earth system, agricultural systems or catchment processes, on the other. Boxes represent typical system boundaries and modeling focus, and lines represent the principal environmental impacts modeled (from Nelson et al., 2018).

cycle for palm methyl ester or biodiesel, including all processes ranging from background input production (e.g., fertilizer manufacture) up to the vehicle tank. They took into account total combustion or engine efficiency in order to calculate the final energy and GHG indicators.

LCAs on palm oil revealed that the agricultural phase of the cycle is the major contributor to the impact, except for human toxicity or respiratory impacts, to which boiler emissions are the main contributor (Stichnothe and Schuchardt, 2011). The impact of palm-oil production on climate change is strongly governed by the use of fertilizers, especially nitrogen, and by changes in carbon stock in the case of land use change or peat drainage (Bessou et al., 2014). The impact on eutrophication is driven by the emissions of nitrogen- and phosphorus-compound, although mill emissions can also contribute. At the agricultural stage, the major eutrophication factors were found to be nitrate leaching, and phosphorus and nitrate run-off. The use of fertilizers influenced both acidification and photochemical ozone emissions, especially due to their role impact on nitrogen compound emissions. The largest uncertainties relate to cultivation of peat and emissions of N_2O (Schmidt, 2015).

Impacts originating from outside the oil-palm grove are also influenced by in-field management practices, in particular the use of fertilizers. Indeed, significant GHG emissions are generated by the production of nitrogen fertilizer (10–30% of total emissions from fertilizers, compared to 70–90% generated in the field), and by palm-oil mill effluent treatment ponds, especially CH_4 (Wicke et al., 2008; Bessou et al., 2014).

To date, Several LCA impact indicators still need to be more intensively studied in palm-oil production systems. Many indicators are still little studied: these include the impacts of herbicides on terrestrial or freshwater ecotoxicity and the impact of irrigation systems on water depletion.

3. Conclusion and Prospects

Forest conversion to oil-palm plantations directly affects climatic resilience of oil-palm plantations through ecosystem functions. Among others, such affected functions include carbon storage, nutrient cycles, soil regeneration and air and water purification

In order to build on climate-smart approaches, palm-oil production will certainly need to address sustainability issues at the landscape level. Indeed, mosaic landscapes are of great potential—while land-sparing approaches proved useful in preserving biodiversity and ecosystem services, land-sharing strategies are often effective complementary tools, especially in degraded landscapes (Mertz and Mertens, 2017). The combination of land sparing and land sharing generates mosaic landscapes of interest. These are able to generate substantial gains in biodiversity and climate resilience, with only modest decreases in oil production per hectare (Gerard et al., 2017).

The global interest in diversified, mixed oil-palm systems is increasing, as these complex designs are able to increase the efficiency of the use of land and other resources, reduce farmer risk and decrease greenhouse gas (GHG) emissions. Considerable economic and environmental system improvements appear to be feasible through agroforestry-based oil-palm cultivation and diversification as a pathway to intensification deserves full attention of research and policy development (Kasanah et al., 2020).

It is necessary to maximize the conservation value of degraded lands and use an integrated, landscape-scale approach to reconcile economic development and biodiversity conservation. Improving oil-palm cultivation must also include the promotion of climate-smart management practices that incorporate an explicit economic valuation of ecosystem services in and around the plantation.

Acknowledgments

The author gratefully thanks helpful colleagues Jean Ollivier and Bernard Dubos for valuable discussions, which facilitated the production of the present chapter.

References

Adam, H., Collin, M., Richaud, F. et al. (2011). Environmental regulation of sex determination in oil palm: Ccurrent knowledge and insights from other species. *Ann. Bot.*, 108(8): 1529–1537.

Ancrenaz, M., Oram, F., Nardiyono, N., Silmi, M., Jopony, M.E., Voigt, M., Seaman, D.J., Sherman, J., Lackman, I., Traeholt, C. and Wich, S.A. (2021). Importance of small forest fragments in agricultural landscapes for maintaining orangutan metapopulations. *Frontiers in Forests and Global Change*, 4: 5.

Ashraf, M., Zulkifli, R., Sanusi, R., Tohiran, K.A., Terhem, R., Moslim, R., Norhisham, A.R., Ashton-Butt, A. and Azhar, B. (2018). Alley-cropping system can boost arthropod biodiversity and ecosystem functions in oil-palm plantations. *Agriculture, Ecosystems & Environment*, 260: 19–26.

Azhar, B., Saadun, N., Prideaux, M. and Lindenmayer, D.B. (2017). The global palm oil sector must change to save biodiversity and improve food security in the tropics. *Journal of Environmental Management*, 203: 457–466.

Bhagwat, S.A. and Willis, K.J. (2008). Agroforestry as a solution to the oil-palm debate. *Conservation Biology*, 22(6): 1368–1369.

Brandão, F., Schoneveld, G. and Pacheco, P. (2018). *Strengthening Social Inclusion within Oil-palm Contract Farming in the Brazilian Amazon*. vol. 206, CIFOR.

Bessou, C., Basset-Mens, C., Tran, T. and Benoist, A. (2013). LCA applied to perennial cropping systems: a review focused on the farm stage. *The International Journal of Life Cycle Assessment*, 18(2): 340–361.

Bessou, C., Chase, L.D., Henson, I.E., Abdul-Manan, A.F., i Canals, L.M., Agus, F., Sharma, M. and Chin, M. (2014). Pilot application of PalmGHG, the roundtable on sustainable palm oil greenhouse gas calculator for oil palm products. *Journal of Cleaner Production*, 73: 136–145.

Bessou, C., Verwilghen, A., Beaudoin-Ollivier, L., Marichal, R., Ollivier, J., Baron, V., Bonneau, X., Carron, M.P., Snoeck, D., Naim, M. and Aryawan, A.A.K. (2017). Agroecological practices in oil-palm plantations: Examples from the field. *OCL-Oilseeds and Fats, Crops and Lipids*, 24(3).

Cochard, B., Amblard, P. and Durand-Gasselin, T. (2005). Oil-palm genetic improvement and sustainable development. *OCL*, 12(2): 141–147. Doi: 10.1051/ocl.2005.0141.

Corley, R.H.V. (1976). Sex differentiation in oil palm: Effects of growth regulators. *Journal of Experimental Botany*, 27: 553–558.

Corley, R.H.V. and Tinker, P.B. (2015). *The Oil Palm*. 5th ed., Chichester, UK: John Wiley & Sons, Ltd. Doi: 10.1002/9781118953297.

Cornaire, B., Daniel, C., Zuily-Fodil, Y. and Lamade, E. (1994). *Le comportement du palmier sous stress hydrique, Données du problème, premiers résultats et voies de recherche. Oléagineux*, 49: 1–12.

De Franqueville, H. (2003). Oil palm bud rot in Latin America. *Experimental Agriculture*, 39(3): 225–240.

Dislich, C., Keyel, A.C., Salecker, J., Kisel, Y., Meyer, K.M., Auliya, M., Barnes, A.D., Corre, M.D., Darras, K., Faust, H. and Hess, B. (2017). A review of the ecosystem functions in oil palm plantations, using forests as a reference system. *Biological Reviews*, 92(3): 1539–1569.

Eycott, A.E., Advento, A.D., Waters, H.S., Luke, S.H., Aryawan, A.A.K., Hood, A.S., Naim, M., Purnomo, D., Rambe, T.D.S., Tarigan, R.S. and Turner, E. C. (2019). Resilience of ecological functions to drought in an oil palm agroecosystem. *Environmental Research Communications*, 1(10): 101004.

Fitzherbert, E.B., Struebig, M.J., Morel, A., Danielsen, F., Brühl, C.A., Donald, P.F. and Phalan, B. (2008). How will oil palm expansion affect biodiversity? *Trends in Ecology & Evolution*, 23(10): 538–545.

Foster, W.A., Snaddon, J.L., Turner, E.C. et al. (2011). Establishing the evidence base for maintaining biodiversity and ecosystem function in the oil palm landscapes of Southeast Asia. *Philosophical Transactions of the Royal Society B, Biol Sci.*, 366(1582): 3277–3291.

Gerard, A., Wollni, M., Hölscher, D., Irawan, B., Sundawati, L. et al. (2017). Oil-palm yields in diversified plantations: initial results from a biodiversity enrichment experiment in Sumatra, Indonesia. *Agric. Ecosyst. Environ.*, 240: 253–60.

Haran, J., Abanda, R.F.X.N., Benoit, L., Bakoumé, C. and Beaudoin-Ollivier, L. (2020). Multilocus phylogeography of the world populations of *Elaeidobius kamerunicus* (Coleoptera, Curculionidae), pollinator of the palm *Elaeis guineensis*. *Bulletin of Entomological Research*, pp. 1–9.

Heeb, L., Jenner, E. and Cock, M.J. (2019). Climate-smart pest management: building resilience of farms and landscapes to changing pest threats. *Journal of Pest Science*, 92(3): 951–969.

Horton, A.J., Lazarus, E.D., Hales, T.C., Constantine, J.A., Bruford, M.W. and Goossens, B. (2018). Can riparian forest buffers increase yields from oil palm plantations? *Earth's Future*, 6(8): 1082–1096.

Jaligot, E., Hooi, W.Y., Debladis, E., Richaud, F., Beulé, T., Collin, M. and Rival, A. (2014). DNA methylation and expression of the EgDEF1 gene and neighboring retrotransposons in mantled somaclonal variants of oil palm. *PloS One*, 9(3): e91896.

Khasanah, N., Van Noordwijk, M., Slingerland, M., Sofiyudin, M., Stomph, D., Migeon, A.F. and Hairiah, K. (2020). Oil palm agroforestry can achieve economic and environmental gains as indicated by multifunctional land equivalent ratios. *Frontiers in Sustainable Food Systems*, 3: 122.

Kleijn, D., Bommarco, R., Fijen, T.P., Garibaldi, L.A., Potts, S.G. and van der Putten, W.H. (2019). Ecological intensification: bridging the gap between science and practice. *Trends in Ecology & Evolution*, 34(2): 154–166.

Korol, Y., Khokthong, W., Zemp, D.C., Irawan, B., Kreft, H. and Hölscher, D. (2021). Scattered trees in an oil palm landscape: density, size and distribution. *Global Ecology and Conservation*, p.e01688.

Lee, A.T.K., Carr, J.A., Ahmad, B., Arbainsyah, Ferisa, A., Handoko, Y., Harsono, R., Graham, L.L.B., Kabangnga, L., Kurniawan, N.P., Keßler, P.J.A., Kuncoro, P., Prayunita, D., Priadiati, A., Purwanto, E., Russon, A.E., Sheil, D., Sylva, N., Wahyudi, A. and Foden, W.B. (2019). Reforesting for the climate of tomorrow: Recommendations for strengthening orangutan conservation and climate change resilience in Kutai National Park, Indonesia. *Gland*, Switzerland: IUCN, viii + 70pp.

Legros, S., Mialet-Serra, I., Caliman, J.P., Siregar, F.A., Clément-Vidal, A. and Dingkuhn, M. (2009). Phenology and growth adjustments of oil palm (*Elaeis guineensis*) to photoperiod and climate variability. *Annals of Botany*, 104: 1171–1182.

Lipper, L., Thornton, P., Campbell, B.M., Baedeker, T., Braimoh, A., Bwalya, M., Caron, P., Cattaneo, A., Garrity, D., Henry, K. and Hottle, R. (2014). Climate-smart agriculture for food security. *Nature in Climate Change*, 4(12): 1068–1072.

Maillard, G., Daniel, C. and Ochs, R. (1974). *Analyse des effets de la sécheresse sur le palmier à huile. Oléagineux*, 29: 8–9.

Manik, Y. and Halog, A. (2013). A meta-analytic review of life cycle assessment and flow analyses studies of palm oil biodiesel. *Integrated Environmental Assessment and Management*, 9(1): 134–141.

Mariau, D. and Genty, P. (1988). IRHO contribution to the study of oil palm insect pollinators in Africa, South America and Indonesia, IRHO contribution to the study of oil palm insect pollinators in Africa, South America and Indonesia. *Oléagineux*, 43(6): 233–240.

Mertz, O. and Mertens, C.F. (2017). Land sparing and land sharing policies in developing countries—Drivers and linkages to scientific debates. *World Dev.*, 98: 523–35.

Messier, C., Bauhus, J., Sousa-Silva, R., Auge, H., Baeten, L., Barsoum, N. and Zemp, D.C. (2022). For the sake of resilience and multifunctionality, let's diversify planted forests! *Conservation Letters*, 15(1): e12829.

Miccolis, A., Robiglio, V., Cornelius, J.P., Blare, T. and Castellani, D. (2019). Oil palm agroforestry: Fostering socially inclusive and sustainable production in Brazil. Exploring inclusive palm oil production. *EFTERN News*, 59.

Nchanji, Y.K., Nkongho, R.N., Mala, W.A. and Levang, P. (2016). Efficacy of oil palm intercropping by smallholders. Case study in South-west Cameroon. *Agroforestry Systems*, 90(3): 509–519.

Nelson, P.N., Huth, N., Sheaves, M., Bessou, C., Pardon, L., Lim, H. and Kookana, R.S. (2018). *Modeling Environmental Impacts of Agriculture, Focusing on Oil Palm*. Burleigh Dodd Science Publishing.

Ong-Abdullah, M., Ordway, J.M., Jiang, N., Ooi, S.E., Kok, S.Y., Sarpan, N. and Martienssen, R.A. (2015). Loss of Karma transposon methylation underlies the mantled somaclonal variant of oil palm. *Nature*, 525(7570): 533–537.

Paterson, R.R.M., Sariah, M. and Lima, N. (2013). How will climate change affect oil palm fungal diseases? *Crop Protection*, 46: 113–120.

Paterson, R.R.M. and Lima, N. (2018). Climate change affecting oil palm agronomy, and oil palm cultivation increasing climate change, require amelioration. *Ecology and Evolution*, 8(1): 452–461.

Paterson, R.R.M. (2019). Ganoderma boninense disease of oil palm to significantly reduce production after 2050 in sumatra if projected climate change occurs. *Microorganisms*, 7(1): 24.

Rival, A. (2017). Breeding the oil palm (*Elaeis guineensis* Jacq.) for climate change. *OCL*, 24(1): D107.

Schmidt, J.H. (2010). Comparative life cycle assessment of rapeseed oil and palm oil. *The International Journal of Life Cycle Assessment*, 15(2): 183–197.

Schmidt, J.H. (2015). Life cycle assessment of five vegetable oils. *Journal of Cleaner Production*, 87: 130–138.

Singh, R., Ong-Abdullah, M., Low, E.T.L., Manaf, M.A.A., Rosli, R., Nookiah, R. and Sambanthamurthi, R. (2013). Oil palm genome sequence reveals divergence of interfertile species in Old and New worlds. *Nature*, 500(7462): 335–339.

Somyong, S., Poopear, S., Sunner, S.K., Wanlayaporn, K., Jomchai, N., Yoocha, T. and Tragoonrung, S. (2016). ACC oxidase and miRNA 159a, and their involvement in fresh fruit bunch yield (FFB) via sex ratio determination in oil palm. *Molecular Genetics and Genomics*, 291(3): 1243–1257.

Stichnothe, H. and Schuchardt, F. (2011). Life cycle assessment of two palm oil production systems. *Biomass and Bioenergy*, 35(9): 3976–3984.

Stiegler, C., Meijide, A., Fan, Y., Ashween Ali, A., June, T. and Knohl, A. (2019). El Niño-Southern Oscillation (ENSO) event reduces CO_2 uptake of an Indonesian oil palm plantation. *Biogeosciences*, 16(14): 2873–2890.

Wicke, B., Dornburg, V., Junginger, M. and Faaij, A. (2008). Different palm oil production systems for energy purposes and their greenhouse gas implications. *Biomass and Bioenergy*, 32(12): 1322–1337.

Zemp, D.C., Ehbrecht, M., Seidel, D., Ammer, C., Craven, D., Erkelenz, J., Irawan, B., Sundawati, L., Hölscher, D. and Kreft, H. (2019). Mixed-species tree plantings enhance structural complexity in oil palm plantations. *Agriculture, Ecosystems & Environment*, 283: 106564.

6

Papaya Resilience to Global Warming and Climate Change

Daniel Chebet,[1] *Isaac Savini*[2] and *Fredah K Rimberia*[3,*]

1. Introduction

1.1 Background Information

The papaya plant is a semi-woody, latex-producing, usually single-stemmed, short-lived perennial herb (Fig. 1). Papaya (*Carica papaya* L.) is commercially produced in all tropical and subtropical parts of the world. Papayas are produced in about 60 countries, with the bulk of production occurring in developing economies. Asia is the largest papaya-producing continent, providing 55.5% of the total production, followed by South America (23.0%) and Africa (13.2%).

Papaya is ranked third with 11.22 metric tons (mt) (15.36%) of the total tropical fruit production, behind mango with 38.6 mt (52.86%) and pineapple with 19.41 mt (26.58%) (Evans Ballen, 2012). Papaya produces nutrient-rich fruits for a balanced diet and also as a source of papain—a commercially valuable proteolytic enzyme that is produced in the milky latex of green, unripe fruit for industrial applications (Dunne and Horgan, 1992).

[1] Department of Seeds, Crops and Horticultural, University of Eldoret, Kenya.
[2] A private consultant in agriculture and plant nutrition (formerly, International Center for Tropical Agriculture (CIAT) Staff, Nairobi, Kenya.
[3] Department of Horticulture and Food Security, Jomo Kenyatta University of Agriculture and Technology, Kenya.
Emails: daniel.chebet@uoeld.ac.ke; wasavini@gmail.com
* Corresponding author: frenda@agr.jkuat.ac.ke

Fig. 1. A papaya orchard near the Indian Ocean coast of Kenya (Fredah K. Rimberia, personal communication).

Papaya fruits are rich sources of carbohydrates, natural vitamins A and C and minerals, like calcium (Nakasone and Paull, 1998). Papaya seeds have commonly been used in traditional medicine, while papain, a protein-digesting enzyme obtained from unripe papaya fruits has been used to tenderize meat and as a food supplement to aid digestion (Milind and Gurditta, 2011). The fruits and seeds have anthelminthic and anti-amoebic activities (Teixeira da Silva, 2007). Studies on the effects of papaya preparations document positive outcomes for patients with constipation, heartburn and symptoms of irritable bowel syndrome (Muss et al., 2012). Thus, papaya is considered a nutraceutical fruit. Papaya leaves and immature fruits produce several proteins and alkaloids with important pharmaceutical and industrial applications (El Moussaoui et al., 2001).

1.2 Climatic Adaptation

Different papaya fruit-bearing sex types are produced under both tropical and subtropical conditions. Hermaphrodite cultivars are commercially cultivated in the tropics, while female cultivars are predominantly used in the subtropical regions (Aquilizan, 1987; Drew, 1988). In a study carried out in the subtropical Okinawa Islands of Japan (Table 1), the hermaphrodites were liable to change their flower sex and consequently their fruit yield decreased. On the other hand, the females showed stronger parthenocarpic

Table 1. Fruit setting rates between female and hermaphrodite papaya in pollination and non-pollination treatments (Source: Rimberia et al., 2018).

Sex type[1]	**Pollination**			
	No. flowers used	**No. fruits developed**	**No. flowers abscised**	**Fruit setting rate (%)****
Female	95	79	16	83.2
Hermaphrodite	145	95	50	65.5
Sex type[1]	**Non-pollination**			
	No. flowers used	**No. fruits developed**	**No. flowers abscised**	**Fruit setting rate (%)****
Female	90	65	25	72.2
Hermaphrodite	92	45	47	48.9

[1] Data from female or hermaphrodite cultivars were summed up. ** Significant at $P < 0.0028$

ability than the hermaphrodites and exceeded the hermaphrodites in fruit yield (Rimberia et al., 2018). Thus, the warmer climate that is emerging in subtropical regions is expanding areas that are suitable for production of papaya (Subedi, 2019).

2. Ecological Requirements for Papaya

2.1 Light

Papaya is classified as a plant with C3 metabolism (Campostrini and Glen, 2007) with characteristic C3 leaf anatomy. It is a light-demanding plant and prolonged low light intensity can cause significant alterations in the leaf anatomy and morphology (Buisson and Lee, 1993). Buisson and Lee (1993) showed that papaya grown under reduced light intensity had reduced leaf thickness, specific leaf weight, stomatal density, leaf area and petiole length, as well as increased chlorophyll content per unit leaf area and the amount of air space into the leaf mesophyll. The reduction in light intensity also greatly reduces the number of lobules on the leaf.

Maximum net carbon assimilation rates of 25–30 $\mu mol\ m^{-2}\ s^{-1}$ were achieved at 2000 $\mu mol\ m^{-2}\ s^{-1}$ photosynthetic photo flux density (PPFD) (Campostrini and Glen, 2007). Reduced light intensity can cause a significant decrease in photosynthetic carbon assimilation in papaya (Marler and Mickelbart, 1998). However, excess solar radiation can also decrease carbon assimilation by the leaf due to the interaction of stomatal and non-stomatal factors. High solar radiation can cause leaf heating, increase the vapor-pressure deficit between the leaf surface and surrounding air (VPDleaf-air) and promote stomatal closing, reducing CO_2 entry into the leaf mesophyll, thus diminishing the concentration of

CO_2 at the Rubisco carboxylation sites. Photo-inhibition, due to damage of photosystem II (PSII) and associated damage to ATP and NADPH synthesis, are the main non-stomatal effects which lower photosynthetic carbon assimilation under excess solar radiation (Vass, 2012).

While photorespiration in C3 plants can decrease the net efficiency of carbon assimilation by 25–30% (Lawlor, 1993), papaya can maintain high net CO_2 assimilation rates under well-watered and PPFD saturating conditions, suggesting minimal photorespiration losses and adaptation to high light intensities (Campostrini and Glen, 2007). On the other hand, papaya can adjust to as low as 30% reduction in PPFD, making commercial greenhouse production feasible (Campostrini and Glen, 2007).

Cultivar also influences maximum net CO_2 assimilation rates. Campostrini et al. (2001) found maximum net CO_2 assimilation rates of 25 μmol m^{-2} s^{-1} for cv. 'Baixinho de Santa Amália' and 20 μmol m^{-2} s^{-1} for cvs. 'Sunrise Solo 72/12', 'Sunrise Solo TJ' and 'Know-You' (Campostrini and Glen, 2007). On the other hand, Jeyakumar et al. (2007), working with Indian cultivars, demonstrated that in field-cultivated papaya, PPFD light saturation was 1250 μmol m^{-2} s^{-1} with net CO_2 assimilation at 12 μol m^{-2} s^{-1}, but at PPFD levels above 1250 μmol m^{-2} s^{-1}, net CO_2 assimilation rates fell sharply to values of 5 μmol m^{-2} s^{-1} (Fig. 2).

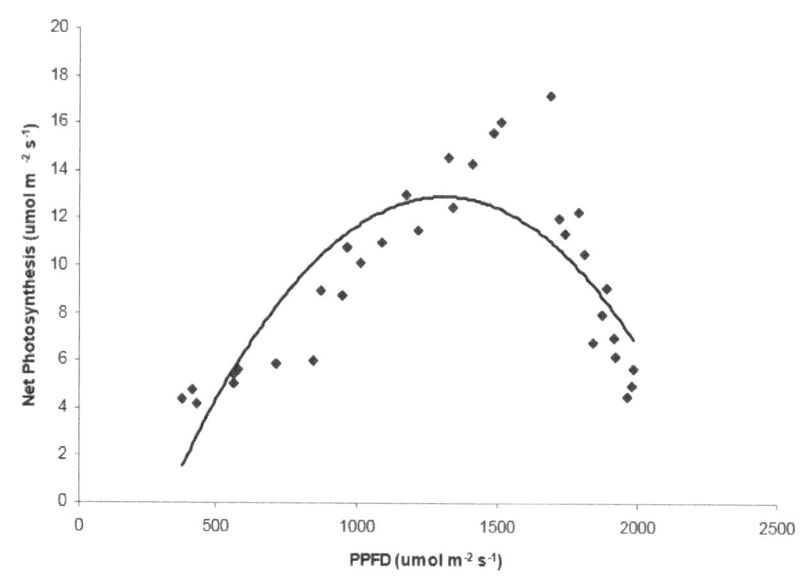

Fig. 2. Effect of photosynthetic photo flux density on net photosynthesis in papaya plants (Source: Jeyakumar et al., 2007).

2.2 *Temperature*

Papaya has optimal growth and development at air temperatures ranging between 21–33°C (Nishina et al., 2000; OECD, 2003). Allan and Jager (1978) reported that net carbon assimilation increased when air temperature rose from 16 to 30°C, and then it decreased linearly at temperatures above 30°C, the value at 41°C being half that at 30°C. Jeyakumar et al. (2007), while working in India, reported that leaf net CO_2 assimilation increased gradually from 27°C to 35°C but reduced considerably when the temperature exceeded 35°C (Fig. 3).

Due to its origin in tropical environments, papaya is classified as a species sensitive to low temperatures (Campostrini and Glen, 2007). Temperatures below 12–14°C strongly retard fruit maturation and adversely affect fruit production (Nakasone and Paull, 1998). Generally in most fruit crops, like mango, papaya, guava and others, flower drop occurs when extreme low temperature conditions prevail during flowering and ultimately hamper fruit yield (Dinesh and Reddy, 2012). Papaya is extremely sensitive to frost, which can kill the plant. Papaya fruits become insipid when they ripen in periods when the temperature is at sub-optimal level. Temperatures below 20°C result in other problems, such as carpelloidy, gender changes, reduced pollen viability, and low-sugar content of fruits (Galán-Saúco and Rodríguez-Pastor, 2007; Campostrini and Glen, 2007).

Air temperature influences papaya plant gender expression because there is a tendency to produce male flowers at high temperature. High

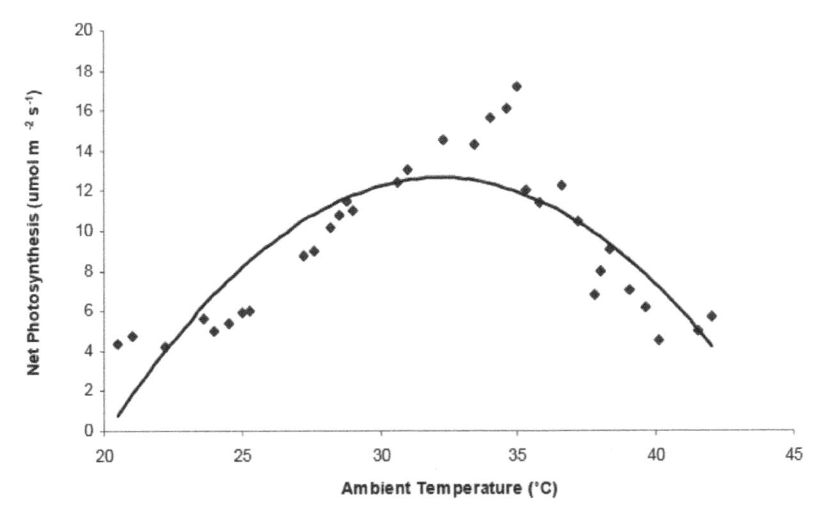

Fig. 3. Effect of ambient temperature on net photosynthesis in papaya (Source: Jeyakumar et al., 2007).

temperatures (30–35°C) induce 'female sterility', in which case normally hermaphroditic papaya plants produce male flowers, resulting in poor fruit set and production (Nishina et al., 2000). The female papaya is stable in terms of sex expression.

2.3 *Water*

A minimum monthly rainfall of 100 mm, minimum relative humidity of 66% (Nishina et al., 2000) is suggested as 'ideal' for papaya growth and production. Below this, irrigation is necessary. Net CO_2 assimilation is negatively affected by water deficit. Jeyakumar et al. (2007) reported that the net CO_2 assimilation was more than 16 $\mu mol\ m^{-2}\ s^{-1}$ when the leaf relative water content was 90.6% but reduced to less than 7 $\mu mol\ m^{-2}\ s^{-1}$ when the relative water content dropped to 77.4%.

Papaya requires good soil drainage. Where soil drainage is restricted, papaya is susceptible to fungal root diseases (Nishina et al., 2000). The plants are severely affected by exposure to waterlogging even for a short length of time. Papaya is sensitive to low oxygen availability in the soil (hypoxia), which is commonly caused by waterlogging. A completely flooded soil can cause death to papaya plants in two to four days (Campostrini and Glen, 2007; Fredah K. Rimberia, personal communication) (Fig. 4). These studies indicate that papaya is sensitive to small reductions in soil oxygen content. Consequently, a well-drained soil is essential for high productivity (Campostrini and Glen, 2007). Papaya exhibits both stomatal and non-stomatal response to soil water deficits and the sources of response signals are both hydraulic and non-hydraulic in nature (Thani et al., 2016) (Fig. 5).

Fig. 4. Papaya plants destroyed by flooding for about four days in a plastic tunnel at Jomo Kenyatta University of Agriculture and Technology (photo by Fredah K. Rimberia in 2017).

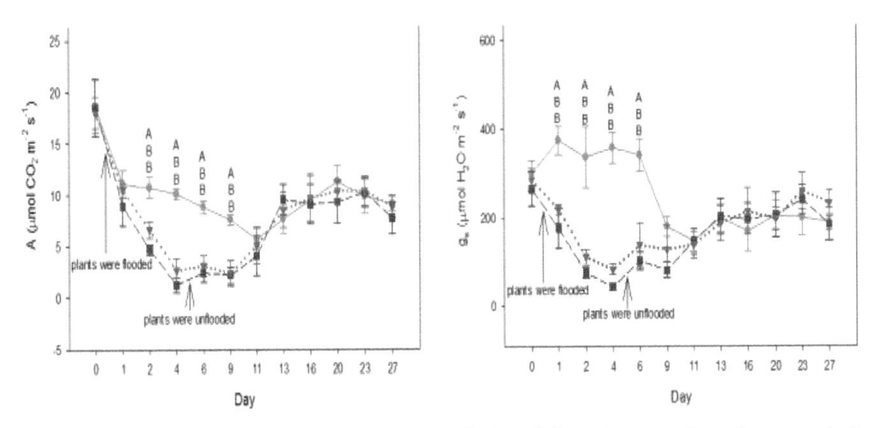

Fig. 5. Effect of flooding on the net CO_2 assimilation (*left*) and stomatal conductance (gs) (*right*) of papaya in Florida (Thani et al., 2016).

Marler et al. (1994) proposed that it is highly unlikely that stomata of drought-stressed papaya plants closed due to hydraulic signals from leaf dehydration since leaf relative water content and pre-dawn xylem potential were unrelated to stomatal conductance at mild and moderate soil water deficits. They proposed that other non-hydraulic plant signals control stomatal behavior. Marler et al. (1994) suggested that delaying dehydration appears to be the adaptation that papaya uses in response to drought. However, Mahouachi et al. (2006) found that osmotic adjustment is a contributing factor in drought adaptation in 'Baixinho de Santa Amália' papaya. In any case, Marler et al. (1994) and Torres-Netto (2005) demonstrated that there is genetic variability in papaya cultivar response to soil water deficits providing clues to the mechanisms of drought adaptation (Campostrini and Glen, 2007).

2.4 *Wind*

Adequate air movement is important for ventilation, to avoid build-up of high humidity pockets that encourage fungal diseases, such as phytophthora and anthracnose (Nishina et al., 2000). On the other hand, papaya is soft stemmed and requires protection from strong wind. Excessive wind causes reduced growth, fruit set, fruit quality and productivity.

Papaya plants are vulnerable to wind damage; especially when the soil is wet and the plants are bearing a heavy load of fruits. The tree can be uprooted by winds of 64 km/h, especially if the soil is shallow and softened by rain and/or in peat soil because of poor anchorage. Strong winds also cause loss of leaf area which leads to flower and young fruit abscission, as well as low total soluble solids in the mature fruits on the column. Full recovery of leaves from wind damage can take from four to eight weeks. Clemente and Marler (2001) demonstrated the sensitivity of papaya seedlings to exposure to wind at speeds ranging between 0–2.5 m s^{-1}. The rate of dry mass gain by the young plants (three to six weeks old) declined continuously with increase in wind speed from an average of 1900 mg $plant^{-1}$ to 900 mg $plant^{-1}$. Close spacing of papaya plants results in tall plants with thin stems, making them more susceptible to wind damage than plants grown at a wider spacing (Zimmerman, 2008).

2.5 Soil

Papaya can be grown on a variety of soil types; however, soils must be fertile, well drained and aerated for satisfactory performance. Prolonged waterlogging results in rotting of stem and roots, yellow foliage, leaf dropping and growth reduction. Being a shallow-rooted plant, papaya can be grown in soils about 45 cm deep. The desirable soil pH range is between 5.5 and 6.5 (Awada, 1975). At pH levels below 5.0, seedling growth is poor and mortality is high. In soils with a pH range of 5.0–5.5, liming is recommended for optimal growth and yield.

Soil compaction is common in commercial enterprises. It is caused by use of heavy machines on wet soils. Campostrini and Yamanishi (2001), while evaluating cultivar adaptation to soil compaction and root restriction, reported that all cultivars had reduced total leaf number, average leaf area, length of leaf central vein, total leaf area, trunk diameter and tree height compared to non-restricted plants. Restricted root growth also induces reduced photosynthetic rates and stomatal conductance. Campostrini and Yamanishi (1998) concluded that rooting volume restriction induced senescence as a general physiological response.

3. Symptoms of Climate Change Impacts on Papaya

3.1 Nature of the Papaya Plant

Papaya is an herbaceous perennial with three sex forms: male, female and hermaphrodite. Both females and hermaphrodites are commercially useful for fruit production, while the males are solely used as a pollen source. Hermaphroditic cultivars are commercially cultivated in the tropics, while female cultivars are predominantly used in the subtropical regions (Aquilizan, 1987; Drew, 1988). This difference in cultural adaptation between both types of papaya may be due to sex reversal. Hermaphrodite and male papaya plants seasonally reverse their flower sex (Nakasone and Paull, 1998; Ray, 2002). Sex reversal in hermaphrodites is usually accompanied by stamen carpellody and female sterility (Nakasone and Paull, 1998), resulting in poor quality and low yield of fruits. On the other hand, the female is stable with respect to sex expression throughout the year (Nakasone and Paull, 1998; Ray, 2002).

3.2 *Impact of Climate Change*

The main effect of climate change that affects fruit crop production is shifting of climatic zone and variation in the seasonal distribution of rainfall, leading to more periodic droughts and floods, rise in temperature, disease and pest outbreak among others (Subedi, 2019). In papaya, high temperatures resulting from global warming and climate change resulted in floral abscission in female and hermaphrodite plants, sex reversal in hermaphrodite plants as well as promotion of stigma and stamen sterility (Reddy et al., 2017; Subedi, 2019). Additionally, high temperatures coupled with high rainfall and humidity result in fruits with higher TSS and create ideal conditions for proliferation of disease pathogens, like those that cause powdery mildew (Reddy et al., 2017).

Diseases are responsible for losses of at least 10% of global food production, representing a threat to food security (Strange and Scott, 2005). It is estimated that the damage caused by Papaya Ring Spot Virus (PRSV-P) will increase in severity with the expected increase in temperature (Ghini et al., 2011). Mangrauthia et al. (2009) observed that at temperatures at 26–31°C, symptoms were more severe. Jesus Junior et al. (2007) evaluated the impact of climate change on leaf lesions (*Asperisporium caricae*) of papaya in Brazil and found that in the future, there will be a reduction in the area favorable to this disease, though large areas, particularly in Espirito Santo State, a leading producer of papaya in Brazil will still be favorable to it.

Other effects of global warming and climate change include strong winds (cyclones) and drought. Papaya is soft wooded, so cyclones create significant destruction of both trees and fruits (Fakava, 2012). Severe drought presents problems for rain-fed agriculture as well as lack of irrigation water (Fakava, 2012).

4. Papaya Management under Changed Climate Conditions

4.1 *Sustainable Production*

Sustainable management of natural resources and climate change adaptation is a must for the agricultural sectors to achieve food and nutrition security. Improved sustainable farming practices that have been developed and are successful in Pacific and Asian developing countries include improved farm practices and management, like irrigated cropping systems, use of higher yielding, drought and disease tolerance varieties, integrated pest management, conservation agriculture, organic agriculture, agro-forestry system as well as integrated crops and livestock in aquaponic production system management (Fakava, 2012).

To sustain the productivity, modification of present horticultural practices and greater use of greenhouse technology are some of solutions to minimize the effect of climate change. Development of new hybrids, varieties of fruit crops tolerant to abiotic and biotic factors, producing good yield under stress conditions, as well as adoption of hi-tech horticulture and wise management of resources will be the main strategies to meet this challenge (Reddy et al., 2016). Greenhouse production of papaya requires varieties with both dwarfness and parthenocarpy traits, like Kamiya, Co 1 and Co 2, considering the limitation in space (Rimberia et al., 2018).

4.2 Response to Carbon Dioxide Enrichment

Among the principal abiotic requirements for plant growth, namely, light, water, nutrients and gases, carbon dioxide (CO_2) is an anthropogenic gas associated with potential global warming. The current annual rate of increase in CO_2 (0.5%) is expected to continue with concentrations exceeding 600 ppm by the end of this century from the current 380 ppm (Houghton et al., 2001). Such an increase in the CO_2 levels will certainly affect the globally important process of photosynthesis, which sustains life on this planet. However, the increased CO_2 concentration in the atmosphere is believed to provide enhanced fertilization to plant growth, especially for C3 crops (Derner et al., 2003).

The primary effects of rising CO_2 on plants include reduction in stomatal conductance and transpiration, improved water-use efficiency, higher rates of photosynthesis and increased light-use efficiency. Cruz et al. (2016) reported that compared to ambient levels of CO_2, elevated CO_2 generally increased photosynthesis irrespective of rates of N applied (though the rate was higher in the high N treatments compared to those that received low N). They also observed a significant enhancement in plant height, stem diameter and leaf area in the high N treatment. Elevated CO_2 also increased the biomass of leaf, stem plus petiole, and root dry mass of papaya plants regardless of N treatment, leading to total dry mass enhancements of 56.6% in the high N treatments and 64.1% in the low N treatments (Fig. 6). Importantly, their data indicate that elevated CO_2 'alleviated the effect of low N on dry matter accumulation in papaya', which they concluded was at least partially explained by a larger leaf area and higher rate of photosynthesis per leaf area unit observed under elevated CO_2.

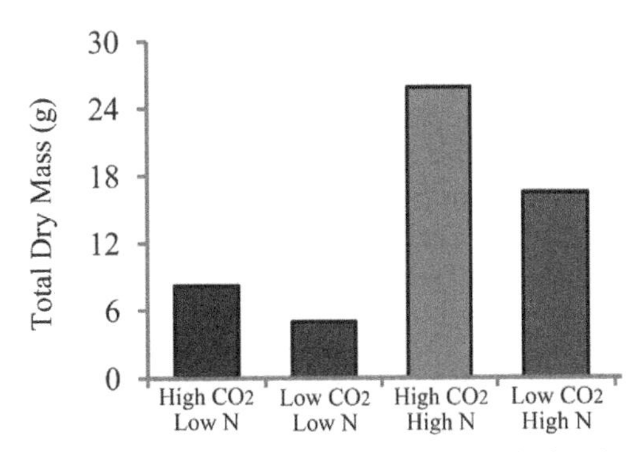

Fig. 6. Total dry mass of papaya plants grown in controlled chambers at two different CO_2 concentrations (high and low; 750 and 390 ppm) and two different N treatments high and low; 8 mM NO_3^- or 3 mM NO_3^-) (Source: Cruz et al., 2016).

5. Natural Adaptation of Papaya Genetic Resources

Jeyakumar et al. (2007) evaluated five papaya cultivars in India: 'Co. 2', 'Co. 5', 'Co. 7', 'Surya' and 'Sun Rise Solo' and found 'Co. 7' having significant abiotic stress tolerance as evidenced by higher leaf gas exchange characters, water use efficiency, cell membrane integrity, chlorophyll stability index and chlorophyll fluorescence under water-limited conditions.

6. Mitigation Strategies against Biotic and Abiotic Stresses on Papaya

6.1 *Cultural Techniques*

Science-based management can meet the water demands of the plant in a more efficient manner through improved irrigation technology, while temperature can be reduced through overhead cooling systems (Campostrini and Glen, 2007) or reflective materials for growing structures. The papaya crop needs more irrigation during dry periods to maintain its growth and fruit production (Malo and Campbell, 1986). Water-deficit stress in field-grown papaya is reported to have caused 50% reduction in the number of leaves, 86% reduction in the number of flowers and 58% reduction in fruit set (Laxman and Bhatt, 2017). Additionally, the growth and development of the papaya fruits was retarded (Masri et al., 1990).

6.2 *Greenhouse Cultivation*

Greenhouse cultivation of papaya is useful for controlling adverse environmental factors, like light and temperature as well as pests that transmit disease-causing micro-organisms, like bacteria and viruses.

Campostrini and Glen (2007) quoting Reis et al. (2005) demonstrated that the photosynthetic apparatus of papaya may adapt to changes in light intensity and quality using 'Baixinho de Santa Amália' papaya that is cultivated in greenhouses (at 30% sunlight interception). Compared to plants grown in full sunlight, greenhouse cultivated plants of this variety had higher carbon assimilation and greater stomatal conductance at midday, which was attributable to reductions in the leaf-to-air vapor pressure deficit under greenhouse conditions.

Galán Saúco and Rodríguez-Pastor (2007) showed that papaya benefited from the climatic modifications of the greenhouse through improved yields (both in quantity and quality), reduced water consumption as well as exclusion of Papaya Ring Spot Virus (PRV). Consequently, Campostrini and Glen (2007) proposed the greenhouse papaya cultivation as an alternative production system in the subtropics.

Greenhouse production requires varieties with both dwarfness and parthenocarpy traits, like Kamiya, Co 1 and Co 2, considering the limitation in space for inclusion of pollen sources as well as pollinators. Rimberia et al. (2018) recommended that the approach to developing varieties that combine dwarfness, parthenocarpy and PRSV tolerance needs to be adopted and encouraged to take advantage of greenhouse production as well as decreasing sizes of land available for production.

6.3 *Inoculation with Mycorrhiza*

Land degradation and soil fertility depletion are considered the major threats to food security and natural resource conservation in sub-Saharan Africa (Cardoso and Kuyper, 2006). Mycorrhizal fungi are specialized organisms that live on plant roots in a symbiotic relationship. The host plant supplies the fungus with carbohydrates produced during photosynthesis (Le Tacon et al., 2013). In return, the fungi use its extensive network of hyphae in the soil to transfer water and nutrients to the roots (Le Tacon et al., 2013).

Studies have also shown that mycorrhiza fungi also enhanced tolerance to drought stress (Kohler et al., 2008), caused faster recovery after moisture stress (Qiang-Sheng et al., 2007), increased the transpiration and photosynthetic rates (Smith and Smith, 2011) and conferred tolerance to flooding (Muok and Ishii, 2006) and high soil salinity (Muok and Ishii, 2006). Arbuscular Mycorrhiza (AM) inoculation also antagonized parasitic soil-borne pathogens and pests (Elsen et al., 2003), improved the plant's ability to withstand transplanting shock and improved general plant performance after transplanting (Raviv, 2010).

Mamatha et al. (2002) demonstrated that field-planted, 1.5-year-old plants (var. 'Solo') had increased fruit yield when inoculated with *Glomus mosseae* and *G. caledonium* with or without the addition of *Bacillus coagulans*, which increases AM colonization. The species of AM used affects plant productivity. Effective species include *G. mosseae, G. claroideum* and *G. fasciculatum* (Campostrini and Glen, 2007). Colonization rates and spore density were positively correlated with soil organic matter and coarse sand fractions and negatively correlated with fine sand.

The mycorrhizal network is a key to improving the acquisition of nutrients and water in papaya production. Management factors that increase colonization of effective fungi can be expected to improve nutrient and water-use efficiencies. Soil ethylene levels and ACC activity were reduced by AM under these water-deficit conditions, further supporting a reduced water-stress severity in AM-treated plants. Such a reduction occurred despite an increase in above-ground mass and leaf area and was due largely to a significant increase in root mass in the AM treatment that was more effective in water uptake than the non-AM treatment (Campostrini and Glen, 2007).

A study by Chebet et al. 2008 found that inoculation with AM fungi increased the plant height, leaf number and stem girth in papaya variety, 'mountain' (Fig. 7).

Arbuscular mycorrhiza also increased the uptake of phosphorus, and potassium in the leaf tissues. Inoculation also favored mycorrhizal infectivity of roots and increased the root mass. This study indicated that AM fungi improved the capacity of papaya seedlings to absorb and utilize plant nutrients possibly by increasing the effective root surface area from which available form of nutrients are absorbed and also by increasing access of roots by bridging the depletion zones (Chebet, 2020). Mycorrhizal-inoculated seedlings had greater root mass compared to un-inoculated seedlings, as indicated by greater root fresh weight (Chebet, 2020). Likewise, the extent of mycorrhizal root infection was significantly greater in inoculated seedlings than in un-inoculated seedlings (Chebet, 2020). It is expected that this greater

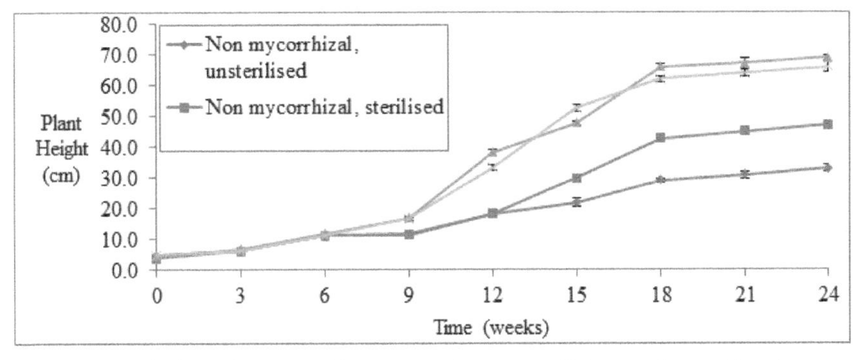

Fig. 7. Effect of arbuscular mycorrhiza fungi and media condition on plant height of papaya (*Carica papaya* var 'mountain') seedlings (Source: Chebet et al., 2008).

mass of mycorrhizal roots corresponded to greater absorptive surface area for nutrients and water.

7. Remote Sensing and Environmental Certification

Site-specific Crop Management (SSCM) is one facet of precision agriculture which is helping increase production with minimal input. SSCM involves spatial referencing, crop and climate monitoring, attribute mapping, decision support systems and differential action. SSCM is carried out with a greater degree of precision through the use of geospatial technologies. Geospatial technology is a combination of four essential tools: remote sensing, geographic information systems (GIS), global positioning systems (GPS), and information technology (data management) (Panda et al., 2010).

The application of aerial or satellite imaging, along with GPS and GIS, is the first step towards the successful application of SSCM for fruit and nut crops (Sevier and Lee, 2005). The delineation of orchards and spatial analysis using geospatial technology can provide additional information, such as the determination of fruit yield, the quantification and scheduling of precise and proper fertilizer, irrigation needs and the application of pesticides for pest and disease management, ultimately improving profits for producers (Panda et al., 2009). Spatial data processing is most essential for SSCM-related model development, such as yield prediction, water-demand forecasting or the determination of soil nutrient availability (Panda et al., 2010).

Baugher et al. (2009) provides an overview of current progress and the potential for future innovations in specialty crops, which include fruit and nut crops. They stated that video, laser and satellite imaging technologies will become pertinent to future improvements in orchard production. To make this possible, tree crop systems that are compatible with imaging technologies, like high density planting and training to a continuous planar fruiting wall (hedgerow), will be mandatory. Baugher et al. (2009) indicated that imaging technology has the potential for crop load assessment, including blossom or green fruit counts, yield estimation, determination of insect presence or disease infection and associated eradication, soil moisture content determination for enhanced irrigation system design, estimation of fertilizer, pesticide and herbicide application rates, and the development of assisted or automated pruning and harvesting strategies.

Geographic Information System (GIS) and remote sensing (RS) can be used to measure growth rate changes caused by stress, like salinity, drought, temperature among others (Meshram et al., 2017). In Thailand, airborne pollen of papaya (*Carica papaya* L.) and its distribution was effectively mapped using GIS and RS, and analyzed using land use data obtained from Quickbird imagery (Sritakae et al., 2011). They reported that pollen grains were densely distributed within 100 m from the papaya plot and dramatically decreased beyond that range, with the farthest distance being 900 m from the plot (Sritakae et al., 2011). Total pollen varied in different land use types, with most pollen found in agricultural land, bare land and harvested areas, and much less in areas sheltered by dense tree lines (Sritakae et al., 2011). This information is very useful for determining isolation distances between genetically modified papaya and the conventional varieties, especially in the seed production industry.

8. Conclusion and Prospects

The soft-wooded nature of papaya makes it very vulnerable to global warming and climate change that are characterized by periodic droughts and flooding, as well as high temperatures. High temperature (30–35°C) induce floral abscission in female and hermaphrodite plants, sex reversal in hermaphrodite plants as well as promotion of stigma and stamen sterility. Besides, high temperature and humidity create ideal conditions for proliferation of fungal disease pathogens and severity of viral diseases, like PRSV-P.

Sustainable crop-management strategies, like irrigated cropping systems, use of higher yielding, drought and disease tolerance varieties, integrated pest management, conservation agriculture have been proposed as useful in minimizing unfavorable conditions caused by global warming and climate change. Varieties that combine dwarfness, parthenocarpy and PRSV-P resistant traits, drip irrigation and carbon dioxide-enrichment are quite promising in greenhouse production of papaya. The modern genetic engineering technologies, like the Next Generation Sequencing (NGS) technologies will help to speed up breeding of new papaya varieties for overcoming biotic or abiotic stresses.

Other strategies for overcoming the rigorous effects of global warming and climate change are inoculation with arbuscular mycorrhiza fungi. The mycorrhizal network is a key to improving the acquisition of nutrients and water in papaya production. Mycorrhiza fungi have been shown to enhance tolerance to drought stress, increase the transpiration and photosynthetic rates and confer tolerance to flooding and high soil salinity.

Geographic Information System (GIS), remote sensing (RS) and satellite imaging technologies have come in handy for improving the efficiency of site-specific crop management. They have been applied in developing models to measure growth rate changes caused by stress, like salinity, drought and temperature among others, yield prediction, water-demand forecasting or the determination of soil nutrient availability among others. This technology has also been applied in the determination of isolation distances for genetically modified papaya.

Further research is required in application of modern technologies, like genetic engineering for breeding new varieties, as well as GIS, remote sensing and satellite imaging in papaya fruit improvement.

Acknowledgements

The author wishes to thank the German government through the German Academic Exchange Service (*Deutscher Akademischer Austaunsch Dienst*) (DAAD) and the Kenyan government through the Higher Education Loans Board (HELB) as well as the Research Production and Extension Division of the Jomo Kenyatta University of Agriculture and Technology for funding the papaya research.

References

Allan, P. and de Jager, J. (1978). Net photosynthesis in macadamia and papaw and the possible alleviation of heat stress. *Acta Hort.*, 102: 23–30, doi: 10.17660/ActaHortic.1979.102.4.

Aquilizan, F.A. (1987). Breeding system for fixing stable papaya inbred lines with breeding potential for hybrid variety production. pp. 101–106. *In*: *The Breeding of Horticultural Crop*. Taipei: Food and Fertilizer Technology Center for the Asian and Pacific Region.

Awada, A. (1975). *Critical Nitrogen Level in Petioles of Papaya Issue 94 of Technical bulletin* (Hawaii Agricultural Experiment Station) Hawaii Agricultural Experiment Station, University of Hawaii, 20 pp.

Baugher, T.A., Schupp, J.R., Lesser, K.M., Harsh, R.M., Lewis, K.M., Seavert, C.F. and Auvil, T.D. (2009). Mobile platforms increase orchard management efficiency and profitability. *Acta Hort.*, 824: 361–364.

Buisson, D. and Lee, D.W. (1993). The development responses of papaya leaves to simulated canopy shade. *Am. J. Bot.*, 80(8): 947–952.

Campostrini, E. and Yamanishi, O.K. (1998). Influence of root restriction on physiological characteristics of four papaya (*Carica papaya*) genotypes. pp. 3821–3824. *In*: *Proceedings of XIth International Congress on Photosynthesis*. Budapest, Hungary.

Campostrini, E. and Yamanishi, O.K. (2001). Influence of mechanical root restriction on gas-exchange of four papaya genotypes. *Braz. J. Plant Physiol.*, 13: 129–138.

Campostrini, E. and Glenn, D.M. (2007). Ecophysiology of papaya: a review. *Braz. J. Plant Physiol.*, 19(4): 413–424.

Cardoso, I.M. and Kuyper, T.W. (2006). Mycorrhizas and tropical soil fertility: nutrient management in tropical agro-ecosystems. *Agric. Ecosyst. Environ.*, 116: 72–84.

Chebet, D.K., Rutto, L.K., Wamocho, L.S. and Kariuki, W. (2008). Effect of Arbuscular Mycorrhizal (AM) inoculation on the growth, nutrient uptake and root infectivity of tropical fruit seedlings. *Acta Horticulturae*, 773: 253–260.

Chebet, D.K. (2020). *Effect of Arbuscular Mycorrhizal Inoculation on the Growth and Performance of Tropical Fruit Seedlings under Saline, Flooding and Nutrient Stress*. Ph.D Thesis, Jomo Kenyatta University of Agriculture and Technology, Kenya, pp. 55–119.

Clemente, H.S. and Marler, T.E. (2001). Trade winds reduce growth and influence gas exchange patterns in papaya seedlings. *Annals of Botany*, 88: 79–385.

Cruz, J.L., Alves, A.A.C., LeCain, D.R., Ellis, D.D. and Morgan, J.A. (2016). Interactive effects between nitrogen fertilization and elevated CO_2 on growth and gas exchange of papaya seedlings. *Scientia Horticulturae*, 202: 32–40.

Derner, J.D., Johnson, H.B., Kimball, B.A., Pinter, P.J., Polley, H.W., Tischler, C.R., Boutton, T.W., LaMorte, R.L., Wall, G.W., Adam, N.R., Leavitt, S.W., Ottman, M.J., Matthias, A.D. and Brooks T.J. (2003). Above- and below-ground responses of C3–C4 species mixtures to elevated CO_2 and soil water availability. *Global Change Biology*, 9: 452–460.

Dinesh, M.R. and Reddy, B.M.C. (2012). Physiological basis of growth and fruit yield characteristics of tropical and sub-tropical fruits to temperature. *Tropical Fruit Tree Species and Climate Change*, 45.

Drew, R.A. (1988). Rapid clonal propagation of papaya in vitro from mature field grown trees. *HortScience*, 23: 609–611.

Dunne, J. and Horgan, L. (1992). Meat tenderizers. pp. 1745–1751. *In*: Hui, Y.H. (ed.). *Encyclopedia of Food Science and Technology*. Wiley, New York, USA.

El Moussaoui, A., Nijs, M., Paul, C., Wintjens, R., Vincentelli, J., Azarkan, M. and Looze, Y. (2001). Revisiting the enzymes stored in the laticifers of *Carica papaya* in the context of their possible participation in the plant defence mechanism. *Cell and Molecular Life Sciences*, 58: 556–570.

Elsen, A., Baimey, H., Swennen, R. and De Waele, D. (2003). Relative mycorrhizal dependency and mycorrhiza–nematode interaction in banana cultivars (*Musa* spp.) differing in nematode susceptibility. *Plant and Soil*, 256: 303–313.

Evans, E., Ballen, F. and Crane, J. (2012). An overview of global papaya production, trade and consumption. Corpus ID 220727911.

Fakava, V.T. (2012). *Climate Change Impact on Agriculture and Food Security*, Regional Training Workshop on Adaptation for the Pacific Least Developed Countries, 28 September–3 October 2012, Funafuti, Tuvalu.

Galán-Saúco, V.G. and Rodríguez-Pastor, M.C.R. (2007). Greenhouse cultivation of papaya. *Acta Hort.*, 740: 191–195.

Ghini, R., Bettiol, W. and Hamada, E. (2011). Diseases in tropical and plantation crops as affected by climate changes: current knowledge and perspectives. *Plant Pathology*, 60: 122–132.

Houghton, J.T., Ding, Y., Griggs, D.J. et al. (2001). *Climate Change 2001: The Scientific Basis. Intergovernmental Panel on Climate Change* (IPCC), Cambridge University Press, Cambridge, 881 p.

Jesus Junior, W.C., Ceci´lio, R.A. and Valadares Junior, R. et al. (2007). *Aquecimento global e potencial impacto na cultura e doenc¸as do mamoeiro*. pp. 83–100. *In*: Martins, D.S., Costa, N.A. and Costa, A.F.S. (eds.). *Papaya Brasil: Manejo, Qualidade e Mercado do Mama˜o. Vito´ria*, Brazil: Incaper.

Jeyakumar, P., Kavino, M., Kumar, N. and Soorianathasundaram, K. (2007). Physiological performance of papaya cultivars under abiotic stress conditions. Chan, Y.K. and Paull, R.E. (eds.). Proc. Ist IS on Papaya. *Acta Hort.*, 740, ISHS 2007.

Kohler, J., Hernández, J.A., Caravaca, F. and Roldán, A. (2008). Plant-growth-promoting rhizobacteria and arbuscular mycorrhizal fungi modify alleviation biochemical mechanisms in water-stressed plants. *Funct Plant Biol.* 2008 Apr; 35(2): 141–151. doi: 10.1071/FP07218. PMID: 32688765.

Lawlor, D.W. (1993). *Photosynthesis: Molecular, Physiological and Environmental Process.* Longman Scientific & Technical, Hong Kong.

Laxman, R.H. and Bhatt, R.M. (2017). Abiotic stress management in fruit crops. pp. 399–412. *In*: Minhas, P.S. et al. (eds.). *Abiotic Stress Management for Resilient Agriculture*. Doi. 10.1007/978-981-10-5744-1_18.

Le Tacon, F., Zeller, B., Plain, C., Hossann, C., Bréchet, C. and Robin, C. (2013). Carbon transfer from the host to tuber melanosporum mycorrhizas and ascocarps followed using a 13C pulse-labeling technique. *PLoS ONE*, 8: e64626.

Mahouachi, J., Socorro, A.R. and Talon, M. (2006). Responses of papaya seedlings (*Carica papaya* L.) to water stress and rehydration: growth, photosynthesis and mineral nutrient imbalance. *Plant and Soil*, 281(1): 137–146. DOI:10.1007/s11104-005-3935-3.

Malo, S.E. and Campbell, C.W. (1986). The papaya, Univ. Florida (Gainesville) Coop. Ext. Serv. *Fruit Crops*, Fact Sheet FC 11.

Mamatha, G., Bagyaraj, D.J. and Jaganath, S. (2003). Inoculation of field-established mulberry and papaya with arbuscular mycorrhizal fungi and a mycorrhiza helper bacterium. *Mycorrhiza*, 12(6): 313–6.

Mangrauthia, S.K., Shakya, V.P.S., Jain, R.K. and Praveen, S. (2009). Ambient temperature perception in papaya for papaya ringspot virus interaction. *Virus Genes*, 38: 429–34.

Marler, T.E., George, A.P., Nissen, R.J. and Andersen, P.C. (1994). Miscellaneous tropical fruits. pp. 199–224. *In*: Schaffer, B. and Andersen, P.C. (eds.). *Hand Book of Environmental Physiology of Fruit Crops. Vol 2. Sub-Tropical and Tropical Crops*. CRC Press, Boca Raton, Florida.

Marler, T.E. and Mickelbart, M.V. (1998). Drought, leaf gas exchange, and chlorophyll fluorescence of field grown papaya. *J. Amer. Soc. Hort. Sci.*, 123(4): 714–718.

Masri, M., Razak, A.S. and Ghazalli, M.Z. (1990). Response of papaya (*Carica papaya* L.) to limited soil moisture at reproductive stage. *Mardi Res. J.*, 18: 191–196.

Meshram, A., Bhagyawant, S.S. and Srivastava, N. (2017). Environment and biodiversity conservation studies with remote sensing and GIS. *MOJ Proteomics Bioinform.*, 5(2): 36–38.

Milind, P. and Gurditta. (2011). Basketful benefits of Papaya. *Intl. Res. J. Phar.*, 2(27): 6–12.

Muok, O.B. and Ishii, T. (2006). Effect of arbuscular mycorrhiza fungi on tree growth and nutrient uptake of *Sclerocarya birrea* under water stress, salt stress and flooding. *Journal of Japan Society of Horticultural Science*, 75: 26–31.

Muss, C., Mosgoeller, W. and Endler, T. (2012). Papaya preparation (Caricol®) in digestive disorders. *Biogenic Amines*, 26: 1–17.

Nakasone, H.Y. and Paull, R.E. (1998). *Tropical Fruits*. CAB International, Wallingford.

Nishina, M., Zee, F., Ebesu, R., Arakaki, A., Hamasaki, R., Fukuda, S., Nagata, N., Chia, C.L., Nishijima, W., Mau, R. and Uchida, R. (2000). Papaya production in Hawaii. *Fruits and Nuts*, June 2000, F&N-3. Cooperative extension service, University of Hawaii, Manoa.

OECD (Organisation for Economic Co-operation and Development). (2003). Draft Consensus Document on the Biology of *Carica papaya* (L.) (Papaya), *Report No. 5*, February 2003, OECD, France.

Panda, S.S., Hoogenboom, G. and Paz, J. (2009). Distinguishing blueberry bushes from mixed vegetation land-use using high resolution satellite imagery and geospatial techniques. *Comput. Electron. Agr.*, 67: 51–59.

Panda, S.S., Hoogenboom, G. and Paz, J.O. (2010). Remote sensing and geospatial technological applications for site-specific management of fruit and nut crops: a review. *Remote Sensing*, 2(8): 1973–1997.

Qiang-Sheng, W., Ren-Xue, Z. and Ying-Ning, Z. (2007). Osmotic solute responses of mycorrhizal citrus (*Poncirus trifoliata*) seedlings to drought stress. *Acta Physiologia Plantarum*, 29: 543–549.

Raviv, M. (2010). The use of mycorrhiza in organically-grown crops under semi arid conditions: a review of benefits, constraints and future challenges. *Symbiosis*, 52: 65–74. http://dx.doi.org/10.1007/s13199-010-0089-8.

Ray, P.K. (2002). *Breeding Tropical and Subtropical Fruits*. Narosa Publishing House, New Dehli (India) pp. 1–337.

Reddy, C., Sivapriya, T.V.S., Arun Kumar, U. and Ramalingam, C. (2016). Optimization of food acidulant to enhance the organoleptic property in fruit. *J. Food Process Technol.*, 7: 11.

Reddy, A., Gopala Krishna, J., Suresh Kumar, V., Maruthi, K. Venkatasubbaiah and Srinivasa, R.C. (2017). Fruit production under climate changing scenario in India. *Environment & Ecology*, 35(2B): 1010–1017, April–June, 2017.

Reis, F.O., Campostrini, E., Martelleto, L.A.P., Vasconcellos, M.A.S. and Ribeiro, R.L.D. (2005). Trocas gasosas em folhas do mamoeiro 'Baixinho de Santa Amália' sob diferentes ambientes de cultivo. Annals of the X Congresso Brasileiro de Fisiologia Vegetal. Recife, Brazil (on CD ROM).

Rimberia Fredah, K. (2017). Commercial and industrial development of papaya (*Carica papaya* L.): Varietal improvement, production and processing technologies. Research funded by the Government of Kenya through Jomo Kenyatta University of Agriculture and Technology from 2008 to 2013. The extension and additional funding by the Africa Union-african innovation-JKUAT and PAUSTI Network (Africa-ai-JAPAN) project (from 2015 to 2018).

Rimberia, F.K., Ombwara, F.K., Mumo, N.N. and Ateka, E.M. (2018). Genetic improvement of papaya (*Carica papaya* L.). *In*: Al-Khayri, J., Jain, S. and Johnson, D. (eds.). *Advances in Plant Breeding Strategies: Fruits*. Springer, Cham. Doi https://doi.org/10.1007/978-3-319-91944-7_21 ISBN 978-3-319-91944-7.

Sevier, B.J. and Lee, W.S. (2005). Precision farming adoption by florida citrus producers: probit model analysis; Circular # 1461; Institute of Food and Agricultural Sciences, University of Florida: Gainesville, FL, USA, 2005.

Smith, S.E. and Smith, F.A. (2011). Roles of arbuscular mycorrhizas in plant nutrition and growth: new paradigms from cellular to ecosystems scales. *Annual Review of Plant Biology*, 63: 227–250.

Sritakae, A., Praseartkul, P., Cheunban, W., Miphokasap, P., Eiumnoh A., Burns P., Phironrit N., Phuangrat, B., Kitsubun, P. and Meechai, A. (2011). Mapping airborne pollen of papaya (*Carica papaya* L.) and its distribution related to land use using GIS and remote sensing. *Aerobiol.*, 47(1): 291–300.

Strange, R.N. and Scott, P.R. (2005). Plant disease: A threat to global food security. *Annual Review of Phytopathology*, 43: 83–116.

Subedi, S. (2019). Climate change effects of Nepalese fruit production. *Adv Plants Agric Res.*, 9(1): 141–145.

Teixeira da Silva, J.A., Rashid, Z., Tan Nhut, D. et al. (2007). Papaya (*Carica papaya* L.) biology and biotechnology. *Tree For Sci Biotech.*, 1(1): 47–73, Global Science Books.

Thani, Q.A., Vargas, A.I., Schaffer, B., Liu, G. and Crane, J.H. (2016). Krome memorial section responses of papaya plants in a potting medium in containers to flooding and solid oxygen fertilization. *Proc. Fla. State Hort. Soc.*, 129: 27–34.

Torres Netto, A. (2005). *Physiological characterization and water relations in papaya genotypes* (*Carica papaya* L.). DSc thesis, Campos dos Goytacazes-RJ, Universidade Estadual do Norte Fluminense Darcy Ribeiro-UENF, 116 p.

Vass, I. (2012). Molecular mechanisms of photo-damage in the photosystem II complex. *Biochem. Biophys. Acta*, 1817(1): 209–217.

Zimmerman, Thomas W. (2008). Papaya Growth in Double-Row Systems Established During the Dry Season. 44th Annual Meeting, July 13–17, 2008, Miami, Florida, USA 256616, Caribbean Food Crops Society.

7

Passion Fruits Resilience to Global Warming and Climate Change

Daniel Chebet,[1] *Isaac Savini*[2] and *Fredah K Rimberia*[3,*]

1. Introduction

1.1 Background Information

Passion fruit, a perennial woody fruit vine belongs to the family Passifloraceae, is native to South America, specifically in Colombia and Brazil (Ocampo et al., 2010). There are two types of edible passion fruit: purple (*Passiflora edulis* Sims) and yellow (*P. edulis* var flavicarpa Deg.). The purple passion fruit is originally native of tropical America, whereas yellow passion fruit is either a mutation of the purple variety or a natural hybrid between purple and another related species of passion fruit (Alkamin and Girolami, 1959). *Passiflora quadrangularis* L., the giant granadilla, is also cultivated to a limited extent for local consumption. The major world producers of passion fruit are Australia, Hawaii, South Africa and Brazil (Joy, 2010). In Brazil, yellow passion fruit accounts for 95% of production, while the purple passion fruit accounts for the remaining

[1] Department of Seeds, Crops and Horticultural, University of Eldoret, Kenya.
[2] A private consultant in agriculture and plant nutrition (formerly, International Center for Tropical Agriculture (CIAT) Staff, Nairobi, Kenya.
[3] Department of Horticulture and Food Security, Jomo Kenyatta University of Agriculture and Technology, Kenya.
Emails: daniel.chebet@uoeld.ac.ke; wasavini@gmail.com
* Corresponding author: frenda@agr.jkuat.ac.ke

5%. Other passion fruit producers include Ecuador, Indonesia, Colombia, Kenya and South Africa (Othman, 2016).

The fruit has high nutritional and medicinal value. It is a rich source of Vitamins A and C and contains fair amounts of iron, potassium, sodium, magnesium, sulphur and chlorides and has dietary fiber and protein. Fruits are eaten fresh or processed into products like jams, squash, juice, cakes, pies and ice-cream (Thokchom and Mandal, 2017). *Passiflora edulis* plant contains anti-inflammatory, anticonvulsant, antimicrobial, anticancer, anti-diabetic, antihypertensive, anti-sedative, antioxidant properties and various remedial actions for treating conditions like osteoarthritis, asthma and acts as colon cleanser (Zas and John, 2012). The different parts of the plants have been used for treatment of ulcers, haemorrhoids, as sedatives, remedy for insomnia, digestive stimulant and remedy for gastri-carcinoma (Zas and John, 2012).

1.2 *Climatic Adaptation*

Passion fruits are adapted to climates ranging from cool subtropical (purple variety) to warm tropical (yellow variety), but are sensitive to extremes of both low and high temperatures. Menzel et al. (1987) observed that the potential yield was reduced by either low temperatures which restricted vegetative growth or by high temperatures which prevented flower production. They concluded that selection criteria for new cultivars should include both cold- and heat-tolerance to expand passion fruit production.

2. Ecological Requirements of Passion Fruits

2.1 *Agro-ecological Zones and Altitude*

Purple passion fruit does well in the upper midland to upper highland zones. Yellow passion fruits do well in the lower midland and lowland zones. The most suitable altitude for passion fruit production in Kenya ranges from 1200–2000 m above sea level. In South Africa, at altitudes between 1,200–1,400 m, plantations may be productive for eight years, owing to the longer cycles, implicating a greater longevity (Joy, 2010).

2.2 *Temperature*

According to USAID (2018), there was increased average temperature of 0.34°C per decade in Kenya from 1985–2015. The greatest increases occurred in March to May season and in arid and semi-arid regions. According to the World Bank (2020), the projected increase in temperature in 2020–2039 will be 1.15°C, in 2040–2059 it will be 2.1°C and from 2080–2099 it will be 4.31°C (Fig. 1). The projected days of heat waves

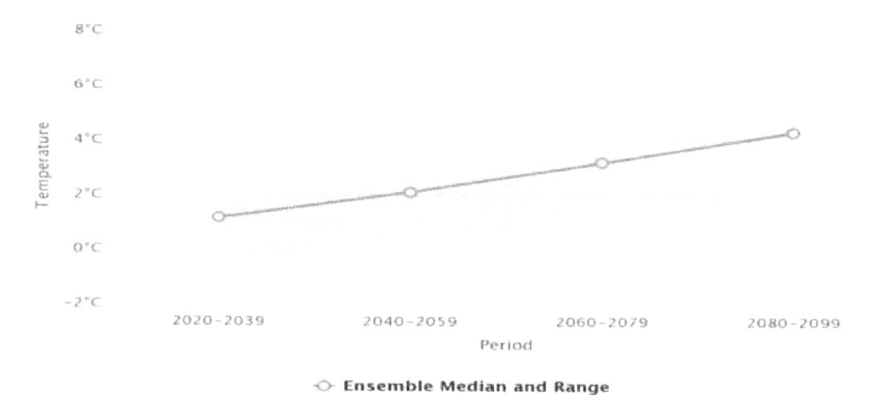

Fig. 1. Projected change in minimum temperature for Kenya from 2020–2099 (*Source*: World Bank, 2020).

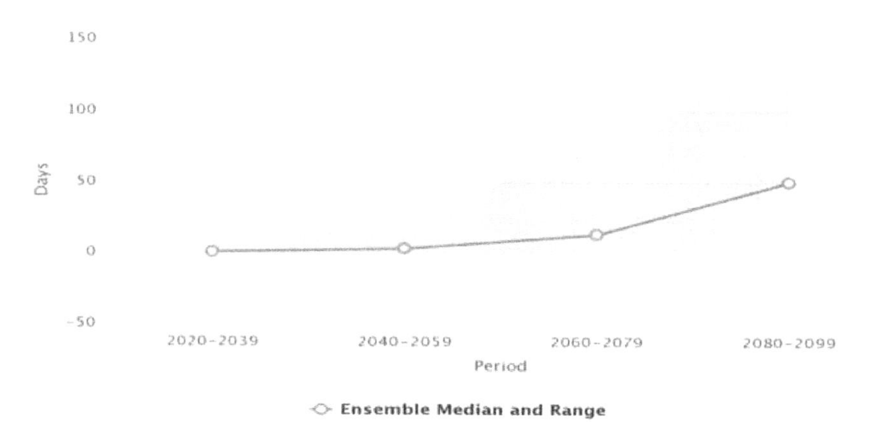

Fig. 2. Projected increase in number of days of heat waves (> 35°C) in Kenya (*Source*: World Bank, 2020).

(> 35°C) will increase from 0.17 in 2020–2039 to 2.59 days in 2040–2059, 12.56 days in 2060–2079 and 49.44 from 2080–2099 (Fig. 2). The optimum temperature for production of purple and yellow passion fruit is between 18°C–25°C and 25°C–30°C respectively (Joy, 2010). Higher temperatures also reduce plant dry weight by reducing the leaf and stem weights and causing a smaller increase in root weight (Menzel et al., 1987).

2.3 *Light*

Successful flower development requires certain environmental conditions. In many species, photoperiod influences flower induction. In passion fruit, long days (LDs) are necessary for intact flower development but not for flower induction (Chayut et al., 2014). Flower primordia of most varieties abort at an early stage of development if exposed to photoperiods shorter than 11 hours (Chayut et al., 2014). In Israel, flowers of the purple 'Passion Dream' (PD) cultivar reach anthesis only during spring and autumn and abort during winter, due to short photoperiods (Nave et al., 2010).

2.4 *Rainfall*

Passion fruits develop continuously in the tropics and require a constant supply of water to enable continuous flowering and fruiting. Water requirement is highest when fruits are approaching maturity. Well distributed rainfall of 900–2000 mm throughout the year is suitable, preferably with 60–120 mm of water every month. Productivity of around 40 t/ha has been obtained with a total water supply (rain and irrigation) of 1,300–1,470 mm. Irrigation is needed in areas where the annual rainfall is below 1,200 mm per year (Joy, 2010; Infonet-Biovision, 2019).

On an average, there has been little change in precipitation in Kenya since 1970s, though a decline in the long rains has been observed in central Kenya. The future projections show a likely increase in average rainfall (projections range from –3 to +28%), in the coast and highlands mainly from October to May (Opondo, 2013). According to World Bank (2020), the projected annual increase in rainfall for Kenya will be 60.7 mm from 2020–2039, 37.09 mm in 2040–2059, 65.18 mm in 2060–2079 and 90.06 mm from 2080–2099 (Fig. 3).

While the increase in rainfall favours passion fruits, the challenge is by the rainfall intensity. Excessive rainfall causes poor fruit set in passion fruits. The main effect is on pollination. If rain occurs for 1–1/2 hrs after pollination, there will be no fruit set, but if 2 hrs pass before rain falls, it

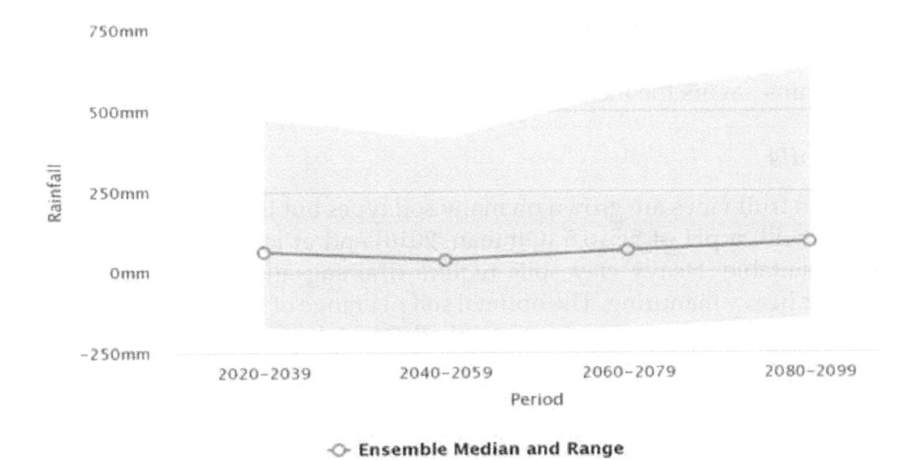

Fig. 3. Projected change in annual rainfall range for Kenya 2020–2099 (*Source*: World Bank, 2020).

will have no detrimental effect (Morton, 1987). Excess rainfall also inhibits pollination by diminishing the activity of pollinating insects and causing pollen grains to burst. Excess rainfall also encourages diseases (Joy, 2010).

Passion fruits are highly sensitive to flooding stress. Climate-related floods and droughts have doubled in sub-Saharan Africa within the last quarter century and countries like Mozambique, Malawi, Kenya, Madagascar and Ethiopia are projected as likely to keep experiencing such unexpected and extreme climatic events (Opondo, 2013). Passion fruit seedlings, subjected to flooding stress for 28 days, led to a reduction in leaf number (due to leaf abscission), total chlorophyll content, root fresh and dry weights and leaf nitrogen and phosphorus content in passion fruit seedlings. Flooding, however, caused an increase in leaf proline and carotenoid content and total soluble solids (Chebet et al., 2020).

Passion fruits withstand droughts relatively well. Prolonged droughts, however, inhibit vegetative development, and in severe cases, cause leaf fall and the formation of smaller and lighter fruits which may shrivel and fall prematurely (Joy, 2010). Drought stress cause a reduction in leaf, stem and root dry weight, leaf area, vine extension and nodes and number of flower buds and open flowers. The reduction is associated with reduced nutrient uptake and leaf water potential, but not leaf conductance (Menzel et al., 1986).

2.5 *Relative Humidity*

Relative humidity has a great influence on vegetative development and the phytosanitary state of the passion fruit. Air relative humidity of around 60% is most favorable for the passion fruit. Elevated temperature, associated with constant wind and low relative humidity, causes a drying out of the tissues by excessive transpiration and impedes the development of passion fruits. Relative humidity of greater than 60%, when associated with rains, favors the incidence of foliar diseases (Joy, 2010).

2.6 *Soils*

Passion fruit vines are grown on many soil types but light to heavy sandy loams with a pH of 5.5–6.5 (Othman, 2016) and at least 60 cm deep are most suitable. Heavy clay soils require draining and very sandy soils require heavy manuring. The optimal soil pH range of 5.5–7 is preferred. If the soil is too acidic, lime must be applied. Good drainage and aeration are essential to minimize the incidence of diseases (Infonet-Biovision, 2019).

One of the challenges affecting passion fruit production is soil salinity. The soil salinity component that is related to climate change is that caused by sea-water extrusion into low-lying coastal soils. According to USAID (2018), Kenya's sea level rose by 5.8 cm in 1932–2001. However, the future projection of sea level rise is 16–42 cm by 2050. A study by Chebet (2020) showed that soil salinity caused a reduction in plant height, leaf number, leaf area and chlorophyll content in passion fruits seedlings. Soil salinity also reduced the leaf and root fresh and dry weights, causing a decrease in the leaf potassium and magnesium contents, but the nitrogen, phosphorus and calcium contents were unaffected (Chebet, 2020).

3. Symptoms of Climate Change Impacts on Passion Fruits

3.1 *Abiotic Factors*

Primordia of most *Passiflora edulis* varieties are sensitive to hot ambient temperatures and abort during warm summers (Chayut et al., 2014). Temperature above 25°C was detrimental to vegetative growth in purple passion fruits. Purple passion grown at maximum and minimum diurnal temperature of 25 and 20°C respectively, had higher node number and higher internode extension and also had higher leaf area compared to those grown at maximum and minimum of 30 and 25°C, respectively. Higher temperature also reduced plant dry weight by reducing the leaf and stem weights and causing a smaller increase in root weight (Menzel et al., 1987).

Increased temperature has detrimental effects on flowering and fruiting. The greatest numbers of floral buds and open flowers were obtained at 20/15°C compared to higher temperatures of 25/20°C and 30/25°C (Menzel et al., 1987). At 23–28°C, the fruit growth period in purple passion was 60 days. However, when the temperature was lower than 23°C or higher than 33°C, the fruit growth period increased to 75 days (Joy, 2010).

The greatest uptake of nutrients into the shoots occurred at 25/20°C. Temperature rise to 30/25°C caused a reduction in the uptake of minerals, like phosphorus, calcium, sodium, manganese and aluminium but did not affect the uptake of nitrogen, potassium, sulphur, magnesium, zinc, copper and boron (Menzel et al., 1987).

3.2 *Biotic Factors*

Climate change-inspired increase in temperature is one factor driving the spread of pests and diseases. Temperature affects the population size, survival rate and geographical distribution of pests, as well as the intensity, development and geographical distribution of diseases. In general, increase in temperature and precipitation levels favors the growth and distribution of most pest species by providing a warm and humid environment and the necessary moisture for their growth (Doody, 2020).

3.2.1 Fusarium Wilt (Fusarium oxysporum)

Fusarium oxysporum f.sp. is a soil-inhabiting pathogen responsible for wilt and cortical rot diseases in more than a hundred crops, including passion fruits (Kristler et al., 1998; Venditti et al., 1988). The disease commonly occurs on adult plants from the beginning of the yield.

The first symptom (Fig. 4A) is a slight wilt of the branch tips, followed by sudden death of the plant within four to 14 days. Before the wilt of the whole plant, a partial wilt (one side of the plant) can occur. Usually, the wilted leaves do not turn yellow and are still attached to the plant for a few days. The collar region of affected plants turns brownish at soil level and vertically cracks, followed by a complete collapse of the plant. On dissection of affected stems, vascular tissues show brownish discolouration (Fig. 4D) (Liberato and Laranjeira, 2005; Infonet-Biovision, 2019). Increase in temperature associated with climate change plays a significant role in increased incidence of fusarium wilt in passion fruits. The most favorable temperature for disease development is 22–26°C (Garibaldi et al., 2006). *Fusarium oxysporum* f.sp. is also sensitive to moisture and disease symptoms appear quicker under water-stress conditions (Ghaemi et al., 2011).

Fig. 4. Symptoms of Fusarium wilt damage in passion fruit leaves (A), vines (B), stem (C) and collar region (D) (*Source*: Infonet-Biovision, 2019).

3.2.2 Passion Fruit Crown and Root Rot Disease

This is a fairly new disease affecting passion fruit production in Kenya. It was first recorded in central Kenya in 2004 and is currently widespread in most passion fruit-producing areas (Mbaka et al., 2006). The disease is also called dieback in Kenya (Amata et al., 2009), collar rot in Fiji, fatal

Fig. 5. Symptoms of crown and root rot damage in passion fruit leaves (A), fruits (B), vines (C) and stem (D) (*Sources*: Infonet-Biovision, 2019; Dr. Chebet, personal communication; Nguyen et al., 2015).

blight or stem and fruit rot in South Africa, root rot in New Zealand and Australia, damping-off and leaf blight in India (Morton, 1987) and stem canker in Vietnam (Nguyen et al., 2015). The disease is highly virulent and in Kenya reduces the orchard lifespan to less than two years (Otipa et al., 2009), while in Fiji, studies have showed reduced lifespan to 30–36 months (Morton, 1987).

Passion fruit crown and root rot is caused by a complex of soil fungi, including *Phytophtora cinnamon, P. nicotiniae, Haematonecria haematococca* and *Fusarium solani* (Joy, 2012; Amata et al., 2009). Infection occurs optimally at higher temperatures but can occur at as low as 15°C. Chlamydospores, the survival spores of the pathogens, persist for up to six years in infected roots, crown and infested soil. Under warm temperatures and wet conditions, the chlamydospores germinate to produce sporangia which then release infectious zoospores. These zoospores then swim to susceptible roots to initiate infections. Once a host is infected, the water flow through the xylem is reduced via wilt-inducing toxins, such as ß-glucans and ß-glucan hydrolases.

Excessive use of nitrogen-based fertilizers further increases susceptibility to disease due to the increased uptake of water from the soil matrix (Coyier and Roane, 1986; Erwin and Ribiero, 1996). The first above-ground symptoms of the disease show mild dieback of the plant followed by changing of leaf colour to pale green, then a light copper colour. Infection on the leaves spreads to the stem, causing initial purpling, then brown areas on the stem. Large, gray/green aqueous spots are visible in the fruits which easily fall down. Wilting, defoliation and plant death occurs, resulting from complete necrotic girdling of the plant collar. The necrosis may reach 2–10 cm above ground and may extend to the roots. Tumescence and fissures in affected collar bark show purple lesion borders and white or reddish mycelia and perithecial structures appearing under high humidity (Joy, 2012).

4. Passion Fruit Management under Changed Climate Conditions

4.1 *Use of Beneficial Organisms*

a) Trichoderma

Trichoderma are fungal organisms that are widespread in soil and root ecosystems. Several strains of *Trichoderma* have been developed as biocontrol agents against fungal diseases of plants with the most common being *T. harzianum, T. viride, T. koningii* and *T. Hamatum. Trichoderma* counteracts the spread of soil-borne diseases brought about by climate change.

The mechanism of action of *Trichoderma* is multifaceted, mainly related to antagonism against plant pathogens (Verma et al., 2007) or activation of plant defences (Hermosa et al., 2012). Several strains produce hydrolases

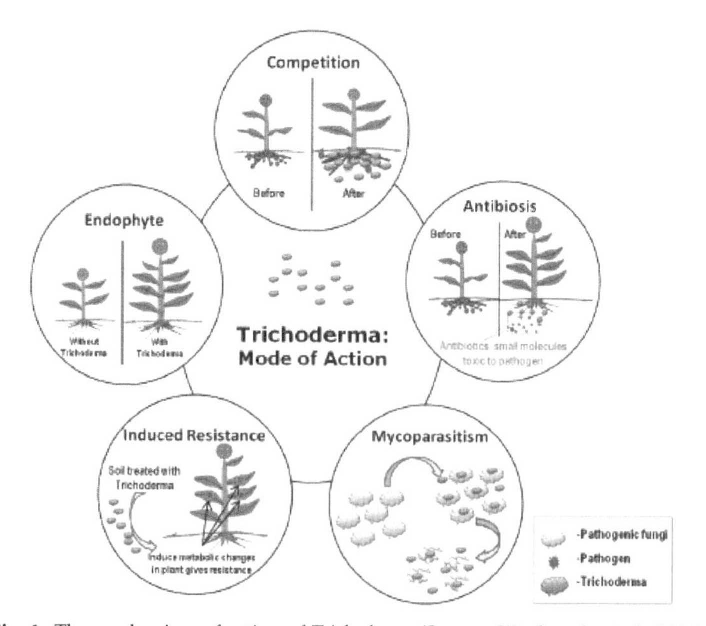

Fig. 6. The mechanism of action of *Trichoderma* (Source: Waghunde et al., 2016).

and proteins which weaken fungal cell walls and act as microbial elicitors (Gomes et al., 2015). The activation of plant defence following *Trichoderma* treatments goes beyond reducing plant diseases. *Trichoderma* application increases the plant tolerance to abiotic stress related to climate change, such as drought and salinity stress (Mona et al., 2017).

There are various methods of application of *Trichoderma*. According to Dudutech (2020), the methods include seed treatment by mixing 10 gms of *Trichoderma* formulation in one liter of cowdung slurry to treat 1 kg of seeds before planting. Nursery treatment is by drenching nursery soils with 5 gms *Trichoderma* per liter of water before planting. In the soil, mix one kg of *Trichoderma* formulation in 100 kg manure and cover for seven days with polythene. Turn the manure in the mixture every three to four days and broadcast in the field. Cuttings and seedlings root tip application is done by mixing 10 gms of *Trichoderma* in one liter and water and dip cutting for 10 minutes before rooting.

To ensure complete protection of passion fruits against soil-borne diseases: (i) the planting hole should be treated with *Trichoderma* product before transplanting. Repeat after 45 days; (ii) thereafter monitor the population of pathogenic *Fusarium oxysporum* f. sp. *Passiflorae,* macroconidia cultured per gram of substrate should be monitored. When it is more than 1.8×10^6/gm, application of *Trichoderma* is recommended; (iii) in cases where the inoculum level cannot be monitored, apply *Trichoderma* when there is a 10-day dry spell and if the dry spell continues, a repeat application should be done every 45 days. Alternating the use of the two drenches is recommended (Dudutech, 2020).

b) Arbuscular Mycorrhizal Inoculation

Mycorrhiza refers to the mutualistic symbiosis between soil-borne fungi with the roots of plants. Arbuscular Mycorrhiza (AM) (also called vesicular arbuscular mycorrhiza VAM or endomycorrhiza) are the most common types of mycorrhizas with over 90% of all plant families believed to contain AM species. The beneficial effects of AM inoculation are greatest under adverse soil and crop conditions. Mineral nutrient acquisition is considered to be the primary function of mycorrhizas, especially uptake of non-mobile nutrients, such as phosphorus, copper and zinc. This is done by extending the volume of soil accessible to plants, leading to increased plant nutrient supply. Mycorrhizas also play a role in the uptake of potassium, nitrogen, calcium and magnesium, although to a lesser extend (Smith and Smith, 2011).

Arbuscular mycorrhizae help to alleviate salinity stress. Mycorrhiza inoculated passion fruit seedlings showed higher plant height, leaf number and chlorophyll content than non-inoculated seedlings under both salt-stress and unstressed conditions. Mycorrhiza inoculated seedlings also had significantly higher root, stem and leaf fresh and dry weights and had

Fig. 7. Effect of salt stress and mycorrhizal inoculation on passion fruit seedlings: (A) extreme salinity, non mycorrhizal, (B) moderate salinity, mycorrhizal, (C) no salt stress, inoculated with mycorrhizae fungi (*Source*: Chebet, 2020).

higher leaf concentration of P, K and Mg than non-inoculated seedlings under both control and medium salt stress conditions. Mycorrhiza inoculated plants also showed lower shoot Na concentrations than non-inoculated plants grown under saline soil conditions (Chebet, 2020).

Arbuscular mycorrhizae help to alleviate flooding stress. A study in passion fruits showed that mycorrhiza inoculation induced greater root, stem and leaf fresh and dry weights under flooding. Mycorrhiza inoculated seedlings also maintained greater leaf area as opposed to leaf abscission that occurred more rapidly in non-inoculated seedlings under flooding. There were declines in chlorophyll a,b concentration and total chlorophyll, and increase in carotenoid content under flooding, but these were more pronounced in non-inoculated seedlings (Fig. 8) (Chebet et al., 2020).

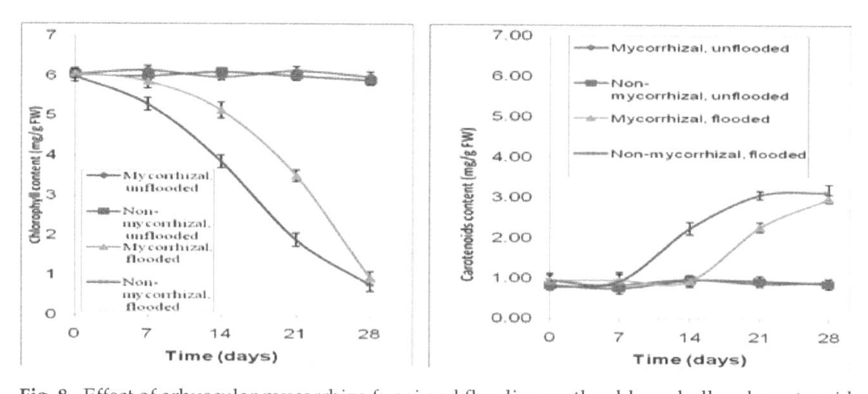

Fig. 8. Effect of arbuscular mycorrhiza fungi and flooding on the chlorophyll and carotenoid content of passion fruit seedlings (*Source*: Chebet et al., 2020).

Flooding induced an increase in leaf proline concentration on 14 and 21 days of flooding, with mycorrhiza inoculated seedlings having the highest proline concentration. Additionally, flooding induced a reduction, but did not completely inhibit mycorrhizal root colonization. The leaf nitrogen and phosphorus contents declined under flooding, especially in non-inoculated seedlings (Chebet et al., 2020).

4.2 *Use of Plant Growth Regulators*

Growth regulators have successfully been used to reduce the sensitivity of passion fruit flowers to hot ambient temperatures, a characteristic of warm summer. Passion fruit flowers at a specific stage showed more resistance in response to hot ambient temperatures after cytokinin application (Sobol et al., 2014). The commercial passion fruit hybrid cultivar 'Passion Dream' ('PD') flowers twice a year under Mediterranean conditions; during spring and early fall (autumn) (Nave et al., 2010). Chayut et al. (2014) managed to extend the anthesis period of the 'PD' vines by three to six weeks through spraying with 200 ppm Uniconazole (an inhibitor of gibberellins production) in mid-August (warm summer), leading to a longer period of passion fruits production.

4.3 *Orchard Floor Management*

Orchard floor management involves managing both tree rows and alleyways. Orchard floor management influences orchard temperatures and moisture retention. Bare ground will absorb more heat during the day and release more heat at night than soil covered in vegetation. Studies in apples show that the soil temperature is inversely related to the amount of mulch applied. Use of plant mulch reduced the minimum air temperature by 0.5–1.0°C while bare, compacted wet soil raised the minimum air temperature by as much as 2°C (Skyroch and Shribbs, 1986).

Fig. 9. Use of ground covers in a passion fruit orchard in Australia (Source: aussiepassionfruit.com.au).

Orchard floor cover also protects the soil from erosion by wind and water, prevents runoff, facilitates water percolation and helps manage dust (Granatstein, 2014). Other benefits include suppressing weeds, stabilizing the soil, maintaining beneficial insect populations and minimizing maintenance inputs (Rowley et al., 2012).

4.4 *Mounds or Berms*

Mounds or berms are raised beds used for planting crops. It is a common practice in landscape or vegetable farming in poorly drained soils.

With flooding expected to increase with climate change, this practice can help to minimize fusarium wilt and root and crown rots in passion fruits. Mounding is also used in shallow soils to raise the effective root area. The beds are raised 20–30 cm high, 1–2 m width be of any length.

4.5 *Genetic Engineering*

The applications of recombinant-DNA technology or genetic engineering in crop improvement are immense and include combating various production constraints, like biotic or abiotic stresses and fruit quality improvement (Parmar et al., 2017).

Detailed information concerning the genomes of passion fruit species is already published (Cutri and Dornelas, 2012; Yotoko et al., 2011). These genomics studies have the potential to enhance the development of new markers (e.g., SSR markers, expressed sequence tags [ESTs] and SNPs) and expand discussions of the structure, organization and evolution of the *Passiflora* genome (Cerqueira-Silva et al., 2016).

Moreover, the development and use of large-scale genotyping should enable the use of genome-wide selection strategies, thereby enhancing the association of molecular diversity data with characteristics of agronomic interest (Cerqueira-Silva et al., 2016). Such techniques will fast track the development of passion fruit varieties that are resistant to abiotic and abiotic stresses.

5. Conclusion and Prospects

Climate change poses serious threat to passion fruit farming worldwide. Increased temperature reduces the number of floral buds and open flowers, reduces the rate of fruit growth and reduces the uptake of minerals, like phosphorus, calcium and manganese. Excess temperature and rainfall also encourage proliferation of diseases, such as fusarium wilt, crown and root rots and brown spots. Increase in rainfall causes poor pollination by diminishing the activity of pollinating insects and causing pollen grains to burst. Other climate change effects on passion fruits are expressed via increased incidences of floods and sea water extrusion due to rise in sea levels. This leads to increased salinity in low lying farmlands which is detrimental to yellow passion fruit production.

Sustainable management strategies proposed to mitigate against the climate change effects in passion fruits include use of beneficial microorganisms, such as *Trichoderma* and Arbuscular mycorrhizae. *Trichoderma* counteracts the spread of soil-borne diseases brought about by climate change. Use of *Trichoderma* is effective by antagonising against plant pathogens, activation of plant defences and increasing the plant tolerance to abiotic stress related to climate change, such as drought and salinity stress. On the other hand, Arbuscular mycorrhiza improves the plant's ability to acquire nutrients and water, enhance tolerance to drought stress and confer tolerance to flooding and high soil salinity. Other strategies include use of plant growth regulators, orchard surface management and soil mounding.

The application of recombinant-DNA technology (genetic engineering) and technologies like the Next Generation Sequencing is to speed up breeding of new passion fruit varieties for overcoming biotic or abiotic stresses. In addition to this, Geographic Information System, remote sensing and satellite imaging technologies have come in handy for improving the efficiency of site-specific crop management. They have been applied in developing models to measure growth rate changes caused by stress, like salinity, drought and temperature among others, yield prediction, water-demand forecasting or the determination of soil nutrient availability among others. A combination of these technologies holds promise in boosting passion fruit production and productivity in the face of challenges caused by global warming and climate change. Further research is required in application of modern technologies, like genetic engineering for breeding new varieties, as well as Geographic Information System, remote sensing and satellite imaging in passion fruit improvement.

Acknowledgements

The author wishes to thank the German government through the German Academic Exchange Service (*Deutscher Akademischer Austaunsch Dienst*) (DAAD) and the Kenyan government through the Higher Education Loans Board (HELB) for funding the passion fruit research.

References

Alkamin, E.K. and Girolami, G. (1959). Pollination and fruit set in the yellow passion fruit. *Hawaii Agri. Stat. Tech. Bull.*, 59: 44.

Amata, R.L. Otipa, M.J., Waiganjo, M., Wabule, M., Thuranira, E.G., Erbaugh, M. and Miller, S. (2009). Incidences, prevalence and severity of passion fruit fungal diseases in major production regions of Kenya. *Journal of Applied Biosciences*, 20: 1146–1152.

Cerqueira-Silva, C.B.M., Faleiro, F.G., Nunes de Jesus, Lisboa dos Santos, E.S. and Pereira, de Souza, A. (2016). The genetic diversity, conservation and use of passion fruit (*Passiflora* spp.). *In*: Ahuja, M.R. and Jain, S.M. (eds.). *Genetic Diversity and Erosion in Plants, Sustainable Development and Biodiversity*, 8. Doi: 10.1007/978-3-319-25954-3_5.

Chayut, N., Sobol, S., Nave, N. and Samach, A. (2014). Shielding flowers developing under stress: translating theory to field application. *Plants*, 3: 304–323. Doi: 10.3390/plants3030304.

Chebet, D.K. (2020). *Effect of Arbuscular Mycorrhizal Inoculation on the Growth and Performance of Tropical Fruit Seedlings under Saline, Flooding and Nutrient Stress.* Ph.D thesis, Jomo Kenyatta University of Agriculture and Technology, Kenya, pp. 55–119.

Chebet, D., Kariuki, W., Wamocho, L. and Rimberia, F. (2020). Effect of Arbuscular Mycorrhizal inoculation on growth, biochemical characteristics and nutrient uptake of passion fruit seedlings under flooding stress. *International Journal of Agronomy and Agricultural Research (IJAAR)*, 16(4): 24–31.

Coyier, D.L. and Roane, M.K. (1986). *Compendium of Rhododendron and Azalea Diseases*. APS Press.

Cutri, L. and Dornelas, M.C. (2012). PASSIOMA: Exploring expressed sequence tags during flower development in *Passiflora* spp. *Comparative and Functional Genomics*, pp. 1–11.

Doody, A. (2020). *Pests and Diseases and Climate Change: Is There a Connection?* https://www.cimmyt.org/news/pests-and-diseases-and-climate-change-is-there-a-connection/.

Dudutech. (2020). *Trichotech*. https://www.dudutech.com/products/trichotech/.

Erwin, D.C. and Ribiero, O.K. (1996). Phytophthora *Diseases Worldwide*, American Phytopathological Society, St Paul, 562p.

Garibaldi, A., Gilardi, G. and Gullino, M.L. (2006). Evidence for an expanded host range of *Fusarium oxysporum* f.sp. raphani. *Phytoparassitica*, 34: 115–121.

Ghaemi, A., Rahimi, A. and Banihashemi, Z. (2011). Effects of water stress and *Fusarium oxysporum* f. sp. lycoperseci on growth (leaf area, plant height, shoot dry matter) and shoot nitrogen content of Tomatoes under Greenhouse Conditions, Iran. *Agricultural Research*, 29: 51–62.

Gomes, E.V., Costa, M.D., de Paula, R.G., de Azevedo, R.R., da Silva, F.L., Noronha, E.F., Ulhoa, C.J., Monteiro, V.N., Cardoza, R.E. and Gutierrez, S. (2015). The Cerato-Platanin protein Epl-1 from *Trichoderma harzianum* is involved in mycoparasitism, plant resistance induction and self cell wall protection. *Science Report*, 5: 13.

Granatstein, D. (2014). Orchard Floor Management, *Washington State University Technical Bulletin*.

Hermosa, R., Viterbo, A., Chet, I. and Monte, E. (2012). Plant-beneficial effects of *Trichoderma* and of its genes. *Microbiology*, 158: 17–25.

Infonet-Biovision. (2019). *Passion Fruits*. https://infonet-biovision.org/PlantHealth/Crops/Passion-fruit.

Joy, P.P. (2010). *Passion Fruit Production Technology*. Kerala Agricultural University.

Joy, P.P. and Sherin, C.G. (2012). *Diseases of Passion Fruits (Passiflora edulis): Pathogen, Symptoms, Infection, Spread & Management*. Kerala Agricultural University.

Kistler, H.C., Alabouvette, C. Baayen, R.P., Bentley, S., Brayford, D. et al. (1998). Systematic numbering of vegetative compatibility groups in the plant pathogenic fungus, *Fusarium oxysporum*. *Phytopathology*, 88: 30–32.

Liberato, J.R. and Laranjeira, F.F. (2005). *Fusarium Wilt of Passionfruit (Fusarium oxysporum* f. sp. *passiflorae)*; updated on 12/21/2007 9:21:11 AM; aailable online: PaDIL – http://www.padil.gov.au.

Mbaka, J.N., Waiganjo, M.N., Chegeh, B.K., Ndungu, B., Njuguna, J.K., Wanderi, S., Njoroge, J. and Arim, M. (2006). A survey of the major passion fruit diseases in Kenya; *10th Biennial KARI Proceedings*.

Menzel, C.M., Simpson, D.R. and Dowling, A.J. (1986). Water relations in passion fruit: Effect of moisture stress on growth, flowering and nutrient uptake. *Scientia Horticulturae*, 29(3): 239–249.

Menzel, C.M., Simpson, D.R. and Winks, C.W. (1987). Effect of temperature on growth, flowering and nutrient uptake of three passion fruit cultivars under low irradiance. *Scientia Horticulturae*, 31: 259–268.

Mona, S.A., Hashem, A., Abd_Allah, E.F., Alqarawi, A.A., Soliman, D.W.K., Wirth, S. and Egamberdieva, D. (2017). Increased resistance of drought by *Trichoderma harzianum* fungal treatment correlates with increased secondary metabolites and proline content. *J. Integrated Agriculture*, 16: 1751–1757.

Morton, J. (1987). Passion fruit. pp. 320–328. *In*: *Fruits of Warm Climates*. Julia F. Morton, Miami, Florida.

Nave, N., Katz, E., Chayut, N., Gazit, S. and Samach, A. (2010). Flower development in the passion fruit *Passiflora edulis* requires a photoperiod-induced systemic graft-transmissible signal. *Plant Cell Environ.*, 33: 2065–2083.

Nguyen, T.D., Burgess, T., Dau, V.T. et al. (2015). Phytophthora stem rot of purple passion fruit in Vietnam. *Australasian Plant Dis. Notes*, 10(35). https://doi.org/10.1007.

Ocampo, J., d'Eeckenbrugge, J.C. and Jarvis, A. (2010). Distribution of the genus *Passiflora* L. diversity in Colombia and its potential as an indicator for biodiversity management in the coffee growing zone. *Diversity*, 2: 1158–1180.

Opondo, D.O. (2013). Loss and damage from flooding in Budalangi district, western Kenya. *Loss and Damage in Vulnerable Countries Initiative, Case Study Report*, Bonn: United Nations University Institute for Environment and Human Security.

Othman, N. (2016). *Cultivation of Passion Fruit under Local Climatic Conditions of Lebanon (Ghazir)*. Lebanese University, Thesis, March 2016. Doi: 10.13140/RG.2.2.18388.30087.

Otipa, M., Amata, R., Waiganjo, M., Ateka, E., Mamati, G., Erbaugh, M. and Miller, S. (2009). Incidences and severity of viruses in passion fruit production systems in Kenya. *First African Biotechnology Congress*, Nairobi, Kenya.

Parmar, N., Singh, K.H., Sharma, D., Singh, L., Pankaj Kumar, Nanjundan, J., Khan, Y.J., Chauhan, D.K. and Thakur, A.K. (2017). Genetic engineering strategies for biotic and abiotic stress tolerance and quality enhancement in horticultural crops: A comprehensive review. *Biotech.*, 7: 239. Doi: 10.1007/s13205-017-0870-y.

Rowley, M., Black, B. and Cardon, G. (2012). Alternative orchard floor management strategies. USU Extension Publication Horticulture/Fruit/2012-01pr.

Shribbs, J.M. and Skroch, W.A. (1986). Influence of 12 ground cover systems on young 'Smoothee Golden Delicious' apple trees: I. Growth. *Journal of American Society for Horticultural Science*, 111: 525–528.

Smith, S.E. and Smith, F.A. (2011). Roles of arbuscular mycorrhizas in plant nutrition and growth: New paradigms from cellular to ecosystems scales. *Annual Review of Plant Biology*, 63: 227–250.

Sobol, S., Chayut, N., Nave, N., Kafle, D., Hegele, M., Kaminetsky, R., Wunsche, J.N. and Samach, A. (2014). Genetic variation in yield under hot ambient temperatures spotlights a role for cytokinin in protection of developing floral primordial. *Plant Cell Environ.*, 37: 643–657.

Thokchom, R. and Mandal, G. (2017). Production preference and importance of passion fruit (*Passiflora edulis*): A review. *Journal of Agricultural Engineering and Food Technology*, 4(1): 27–30.

USAID. (2018). https:///Climate%20change/2018_USAID-ATLAS-Project_Climate-Risk-Profile-Kenya.pdf.

Venditti, M., Micozzi, A., Gentile, G., Polonelli, L., Morace, G. et al. (1988). Invasive *Fusarium solani* infections in patients with acute leukemia. *Rev. Infect. Dis.*, 10: 653–660.

Verma, M., Brar, S.K., Tyagi, R.D., Surampalli, R.Y. and Valero, J.R. (2007). Antagonistic fungi, *Trichoderma* spp.: Panoply of biological control. *Biochemistry Engineering Journal*, 37: 1–20.

Waghunde, R., Shelake, R.M. and Sabalbara, A.N. (2016). *Trichoderma*: A significant fungus for agriculture and environment. *African Journal of Agricultural Research*, 11(22): 195–196.

World Bank. (2020). *Kenya Climate Data.* https://climateknowledgeportal.worldbank.org/download-data.

Yotoko, K.S.C., Dornelas, M.C., Togni, P.D., Fonseca, T.C., Salzano, F.M., Bonatto, S.L. and Freitas, L.B. (2011). Does variation in genome sizes reflect adaptive or neutral processes? New clues from Passiflora. *PloS One*, 6: e18212.

Zas, P. and John, S. (2016). Diabetes and medicinal benefits of *Passiflora edulis*. *Int. J. Food Sci. Nutr. Diet.*, 5(2): 265–269.

Part III
Sub-Tropical Fruits

8

Climate Change Implications on Cultivation of Avocado (*Persea americana* Mill.)

Daniel Fernandes da Silva,[1] *Fabíola Villa*[1,*] and *Glacy Jaqueline da Silva*[2]

1. Introduction

The climatic changes that the planet has been going through since the industrial revolution are now a reality that affect all sectors, especially agribusiness, which is dependent on environmental conditions, reflecting the results directly in the economy.

With climate change, all species are forced to adapt in order to guarantee their existence. One of the ways to guarantee this adaptation is to look for new areas in which the edapho-climatic conditions allow proper development, as well as to cease to exist in areas where the current conditions are different from before. Thus, this new geospatial design of the regions that are conducive to the development of each species shapes the agricultural region for the exploration of species of economic interest.

[1] Centro de Ciências Agrárias, Western Parana State University (Unioeste), Campus de Marechal Cândido Rondon. Rua Pernambuco, 1777, Centro, Marechal Cândido Rondon/PR, Brazil. CEP: 85960-000.

[2] Molecular Biology Department, University of Paraná (Unipar), Umuarama Campus I - Sede. Praça Mascarenhas de Moraes, 4282, Zona III., Umuarama/PR, Brazil. CEP: 87512-000.

Emails: daniel_eafi@yahoo.com.br; glacyjaqueline@prof.unipar.br

* Corresponding author: fvilla2003@hotmail.com

Avocado is a fruit considered subtropical, due to the conditions it requires for its full development. Its origin and cultivation for agricultural production is strongly concentrated in Central and South America, precisely in countries that, according to the projections of scientists, will suffer the greatest damage due to climate change, especially with regard to the increase in temperature and rainfall distribution. This species is widely affected by weather conditions in its various stages, especially during the reproductive period, in which the existence of different breeds and races makes it necessary for cross-pollination, which can be difficult in conditions of excessive rain and high temperatures.

Considering the importance of avocado both for local economies, as in the case of Mexico and the Dominican Republic, and its importance from a nutritional point of view, due to its several essential components to human health, the changes in the avocado production scenario arouse the interest of scientists that use computational tools and mathematical models to predict future scenarios for the production of this fruit.

The following content deals jointly with the edapho-climatic requirements of the avocado, correlating them with the possibilities of fruit cultivation in different regions of the world and how climate change will affect the distribution of this fruit around the globe. This review brings the predictions and behavioral studies of the species based on the work of different specialists in the field in order to assist in the understanding and future planning of avocado cultivation.

1.1 *Avocado's Economic Importance*

Avocado is a fruit of great importance in countless countries around the globe and in recent years it has become one of the most important hot climate fruits in the global market, with consumption growth over the last 150 years intensifying in recent years, mainly for two reasons: the increase in consumption among populations that knew the fruit in the past (such as Mexico and the USA) and the constant expansion in new markets around the world, in countries where avocados were not traditionally consumed before, for example Europe, Japan and China (Kourgialas and Dokou, 2021).

According to data from FAO (2019) the total area occupied with the cultivation of avocado, in production was 918,531 ha, which allowed an average productivity of 7 tons per hectare and a total production of approximately 6.4 million tons of fruit (Table 1, Fig. 1).

Mexico leads the ranking among producing countries, with a production of approximately 2.2 million tons in 2018, followed by the Dominican Republic, Peru, Indonesia, Colombia, Brazil and Kenya. However, Mexican productivity is not the most efficient among the seven countries that head the world production of avocado. In this regard, the

Table 1. Worldwide avocado production in 2017 (FAO, 2019).

Country	Harvested area (ha)	Production (tons)	Productivity (tons ha^{-1})
Mexico	206.389	2.184.663	10.6
Dominican Republic	13.924	644.306	46.3
Peru	40.134	504.517	12.6
Indonesia	33.393	410.094	12.3
Colombia	41.519	326.666	7.9
Brazil	12.940	235.788	16.5
Kenya	14.497	233.933	16.1
World	918.531	6.407.171	7.0

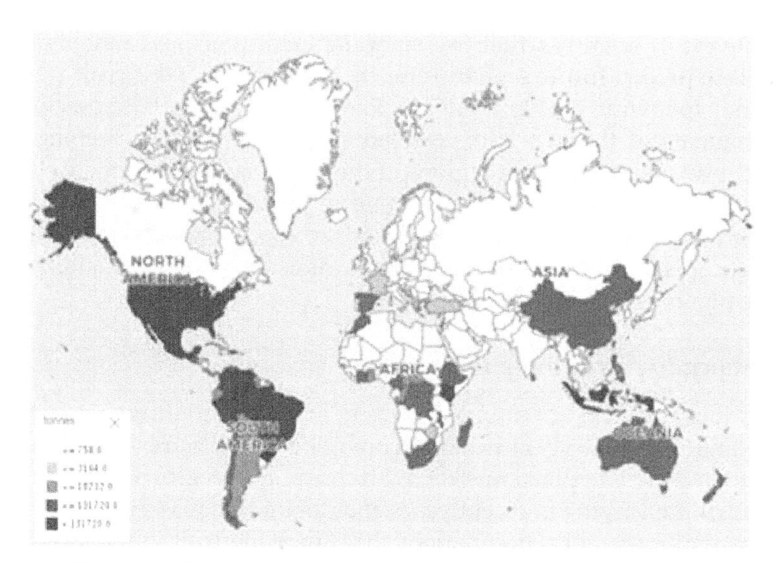

Fig. 1. Distribution of world avocado production in 2018 (FAO, 2019).

only North American country with a prominent position in the production of this fruit occupies the sixth position, with 10.6 tons per hectare. The Dominican Republic has, by far, the highest productivity by cultivation area, with 46.3 tons ha^{-1}, ahead of Kenya, second with 16.1 ton ha^{-1} and Brazil, in third position with 16.5 tons ha^{-1}. Regarding the harvested area, again Mexico was the country with the largest extension, occupying 206,389 ha with productive plantations, followed by Peru with 40,134 ha and Colombia, with 41,519 ha.

The avocado market also shows great movement between countries. In 2017, the total amount of avocado exported in the world was of the order of 2,064.5 thousand tons, generating USD 5,987,508 of income, while the total import was 2,139.3 thousand tons, corresponding to USD 6,290,886. The world's largest avocado exporter that year was Mexico, with 896.6 thousand tons, followed by Peru, the Netherlands, Chile and Spain. At the other end of the Market, as the larger importers, are the United States, the Netherlands, France, the United Kingdom and Spain (FAO, 2019).

Regarding the distribution of avocado cultivation areas today, the overwhelming majority is concentrated in Latin America, with predominance of plantations of the cultivar Hass. Due to the importance that avocado has for Latin America as a whole and part of the United States, whether from a nutritional or economic point of view, an extensive and perhaps the most important study on the areas currently producing Hass avocado and their future behavior in the face of expected global climate changes was carried out by Ramírez-Gil et al. (2019).

Based on data analysis from 12,570 records of Hass avocado production fields distributed throughout America with the exception of Canada, the researchers noted that the potential distribution of the fruit today is concentrated in the tropical and subtropical regions where this variety is already grown. The main suitable areas observed were the high Andes of Colombia, Ecuador and Peru (1400 to 2900 m); coastal valleys of Chile; parts of São Paulo state, Brazil; and mountainous areas of Central America and the Dominican Republic. A large part of Mexico and the California valleys are also among the suitable areas. Some areas according to the study have potential for cultivation today due to their climatic conditions; however, there are no registered plantations. These areas include some inter-Andean valleys, western Venezuela, northeastern and Central Brazil, parts of Uruguay and Argentina, parts of Central America and parts of Baja California, Mexico (Ramírez-Gil et al., 2019) (Fig. 2).

1.2 Botany, Races and Varieties

The avocado tree is an evergreen tree with an average height of 6 meters, but which can reach up to 20 meters according to the genotype (Fig. 3). The tree trunk is cylindrical, corrugated, longitudinally grooved in dark gray, with light and fragile wood, easy to break with the wind, and still has low commercial value. The crown of the tree is dense, upright or spread symmetrically. The leaves are alternated, petiolated with petioles of 15 mm to 5 cm in length. The color of the leaves is reddish or tan when young and green when older, leathery, 10–30 cm long and 3–10 cm wide, oval, lanceolate or elliptical, with an acuminated tip (Maranca, 1980; Koller, 1992).

The inflorescences of the avocado are axillary, appearing at the base of young shoots and grouped at the terminal part of the branches, acquiring the appearance of terminal panicles. The flowers are small, 5–15 mm in diameter, yellowish-green in color, with a characteristic odor and short

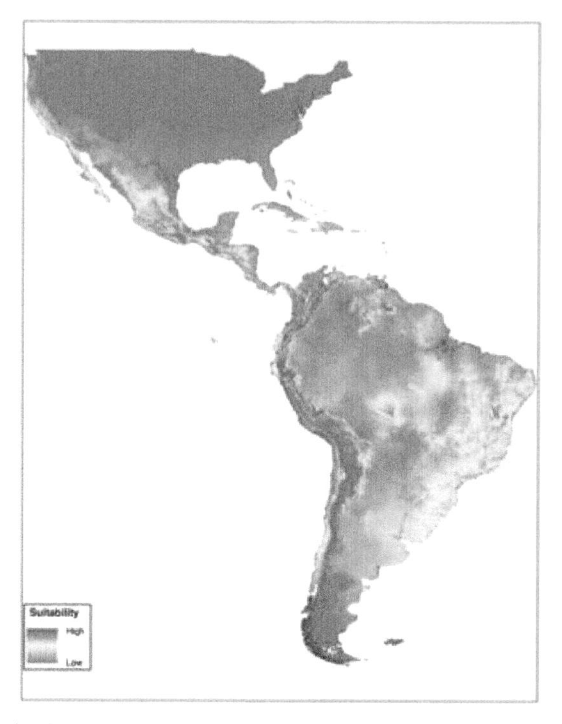

Fig. 2. Areas with edapho-climatic potential for avocado cultivation today (*Source*: Ramírez-Gil et al., 2019).

Fig. 3. Young avocado trees in commercial plantation in full bloom (*Source*: Author: Silva, D.F.).

pedicel. The chalice of the flowers is yellowish-green, with six sepals. The corolla is absent. The stamens are 12, distributed in four groups of three, with nine functional and three asymmetric length staminoids, with the longest internal stamens (Maranca, 1980).

The ovary is unicarpellar and uniovular, with a stigma and a disc-shaped stylus. The fruit is a fleshy monocarp berry formed by membranous epicarp (rind), leathery, mesocarp (pulp) with a buttery consistency and endocarp (structure surrounding the seed). The shape can vary between oval, pyriform or rounded and the external color can vary between green, brown or violet. The seed, also known as 'core', presents itself in a number of one per fruit, of large size, ovoid shape, brown in color and protected by the endocarp (Teixeira et al., 1992).

Systematically the avocado belongs to the Lauraceae botanical family, being included in the genus *Persea*, which covers about 150 species, with a center of origin in Mexico and Central America. This is the reason why the vast majority of species of the genus are found in tropical America and temperate America and only four described species are found in the Amazon (Montenegro, 1951; Maranca, 1980; Koller, 1992).

The genus *Persea* can be divided into two subgenres: *Persea* e *Oreodaphne* (Koller, 1992), however *Persea americana* Mill., which is the species considered most important due to its value in fruit growing. It is found in the *Persea* subgenus (Canto, 1978; Koller, 1992) and *Persea drymifolia*, which for some time was considered a species, and gave rise to the avocado breed known as Mexicana, currently fitting as a botanical variety of *P. americana*.

In addition to the Mexican breed, as mentioned above, avocado varieties can be classified agronomically into two other breeds, totaling three distinct breeds: Mexicana, Guatelmalteca and Antilhana, which botanically correspond to *Persea americana* Mill botanical varieties – var. Drymifolia, *Persea americana* Mill, var. Americana and *Persea nubigena* Mill. var. guatemalensis.

It is possible to differentiate the races by morphological, phenological and sensory parameters. The Mexican breed is originally from the high regions of Mexico and the Cordillera and this makes it more resistant to cold, tolerating even temperatures close to –6°C. The fruit is generally small, with a short peduncle and a thin, smooth skin. The seed is relatively large and the oil content of the pulp is high in the fruit (Koller, 1992), which matures six to eight months after flowering, in the hottest months of the year, between December and February (Campos, 1984). The genotypes of this breed can also be recognized by the characteristic odor of anise when crushing leaves and flowers.

The Guatemalan breed comes from the high regions of Central America and in relation to tolerance to the cold, it is considered intermediate, being more tolerant than Antilhana and less tolerant in relation to Mexicana. Fruit maturation takes longer in relation to Mexicana, occurring 10–15 months after flowering and in the months preceding the hottest period of the year, from May to November. The fruits have a medium to large caliber, a long peduncle, with a thick skin, 1.5–3 mm thick, hard and with a rough surface. The stone occupies the entire cavity of the fruit. The oil content in the pulp is considered to be medium with approximately 12%. Another characteristic of the breed is the deep dark green leaves, stronger than specimens of the Antillean breed. Examples of representatives of this race in Brazil are the cultivars Prince, Wagner and Linda (Campos, 1985; Koller, 1992).

The Antilhana breed corresponds to avocados known popularly known as 'common' or 'butter', coming from the low regions of Central and South America. Its fruits have short peduncles, smooth, leathery skin, about 1.5 mm thick and pulp with low oil content. The seeds are relatively large in size and are usually loose in the cavity. The fruits of the cultivars of this breed usually ripen in the summer. Among the three breeds, this is the most susceptible to cold, damaging itself with temperatures below –2°C (Campos, 1984). This breed includes the Pollock, Simmonds and Princesa cultivars (Canto, 1978; Maranca, 1980; Campos, 1985).

Despite the division into three breeds, this classification has no absolute value, as there are a large number of natural hybrids among the breeds with great potential for economic exploitation. Most of the cultivated varieties originate from these hybrids. Thus, another more simplified classification was carried out by Maranca (1980), who basically divides avocado varieties in two main categories: those of tropical climate, which are generally varieties of the Antilhana breed, of the Guatemalan breed, and/or hybrids of the two; and those of subtropical climate that are of the Mexican or Guatemalan breeds, or even hybrids between the two.

Regarding the varieties, Ramos and Sampaio (2008) claim that it is impossible to compile a complete list with all the existing varieties due to the thousands of genetic lines that arose from the multiplication over thousands of years; however, the adoption of the selection of local hybrids have been used as a tool to obtain new cultivars, which in turn fully meets the requirements of consumers, and, as a rule, are selected for commercial plantations. With the great availability of avocado genetic materials, numerous cultivars have been created; however, despite the large number of cultivars, three of them are more widely cultivated and distributed in the world: 'Hass', 'Fuerte' and 'Nabal' (Pérez-Álvarez et al., 2015).

Hass: Similar to Fuerte, this cultivar is a Guatemalan-Mexican hybrid, which appeared in California/USA in the 1920s and has become the most cultivated and produced cultivar in the world. In relation to climatic conditions, it is a cultivar sensitive to low humidity, especially where there

Fig. 4. (A) Characteristics of Hass avocado fruits. (B) Hass avocado fruit on the right and Fuerte on the left (*Source*: Gonçalves, BHL).

are hot and dry winds, which dry the flowers and young leaves, making them fall and also more susceptible to the cold than Fuerte (Fig. 4A and B).

The Hass cultivar in general is very productive, often tending to excessive fruiting, due to the large amounts of flowers it produces, which reflects negatively on the size of the fruit. It also presents the need for pollination by other cultivars; it is recommended to plant the cultivar Ettinger at distances not exceeding 20 m (Guill and Gazit, 1991). One of the characteristics of the cultivar is the ability to retain the fruit in the plant even after commercial maturation is achieved, allowing the harvest for a long time (Donadio, 1995).

The fruit weighs 180–300 gms, is oval-pyriform, easily peeled, with thick and rough skin, green in the beginning, but it darkens during ripening, reaching a dark violet hue. It presents good resistance to transport. The seed is small, spherical and adherent to the pulp. The pulp production of this cultivar is around 67% (Campos, 1984) with fatty acid composition dominated by oleic acid, followed by palmitic, linoleic, palmitoleic and finally, stearic acid.

Fuerte: It is a hybrid between the Guatemalan and Mexican breeds tuned to the foreign market, with excellent and aromatic buttery pulp (Fig. 4B). Its production requires places with a milder climate and high altitude. Under normal weather conditions, the avocado tree of the Fuerte cultivar does not require the need for orchards interspersed with other varieties. Despite belonging to floral group B, under these conditions, it often manages to self-pollinate, even with alternating production, with years of high production and with years of reduced production (Maranca, 1980).

The fruit is piriform in shape and medium to small in size, weighing 150–450 gms (Donadio, 1995; Medina et al., 1978). The shell is flexible, elastic, green and dull, easily peeled and resistant to transport. The pulp does not have fibers, but it is firm and considered of low yield, being on average 65% and with an average oil content of 22% (Donadio, 1995), reaching up to 35% (Lucchesi and Montenegro, 1975), having as main fatty acids oleic acid, palmitic acid, linoleic acid, palmitoleic acid and stearic acid, in decreasing order (Tango et al., 2004).

The seed is small to medium in size, conical and adherent to the pulp, representing an average 23% of the fruit's mass. This cultivar sometimes produces seedless, small fruit with little or no commercial value (Tango et al., 2004; Donadio, 1995).

This variety is demanding with regard to temperature-environment, especially at the time of flowering and at the beginning of fruiting, when it becomes more sensitive to low temperatures, although the plant resists well to frosts. The harvest is included in the precocious to medium category, with the time from flowering to maturation lasting from five to seven months and its production is spread over several countries in the world, including Mexico, Australia, South Africa and Brazil (Donadio, 1995).

Nabal: Nabal avocado is a Guatemalan cultivar, propagated in California, Florida and Israel since 1927, 1937 and 1934, respectively. Although not as well known as the Hass and Fuerte cultivar, it can be found in crops in the United States, Israel and Australia—places where it produces at different times according to the local climate (Morton, 1987).

Nabal avocado is one of the most sensitive cultivars to frost and, when planted in windy areas, this variety can also be subject to scarring and damage to both the plant and the fruits until they are almost ripe (Morton, 1987).

Avocado can remain on the plant for as long as eight months after flowering. Nabal avocados have a round shape, with a medium to large caliber, weighing up to 480 gms. The fruit's rind is smooth, of medium thickness, dark green, covered with small yellow granules easily removable. They are known for having creamy pulp of exceptionally delicious quality, with a greenish-yellow color and for having a large central cavity in which a generally small seed is found. The cultivar Nabal presents alternation of harvest, producing up to 31 kg of fruit on an average, with about 10–15% of oil in the pulp (Morton, 1987).

1.2.1 Avocado Floral Biology

In addition to the classification in the three races, there is another classification that has nothing to do with the first. The classification of groups A and B, as it is called, is universally adopted in the cultivation of avocado and is only closely related to floral biology, based on the reproductive strategy of protogenic dicogamy occurring in the species (Koller, 1992; Maranca, 1980).

Avocado trees classified in group A are those of cultivars where the first opening of the flower occurs in the morning, ready to receive the pollen (female stage), reopening again in the afternoon of the following day; however, releasing pollen (male stage). In group B cultivars, the first flower opening occurs after noon (female stage), closing at dusk and reopening at dawn in the male stage.

Due to this phenomenon of floral behavior, intercalated planting of groups A and B that bloom contemporary is recommended, to ensure a good pollination rate of the flowers and an economically viable production. In well-organized and fruitful orchards, 25% of the plants are in group A, another 25% in group B and the remaining 50% may be of other varieties. In some cases, this strategy becomes unnecessary, as the variability of brightness can alter the dicogamy, allowing self-pollination in plantations of a single cultivar (Maciel et al., 2008; Maranca, 1980). In general, the ideal temperature conditions for avocado flowering are on an average 26°C during the day and 15°C at night; however, if these limits fall, the floral opening process can change, delaying and modifying the female and male phases. In very cold conditions, the male phase can delay one or more days and the opening can continue at night or until the next day. With changes in temperature, the female phase can also change, especially if the temperature drops below 15°C, as seen in the Fuerte variety. In conditions of high temperatures or less light, the phases overlap when they occur (Bergh, 1976).

Flowering can also vary by race. Guatemalan varieties generally flourish after others (Calabrese, 1989). The total flowering period for each variety is one to two months, which almost always allows a period of coincidence between two varieties, even if they do not flower simultaneously and are of different races. The ideal, however, is to plant varieties that have the same flowering season (Montenegro, 1978).

Pollination of avocados is mainly done by bees, so whenever possible, a good distribution of pollinating plants in the orchard is recommended, preventing pollen from having to be transported over long distances. Knowing how to choose the varieties, it is possible to have avocado throughout the year, as each of them will have its own fruiting season (Pimentel, 2007).

1.3 Edapho-climatic Conditions for Growing Avocado

1.3.1 Soil

As with any other plant, the set of physical, chemical and biological characteristics of the soil must provide adequate conditions for the avocado tree to develop properly, exercising a direct influence on the growth, development and productivity of the plants. Soil is a system composed of a solid part and a porous part; so the solid part that is formed by the minerals that group together has the function of providing nutrients for the plants that grow in it and the porous part, which is made up of air and water, in inversely proportional quantities, guarantees the respiration of the roots and the hydration and conduction of nutrients dissolved in the soil solution, respectively (Stefanoski et al., 2013).

When choosing the area where the avocado orchard will be implanted, attention should be directed to the physical conditions of the soil, as these are difficult to change. Conversely, changes in the chemical attributes of the soil can be more easily obtained through corrections, such as liming and fertilization.

Avocado is one of the most demanding fruits in terms of soil when it comes to physical characteristics, as it requires good drainage and deep soils due to the sensitivity of this species to asphyxiation, which occurs more frequently in clayey or shallow soils with an impermeable layer. Thus, taking into account the principles of good drainage and depth, a series of soils are suitable for the cultivation of avocado—among them being sandy, granitic, volcanic and clayey, but mixed, sandy-clay soils with good depth for root exploration, Donadio (1992).

The lack of oxygenation of the soil can be noticed by the symptoms presented by the plant. The total deprivation (anoxia) or partial (hypoxia) of oxygen in the soil causes the avocado tree to reduce the development of both the root system and the aerial part, moderate to severe withering of the stem and leaves, leaf abscission and root necrosis (Schaffer and Whiley, 2002).

In addition to the difficulty in developing roots in soaked soils, this condition also favors the growth of *Phytophthora cinnamomi* Rans, one of the main disease-causing fungi in avocado living in a humid environment and nourishing itself with organic matter and attacking the plant through smaller roots, until it compromises its conductive system, causing the rotting of the roots, the bark of thick roots and the trunk of the trees in the colon region, making control difficult and expensive (Silva, 2015; Koller et al., 2002).

When there is no previous information about the ground, it is recommended to open drill-holes to a depth of 2, checking that the water-table or rocks have not been reached up to this depth. In this case, if neither situation occurs, the land can be recommended for planting avocado (Koller et al., 2002).

The chemical characteristics of the soil are also important in defining the avocado planting site. Parameters, such as soil acidity (pH) and salinity, directly interfere in the development of plants, as the avocado is very sensitive and demanding.

The appropriate pH range for the avocado tree is about 5–6.5. Below or above these limits, especially in alkaline soils with a pH above 7, the plant is severely damaged, mainly as a result of less iron absorption, although this is not a difficult problem to be corrected and can be solved with liming. On the other hand, the correction of alkaline soils by applying sulfates or sulfur is more difficult (Gayet et al., 1995).

Soil salinity is also another factor that compromises the development of avocado plants. The salinity index, measured by electrical conductivity, of up to 2 mho/cm is considered normal for the crop, while those above 3 mho/cm can cause problems to the plant, such as burning the tip and edges of the leaves, with consequent drop in production (Donadio, 1992).

1.3.2 Climate

Despite the great adaptive capacity of the avocado tree to a series of climatic conditions as varied as possible, just like any other plant, its development is closely correlated with the climatic conditions of the place where it is. This climate in turn is formed by the interaction of several factors. The main influences on the avocado tree are air and soil temperature, solar radiation, rainfall and relative air humidity.

1.3.2.1 Temperature

The air temperature acts in the evapotranspiration process, because the solar radiation absorbed by the atmosphere and the heat emitted by the cultivated surface raise the air temperature, which, being close to the plants, transfers energy to the crop in the form of a sensitive heat flow, increasing evapotranspiratory rates. Temperature also influences the photosynthetic activity of plants due to biochemical reactions involving catalyst enzymes that demand an adequate temperature to express their maximum activity (Larcher, 2000).

Due to its high correlation with plants, temperature is perhaps the most important climate component; so severe winters are limiting in the commercial cultivation of avocado, given the tropical and subtropical origin of the species. However, avocados can be grown in regions of relatively cold winters, which can be explained by the different altitudes of the regions of origin of each breed (Koller et al., 1984). Avocados, belonging to the Guatemalan breed, are the most sensitive to cold, not tolerating temperatures below 6.1°C. The cultivars of the Antilhana breed are described as unable to tolerate temperatures below –4.4°C already being severely injured at –2.8°C and those of Mexican origin are considered very resistant to the cold, withstanding temperatures below –7.7°C (Wolfe et al., 1942).

The temperature, together with the luminosity, exerts a great influence on the reproductive phase of the avocado tree, mainly affecting the flowering in all its phases, from the dicogamy process, to the pollination and germination of the pollen grain (Davenport, 1986; Donadio, 1992). For the floral induction of avocado, it is necessary to have temperatures of 20°C during the day and between 5–15°C at night. However, this need may vary according to the cultivar, based mainly on the genetic load that makes up each cultivar, represented by the three races to which they may belong (Davenport, 1986).

In general, the temperature can affect the flowering of the avocado, altering the floral cycle, providing the coherence of flowers in groups A and B of the male and female phases; In addition, low temperatures can also impair the fertilization of flowers (Lichou and Vogel, 1972). On the other hand, very hot temperatures can induce the abortion of flowers and freshly formed fruits, while temperatures between 40–45°C are capable of causing the fruit to fall in the middle stage of development (La Peña, 1975).

The temperature also has a marked influence on the development of the fruit, so that in regions of higher temperatures, the fruit ripens earlier than in those of colder regions. Flowering and sprouting are also anticipated in regions of warmer temperatures, which interfere with the harvest planning and the choice of cultivars to be planted (Koller, 1984).

Like air temperature, soil temperature also interferes with avocado growth and development by changing parameters, such as root system mass, stem dry mass and plant total in addition to the height and circumference of the stem, and is also related to the absorbed amount of nutrients (Yusof et al., 1969).

1.3.2.2 Solar Radiation

The solar radiation absorbed by the avocado has an effect both on the vegetative cycle of the plant and on the period of development of the fruit, being of great importance for growth, flowering and fruiting, the attributes which are relevant to cultural management, especially to the spacing and care in dense plantings (Duarte Filho et al., 2008).

In regions where the temperature is higher, younger plants may suffer damage to the bark and branches, due to the excess of direct solar radiation. Therefore, in these conditions, the seedlings are produced in a semi-shaded environment; so protection against direct exposure to the sun in the first days after planting in the field is recommended (Koller, 1984).

Avocado trees have a vigorous growth habit, usually with a relatively high percentage of shaded leaves inside the canopy, compared to sunny leaves. These leaves receive low levels of light, decreasing the availability of carbohydrates and consequently causing reductions in growth and production. Among avocado producers, as well as other fruit trees, a cultural treatment recommended to favor greater light penetration is the management of the canopy through pruning, which can provide a significant increase in production and increase in fruit size (Mena, 2005).

The choice of the planting location is also related to the radiation on the avocado orchard. Places of intense light are favorable to photosynthesis and plant productivity, which is why shady areas and regions subject to cloudiness should be avoided as they favor the vertical growth of the plant to the detriment of the lateral opening of the canopy, making harvesting and cultivation difficult. Also, very dense plantations provoke this same effect of vertical growth, in this case, promoting the death of lower branches and compromising fruit production (Koller, 1984).

1.3.2.3 Precipitation

Rainfall is among the main factors that affect the culture of the avocado. Precipitation of the order of 1200 mm per year is sufficient for the crop, as long as there is a reasonable partition throughout the year (Teixeira, 1991), because in summer, plants need much more water than in winter, due to the increase in temperature and the size of the day.

The distribution of rainfall during the months of the year in the different periods of growth of the avocado tree is essential to maintain moisture in the soil during the period of active growth, which includes the emergence of new shoots, flowering and fruit development. If there is a water deficit in the soil during these periods, small flowers and fruits can be aborted, influencing the quantity and quality of the harvested fruits. The occurrence of prolonged droughts can also cause the leaves of the plants to fall, compromising the quality of the final production (Souza and Frizzone, 2008).

Too much water during flowering and fruiting is also detrimental to fruit production and quality (Teixeira, 1991), so it becomes necessary to maintain a balance in the amount of water offered to avocado trees so that there is no water deficit or excess. In regions where there is not enough rainfall, or it is irregularly distributed throughout the year, it is possible to grow avocado trees with the aid of irrigation.

1.3.2.4 Relative Humidity

The relative humidity of the air during the avocado cycle, despite its importance, is not a limiting factor for the species, since it is possible to observe the cultivation of avocado in regions with very different relative humidity conditions, such as in Central America, Israel and California (Koller, 1984).

The most important aspects regarding the relative humidity of the air are linked to other factors, such as the favoring of the appearance of fungal diseases like powdery mildew and anthracnose in conditions of high relative humidity, associated with high temperatures. This can cause economic damage and may even make commercial fruit production unfeasible and also increase the need for water when the plants are in locations of low relative humidity (Duarte Filho, 2008; Koller, 1984). According to Donadio (1992), there is a difference in the adaptability of the avocado breeds in relation to tolerance to the relative humidity levels of the air; thus the Antillean varieties are considered more adapted to places with humidity above 70%, that is, considered high.

1.4 Avocado Climate Change

1.4.1 Climate Change in the Main Avocado-producing Countries

Mexico, Dominican Republic and Peru were the three largest avocado producers in the world respectively in 2016 (FAO, 2019) and both are located in Latin America, which, according to the Working Group of the II Intergovernmental Panel on Climate Change (McCarthy et al., 2001), may represent a risk for this culture because, according to the conclusions of this same group in its *Third Evaluation Report* with 'high confidence', Latin America is highly vulnerable to climate change, given its current low adaptive capacity, particularly for extreme events. In this context, it is highly likely that the yield of the main crops in these countries may decrease significantly, due to a series of consequences that these countries have been suffering as a result of climate change and which are more detailed below (Fig. 5).

Projections show that in Mexico, the occurrence of extreme events will increase with climate change. The temperature will increase, while summer rains tend to decrease (Gay et al., 1996; Gay, 2000). Likewise, in the Dominican Republic, which is also an important avocado producer, the environmental pressure towards the increase in the number of extreme events should grow to strongly affect agriculture, especially due to the geography of the region, which is a small island flanked by two large oceans directly exposed to these phenomena (EuropeAid, 2009).

Peru, the third largest producer of avocado in the world, is naturally a country with characteristics of vulnerability, including low coastal areas, arid and semi-arid areas, areas exposed to floods, droughts and desertification besides a fragile mountain ecosystem (MINAM, 2010). These characteristics, together with the fact that Peru has a predominance of mountain ecosystems, in which hydrological cycles are the main factors affected, and also the occurrence of El Niño and La Niña, will certainly affect the dynamics of agricultural production in the country, depending on the culture of the avocado tree.

1.4.2 Appropriate Climate for Avocado Production

As with all living beings on the planet, the climate is fundamental for maintaining existential conditions, as the climate is defined by the set of

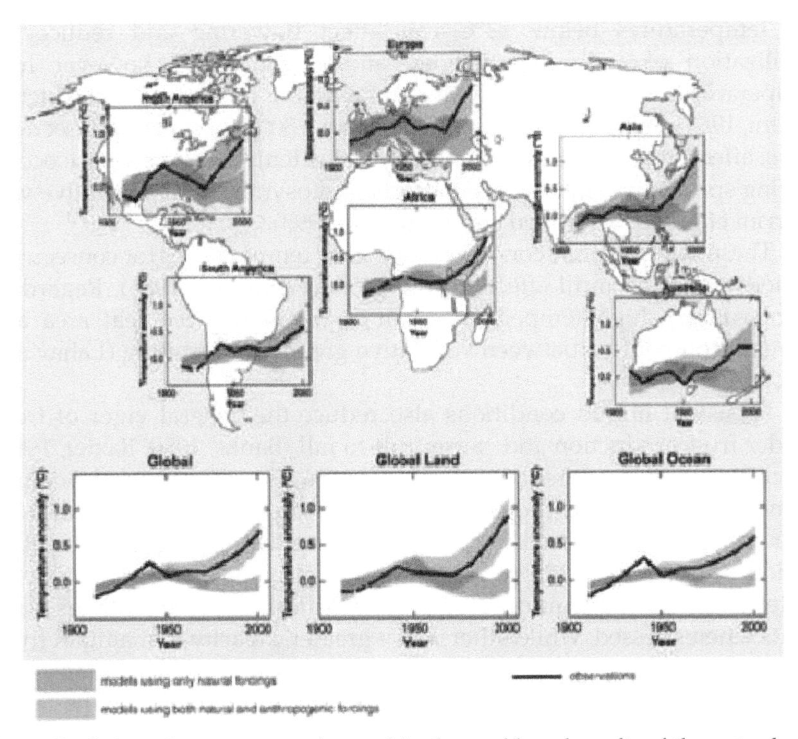

Fig. 5. Evolution of temperatures observed in the world; and results of the natural and anthropogenic models in the period between 1900 and 2000 (*Source*: IPCC, 2007).

atmospheric conditions that characterize a given region, which may or may not favor the survival of a given species in that location.

The climate is composed of elements (temperature, humidity, rain, wind, cloudiness, atmospheric pressure, solar radiation, etc.) and climatic factors (relief, soil type, latitude, altitude, etc.), which together provide the characteristics of a determined place and consequently dictate the rules of behavior for the beings living there. Those unable to adapt to climatic conditions should look for other areas where they do, or they will get extinguished.

Temperature is one of the most important components in the production of avocado because it affects cultivation in several ways, such as influencing the time and intensity of flowering, fruit development, ripening and fruit value at the end of the cycle.

The temperature tolerance of avocado plants is related to the race to which it belongs, since the three existing breeds or ecotypes (Mexican, Guatemalan and West Indies) have markedly different temperature tolerances, mainly due to their areas of origin (Malo et al., 1977). Among the three races, Antilles is undoubtedly the most sensitive to low temperatures (Campos, 1984).

Temperatures below 12°C can affect flowering and reduce the fertilization of avocados (Sedgley and Grant, 1982); however high temperatures, above 42°C are also unfavorable for avocado production (Gafni, 1984). Temperatures maintained above 30°C for several consecutive days, affect fruit production, especially if this temperature increase occurs during spring and summer, as it affects photosynthesis in a negative way (Ferrini et al., 1995; Jackson and Lombard, 1993).

These same climatic conditions of higher temperatures for consecutive periods cause harmful effects to the roots (Whitmore, 1986). Regarding plant canopy, high temperatures can result in reduced leaf area and increase competition between vegetative growth and fruiting (Lahav and Trochoulias, 1982).

Persistent humid conditions also reduce the general vigor of trees, hinder fruit production and cause fruit to fall (Banks, 1980; Baxter, 1981). Precipitation of the order of 1200 mm per year is sufficient for the culture of avocado, as long as there is a reasonable distribution throughout the year (Teixeira, 1991). Excessive rainfall during the flowering and fruiting period causes a drop in production and fruit quality. At the same time, water stress during fruit development is critical, as some cultivars abort fruits when stressed, while others have greater capacity to maintain fruits at different levels of water stress, but at the expense of fruit size and quality (Bower and Cutting, 1988). Prolonged droughts cause the fall of leaves and, consequently, production. Special attention should be paid to young plants, with shallow root system, since the leaves of avocado do not show symptoms of wilt under water stress conditions, evolving directly to necrotic spots with subsequent drying and falling (Koller, 1984).

Solar radiation is an important factor in guaranteeing the production and quality of avocado fruits. This importance starts with the implantation of the orchard when the newly planted seedlings need shading to ensure the setting. Plants in areas that are shaded or subject to cloudiness promote greater vertical growth, making harvesting and cultural treatment difficult, for example. The same can occur in dense plantations, which in addition to forming taller plants, have less production in more internal branches, which can be removed as a way to allow greater light inside the plant. Excessive light can also cause burning of branches and fruits, especially when the plant is defoliated due to the attack of pests or mineral deficiencies (Donadio, 1995).

Air humidity during the avocado cycle is very important, as it favors the appearance of fungal diseases. When high relative humidity values are associated with high temperatures, there is a higher incidence of these diseases, such as powdery mildew and anthracnose, causing economic damage, which may even make commercial fruit production unfeasible. According to Donadio (1992), the Antillean varieties are more adapted to places with high humidity, that is, above 70%.

The occurrence of strong winds is also a limiting factor for the cultivation of avocado because under these conditions, several damaging effects are reported, such as defoliation, fruit fall, branch breaking, difficulty in pollination by insects, increased sweating and predisposition to the effect of drought (Koller, 1984), desiccation of flowers, stains on the fruits because of friction with the branches, etc. (Donadio, 1995).

1.4.3 Effects of Climate Change on Avocado Cultivation

As with all other agricultural crops, climate change will lead to changes in traditional cultivation zones, mainly due to changes in climatic components. Avocado cultivation will be no different and, in this way, areas that are now conducive to cultivation will become inappropriate due to the new environmental conditions, as well as new areas with conditions that currently approach ideal cultivation may be cultivated with avocado trees.

Exemplifying the impact of climate change on avocado cultivation, some studies in regions of high fruit production demonstrate the projections of a future scenario and how the changes will affect the crop. These studies are mainly concentrated in Latin America, since the largest world producers, with the exception of Indonesia, are located on this continent.

Regarding future climate scenarios projected by Ramírez-Gil et al. (2019) for America, what can be observed is that the areas suitable for Hass avocado remain largely stable, without much variation in terms of gains or losses in potential distribution areas (Fig. 6).

It is possible to observe that the changes in the areas with aptitude for the cultivation of Hass avocado were more expressive in the scenario with projections of high emissions of greenhouse gases. In this scenario, the reduction in areas suitable for cultivation will be 19.3% in relation to the current situation, whereas under emissions of greenhouse gases in moderate quantities the reduction will be 14.1%. The areas of loss of fitness in the projected scenarios are concentrated in regions where this list is not planted, such as the interior and northeast of Brazil, inter-Andean valleys (from Peru to Venezuela) and parts of Central America and the Dominican Republic.

At the same time that the loss of areas favorable to avocado cultivation is greater with higher greenhouse gas emissions, this same scenario also favors new areas that will enter the climatic range that allows the cultivation of fruit. The expected gain in new areas will be 30.7%, against 27.3% in new areas in a scenario with moderate gas emissions. These areas are located in the Argentine mid-west; south and northeast of Chile; northwestern Bolivia; southwestern Peru; Andean region of Peru, Ecuador, Colombia and Venezuela; Central Mexico; and the Pacific northwest coast of the United States.

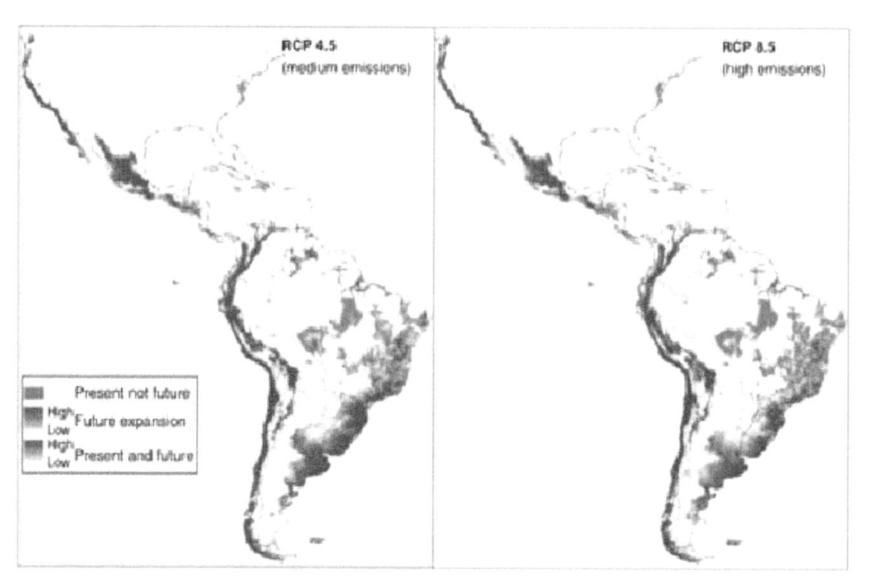

Fig. 6. Patterns of change in potential distribution across America for Hass avocados in two climate change scenarios (RCP 4.5, RCP 8.5). The captions are shaded, corresponding to the expected distribution areas present and possible extensions of future scope: gray area: distribution loss area (present, not future), red area: distributive gain area (future expansion) and blue area: stable area (present and future) (Fonte: Ramírez-Gil et al. (2019)).

The losses and gains in range for potential cultivation seem to be related to temperature increases. Area losses can be seen where the hot areas will become even hotter, and gains were expected where the cold areas will experience an increase in temperature.

Changes in precipitation patterns are also a factor that will contribute to changes in the range with potential for production in the future, although for this factor the pattern of changes was less clear, with similar trends being observed: area gains in dry regions, that got wetter, and losses in areas that became too wet.

Other more localized studies have been released over the years, predicting the impacts of climate change in each country in isolation. These studies were conducted mainly by the local authorities, given the importance of avocado in their respective economies.

Mexico, which is currently the world's largest avocado producer, should be one of the countries most affected by the effects of climate change on this crop. The importance that avocado cultivation has for Mexican agriculture has led the country to expand more and more the areas of fruit cultivation, which have contributed to the deforestation of some regions, as well as changes in the form of land occupation. This is because the climatic requirements of the avocado are very similar to those existing where the Mexican pine and quercus forests exist today. This irregular clearing of the forest, in turn, increasingly contributes to the occurrence of climate change, such as rising temperatures.

In a moderate perspective of greenhouse gas emissions, there is a forecast of an increase in temperature of 2–3°C and a reduction of 20–50 mm of rainfall annually. A study by Charre-Medellín et al. (2019) carried out an assessment of the potential areas for the implantation of avocado crops in Mexico in view of the current scenario and considering three possible levels of greenhouse gas emissions, of which a low gas emission is an intermediate condition with a high concentration of gases of the greenhouse effect in the atmosphere. Of these, in general, it was observed that the greater the emission of gases, in a projection for the year 2050, the smaller the areas suitable for the cultivation of avocado (Fig. 7).

Using 10 global climate models, the study found that, on an average, in the scenario of low greenhouse gas emissions, 20.9% of the area with current potential for avocado cultivation in Mexico would be lost. With the average and high emission of gases, 31.2% and 34.8% of the areas would be lost respectively.

If only the area where there is consensus on the loss of area for the cultivation of avocado in the 10 climatic molasses used in the projection

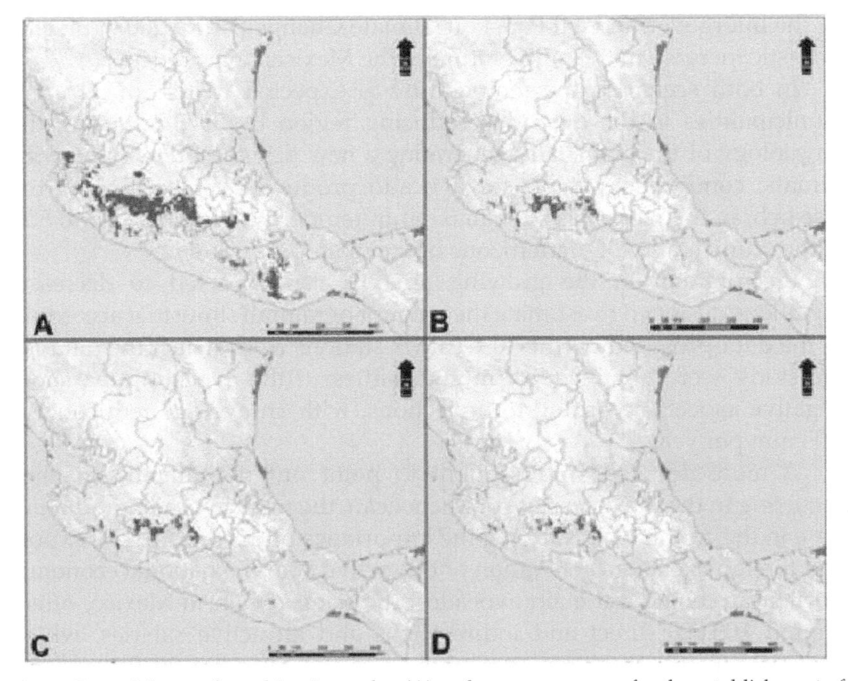

Fig. 7. Potential areas for cultivation today (A) and consensus maps for the establishment of avocado cv. Hass in Mexico against the climate change scenarios for the year 2050. (B) RCP 2.6, (C) RCP 4.5 and (D) RCP 8.5 W m^2.

is considered, for the most optimistic scenario, with low gas emissions, a loss of 79.4% will occur; while for the more pessimistic scenario, the loss will be of the order of 87.6%, concentrated mainly in the states with the greatest damage in the states—Nayarit, Guanajuato, Guerrero, Morelos and Puebla.

Another study, conducted by Paniagua and Gómez (2015), demonstrated the possible climate changes projected for the years 2025, 2050 and 2075, based on climatic data from a historical series obtained at different weather stations located in the region of Mexico, where most of the avocado production is currently concentrated.

In this work, both in the A1B scenario (projection of a planet with rapid economic growth and maximum population growth in the middle of the century, with subsequent reduction and a rapid introduction of newer and more efficient technologies, in addition to having more important distinguishing characteristics, a greater regional, cultural and social equality and being a balanced scenario between the use of different sources of energy), as well as in scenario A2 (a heterogeneous scenario that has distinctive characteristics as self-sufficiency and the conservation of local identities, a world population in continuous growth and economic growth and fragmented and slower technological changes), both described by the Intergovernmental Panel on Climate Change (IPCC, 2007), there is a drastic increase in the temperature of the Mexican avocado area.

In both scenarios the temperature is expected to rise in all eight municipalities in the avocado-producing region by 2075, affecting the physiology of the plants and generating a new design of the region with climatic conditions suitable for avocado production. According to the research, in the worst case, the maximum temperature can rise up to 5°C in the municipality of Madero, one of the main producers.

Precipitation in the growing area is also expected to decrease. Equations designed to estimate the amount of rainfall show that according to the data provided by the 10 weather stations from which the data for the study were obtained, all municipalities in the avocado area show negative aspects according to projections, with an average reduction of 1.05 mm per year.

A more detailed study sought to point out the climatic changes occurring in the Mexican state of Michoacán, the main avocado producing state in the country, considering the importance of the production, export and foreign exchange generation of this activity for the national economy. Michoacán, being the main avocado producing region in Mexico, offers around 100,000 direct and indirect jobs and attractive salaries, which generate wealth for the region (Sommaruga and Eldridge, 2020).

Analyzing data from the historical series between 1963 and 2010, Vargas et al. (2011) concluded that for this state, until the end of the studied period, there was no increase in maximum and minimum temperatures. However, the number of rainy days in the last decade of the study is 8.8 days less compared to the period from 1963 to 1972 and the increase in rainfall in the region was in the order of 345 mm. The most evident indicator of climatic changes for the region is the intensity of rainfall over a 24-hour period, which initially was 12.3 days with precipitation above 30 mm and in the last decade was already 18.2. According to the authors, the combination of lower vegetation covers and greater intensity of rain caused by climate change allows us to infer that more environmental damage can be expected in the areas of cultivation in Michoacán.

Peru, as it also has avocado as an important national agricultural crop, carries out some studies in which future scenarios for cultivation are projected. In this country, during the period 2000–2003, the productivity projection for creole avocado was 5.5–9 t ha^{-1}, but there was already a high availability of areas suitable for cultivation, depending on the regions with ideal climatic conditions to meet the requirements of the avocado and which could serve the market according to the demand of the international market. Despite the favorable climatic conditions in some other Peruvian regions, the slope of the land may somehow present a limitation to fruit cultivation (Varillas et al., 2013).

In a projection for the years 2015–2039, the cultivation of avocado in Peru shows a positive trend, with an increase in productivity, especially in areas where productivity was low in the period 2000–2003. In those areas where productivity was already higher, in future it can easily exceed 11 t ha^{-1} (Varillas et al., 2013). These values, despite appearing low, take into account only crops under rain-fed conditions, without irrigation, which is why climate change can have a great influence.

In Chile, a study shows that in projections for 2050 and 2070 if there is a water restriction, due to reduced rainfall or available water for irrigation, avocado producing properties tend to reduce the profit margin, both in a flow scenario of 2.6–8.5 W m^2. On the other hand, without any reduction in water availability, the expectations for avocado production in the country are promising, with increases in profitability that can reach 90.9% in 2070 under a thermal flow of 8.5 W m^2 (Cavieres, 2015).

In addition to the Latin American countries, other regions also produce avocados and for this reason, they are also concerned with the future effects of climate change for the species. In South Africa, Hunter and Cronin (2020) report that climate change has affected avocado production in different ways, including reduced fertilization and fruit formation and reduced fruit permanence time on the plant without the onset of deterioration, in addition to other ways. More specifically in the Limpopo region, which is an important avocado production area in the country, research has been carried out on climate change and how it would affect fruit production in future scenarios (Randela, 2018). Those responsible for the study observed that in a historical series of data, observed and analyzed between 1970 and 2014, there is an irregularity in the distribution of rain between the locations within the region, causing different points in the avocado region to receive significantly different amounts of rain interfering in their productive capacity. It was also verified that in several years within the studied period, there is a lack of rain, with records below 500 mm per year and that sometimes an amount of rain extremely above average occurs, as, for example, in the year 2000 when the Levebu recorded more than 3000 mm of rain.

Regarding temperature, it was also possible to notice that over the years, there was an average increase in maximum temperature in the four regions with weather stations. Of these four, only the Duiweiskloof station showed a smaller increase in the average maximum temperature; however, the other three regions (Lwamondo, Letaba, Levubu) which showed the greatest increase are where most of the avocado producers in the Limpopo region are concentrated (Randela, 2018).

Climate change in South Africa is noticeable, both from a scientific and empirical point of view. In an interview with avocado producers where they were asked whether they noticed changes in the climate over the years, the vast majority – 87% of respondents—claimed to be observing changes, among which the most pointed out were less rain (78%) followed by a rise in temperature (16%) (Randela, 2018).

In a correlation between the variables, the authors realized that for each unit of temperature that increases, there is a decrease of 1.6 kg ha^{-1} in avocado productivity, but the fluctuation in precipitation did not statistically affect production, which can occur due to the attenuation of the water deficit with irrigation systems (Randela, 2018).

2. Conclusion and Prospects

Avocado is an important fruit with an expanding trade worldwide. In addition to all the nutritional importance that the fruit has for the native peoples of its origin center, contributing to local food security, avocado is present for consumption in Latin America in general, which makes it the main consumer region of the fruit, making avocado cultivation an income-generating activity and consequently one of the pillars of the economy of its main producers.

Therefore, the growth in world consumption of avocados, especially by the European Union and Asian countries, directly implies an increase in the already traditional cultivation areas, from where these fruits are exported until they reach consumer markets. However, the impact of global climate change, through changes mainly in the water regime and in the local temperature, has been redesigning the potential areas for avocado production, making many local farmers, holders of knowledge and technology about the production of avocado, find themselves unable to develop such activity due to the changed natural conditions of the climate.

It is observed that the natural conditions imposed by climate change on avocado-producing regions have forced cultural changes in the consumption of avocados by local populations, as well as agronomic and economic changes in the management of orchards, restricting production due to its cost and the price of the fruit. Thus, there is a prospect that in the coming years, many currently producing areas will become unable to produce, as well as new areas that currently do not meet the edapho-climatic requirements for the avocado crop, will become capable of producing the fruit. In this way, attitudes that contribute to the mitigation of climate change in avocado producing regions can collaborate to sustain production. In addition, studies and research on the impact of these climate changes on avocado cultivation are important to ensure a sustainable chain, capable of guaranteeing production with low environmental impact and, in this way, can be collaborative in food security around the world.

References

Álvarez, S.P., Quezada, G.Á. and Arbelo, O.C. (2015). El aguacatero (*Persea americana* Mill). *Cultivos Tropicales*, 36: 111–123.

Banks, A. (1980). A guide to management of irrigation for avocados. *AGDEX 235/567*, Queensland.

Baxter, P. (1981). *Growing Fruit in Australia*. Thomas Nelson, Melbourne.

Bergh, B.O. (1976). Factors affecting avocado fruitfulness. pp. 83–88. *In*: Sauls, J.W., Phillips, R.L. and Jackson, L.K. (eds.). *Proceedings of the First International Tropical Fruit Short Course: The Avocado*, 5–10 November 1976, Gainesville, USA.

Bower, J. and Cutting, J.G. (1988). Avocado fruit development and ripening physiology. *Horticultural Review*, 10: 229–271.

Calabrese, F. (1989). *Frutticultura Moderna: Avocado. Edizioni Agricole*. Roma, Itália, 217p.

Campos, J.S. (1984). *Abacaticultura paulista, CATI*. Campinas, Brasil, 92pp.

Campos, J.S. (1985). *Cultura racional do abacateiro*. Ícone Editora, São Paulo, Brasil, 150pp.

Canto, W.L., Santos, L.C. and Travaglini, M.M.E. (1978). *Abacate: da cultura ao processamento e comercialização. ITAL*, Campinas, Brasil, 212 pp.

Cavieres, P.C.P. (2015). *Efectos del cambio climático en los cultivos de palto, olivo y uva pisquera insertos dentro de un sistema de agricultura familiar campesina en la provincia del Choapa, Tese de pregrado*. Universidad de Chile, Santiago, Chile.

Charre-Medellín, J., Mas, J. and Chang-Martínez, L. (2019). Áreas potenciales actuales y futuras de los cultivos de aguacate Hass en México utilizando el modelo Maxent en escenarios de cambio climático. *UD Y la Geomática*, 14: 26–33.

Davenport, T.L. (1986). Avocado flowering. *Horticultural Reviews*, 8: 257–289.

Donadio, L.C. (1992). *Abacate para exportação: aspectos técnicos da produção, Denacoop/IICA*. Brasília, Brasil.

Donadio, L.C. (1995). *Abacate para exportação: aspectos técnicos da produção, EMBRAPA*. Brasília Brasil, 53pp.

Duarte Filho, J., Leonel, S., Caproni, C.M. and Grossi, R.S. (2008). Principais variedades de abacateiros. pp. 25–36. *In*: Leonel, S. and Sampaio, A.C. (eds.). *Abacate: Aspectos técnicos da produção*. Cultura Acadêmica, São Paulo, Brasil.

EuropeAid. (2009). *El cambio climático en América Latina*; available at http://euroclimaplus.org/images/Publicaciones/CE/CambioClimaticoAmericaLatina.pdf.

FAO, Food and Agriculture Organization of the United Nations. (2019). FAOSTAT, Food and Agriculture Organization of the United Nations; Statistical Division, Roma; avaliable at http://www.fao.org/faostat/en/#data/QC; accessed on Dec. 2, 2019.

Ferrini, F., Mattii, G.B. and Nicese, F.P. (1995). Effect of temperature on key physiological responses of grapevine leaf. *American Journal of Enology and Viticulture*, 46: 375–379.

Gafni, E. (1984). *Effect of Extreme Temperature Regimes and Different Pollenizers on the Fertilization and Fruit Set Process in Avocado*. MSc thesis, The Hebrew University of Jerusalem, Rehovot, Israel.

Gay, C. (1995). Mexico: Emissions inventory, mitigation scenarios, and vulnerability and adaptation. *In*: Ramos-Mañe, C. and Benioff, R. (eds.). *Interim Report on Climate Change Country Studies*. US Country Studies Program; available at http://www.envirotext.gcrio.org/IR/IRmexico.html.

Gay, C., Ruiz, L.G., Imaz, M., Conde, C. and Mar, B. (1996). *Memorias del Segundo Taller de 'Estudio de País: México', México Ante el Cambio Climático*. INE-SEMARNAP, Cuernavaca, México. 250 pp.

Gay, C. (2000). México: *Una Visión hacia el Siglo XXI, El Cambio Climático en México, Resultados de los Estudios de Vulnerabilidad del País Coordinados por el Instituto Nacional de Ecología con el Apoyo del* U.S. Country Studies Program, SEMARNAP/UNAM/USCSP, México. 220 pp.

Gayet, J.P., Bleinroth, E.W., Matallo, M., Garcia, E.E.C., Garcia, A.E., Ardito, E.F.G. and Bordin, M.R. (1995). *Abacate para exportação: Procedimentos de colheita e pós-colheita*. EMBRAPA, Brasília, Brasil.

Guill, L. and Gazit, S. (1991). Pollination of the Hass avocado cultivars. pp. 241. *In*: Lovatt, C.J. (ed.). *The Proceedings of the 2nd World Avocado Congress*. 21–26 April, 1991, Califórnia, USA.

Hunter, R. and Cronin, K. (2020). Climate change and its impacts on the feasibility and sustainability of small-scale systems of agricultural production, in communal areas and on farms transferred through land reform, *GTAC/CBPEP/EU Project on Employment-intensive Rural Land Reform in South Africa: Policies, Programs and Capacities*. University of Cape Town, South Africa. 10pp.

IPCC, Intergovernmental Panel for Climate Change. (2007). *Climate Change 2007: Synthesis Report*, Contribution of Working Groups I, II and III to the Fourth Assessment Report of the Intergovernmental Panel on Climate Change, IPCC, Geneva, Switzerland, 104 pp.

Jackson, D.I. and Lombard, P.B. (1993). Environmental and management practices affecting grape composition and wine quality—A review. *American Journal of Enology and Viticulture*, 44: 409–430.

Koller, O.C. (1984). *Abacaticultura, Editora da UFRGS*. Porto Alegre, Brasil.

Koller, O.C. (1992). *Abacaticultura, UFRGS*. Porto Alegre, Brasil, 138pp.

Koller, O.C. (2002). *Abacate: produção de mudas, instalação, manejo de pomares, colheita e pós-colheita, Cinco Continentes*. Porto Alegre, Brasil. 154pp.

Kourgialas, N.N. and Dokou, Z. (2021). Water management and salinity adaptation approaches of Avocado trees: A review for hot-summer Mediterranean climate. *Agricultural Water Management*, 252: 106923.

La Peña, F.J.A. (1975). El Aguacate. Ministerio de Agricultura, Madrid, Brasil, 169pp.

Lahav, E. and Trochoulias, T. (1982). The effect of temperature on growth and dry matter production of avocado plants. *Australian Journal of Agricultural Research*, 33: 549–558.

Larcher, W. (2000). *Ecofisiologia Vegetal, Editora RiMa*. São Carlos, Brasil.

Lichou, J. and Vogel, R. (1972). *Biologie florale de l'avocatier en Corse. Fruits*, 27: 705–717.
Lucchesi, A.A. and Montenegro, H.W.S. (1975). *Influência ecológica no desenvolvimento do fruto e no teor de óleo na polpa do abacate (Persea americana* Miller). *Anais da Escola Superior de Agricultura Luiz de Queiroz*, 32: 419–447.
Maciel, M.R.A. (2008). *Botânica e biologia reprodutiva do abacateiro*. pp. 37–64. *In*: Leonel, S. and Sampaio, A.C. (eds.). *Abacate: Aspectos técnicos da produção*. Cultura Acadêmica, São Paulo, Brasil.
Malo, S.E., Orth, P.G. and Brooks, N.P. (1977). *Effects of the 1977 Freeze on Avocados and Limes in South Florida*. Florida State Horticultural Society, 90: 247–251.
Maranca, G. (1980). *Manga e Abacate*. Nobel, São Paulo, Brasil, 138pp.
McCarthy, J.J., Canziani, O.F., Leary, N.A., Dokken, D.J. and White, K.S. (2001). *Climate Change 2001: Impacts, Adaptation and Vulnerability*. Cambridge, United Kingdom, 1032 pp.
Medina, J.C. (1978). *Abacate: Da cultura ao processamento e comercialização*. ITAL, Campinas, Brasil, 212pp.
Mena, F.V. (2005). *Poda en Paltos*. pp. 1–8. *In*: *2nd Seminario Internacional de Paltos*. 29 September–1 Octubre 2004, Quillota, Chile.
MINAM - Ministerio del Ambiente del Perú. (2010). El Perú y el Cambio Climático Segunda Comunicación Nacional del Perú a la Convención Marco de las Naciones Unidas sobre Cambio Climático 2010. Fondo Editorial del MINAM, Lima, 2010.
Montenegro, H.W.S. (1951). *A cultura do abacateiro, Melhoramentos*. São Paulo, Brasil, 102pp.
Montenegro, H.W.S. (1978). *Situação da abacaticultura brasileira*. pp. 40–71. *In*: Donadio, L.C., Pereira, F.M. and Lam-Sanchez, A. (eds.). *Anais do I Simpósio sobre abacaticultura*. 20–25 November 1978, Jaboticabal, Brasil.
Morton, J.F. (1987). *Fruits of Warm Climates*. J.F. Morton, Miami, USA, 505p.
Paniagua, C.F.O. and Gómez, A.M.O. (2015). *Agricultura y cambio climático en la región aguacatera del estado de Michoacán*. pp. 1–29. *In*: *20th Encuentro Nacional sobre Desarrollo Regional en México*. 17–20 Noviembre de 2015, Cuernavaca, Mexico.
Pimentel, G.R. (2007). *Fruticultura Brasileira, Nobel*. São Paulo, Brasil.
Ramírez-Gil, J.G., Cobos, M.E., Jiménez-García, D., Morales-Osorio, J.G. and Peterson, A.T. (2019). Current and potential future distributions of Hass avocados in the face of climate change across the Americas. *Crop and Pasture Science*, 70: 694–708.
Ramos, D.P. and Sampaio, A.C. (2008). *Principais variedades de abacateiro*. pp. 37–64. *In*: Leonel, S. and Sampaio, A.C. (eds.). *Abacate: aspectos técnicos da produção, Cultura Acadêmica*. São Paulo, Brasil.
Randela, M.Q. (2018). *Climate Change and Avocado Production: A Case Study of the Limpopo Province of South Africa*. M.S. thesis, University of Pretoria, Pretoria, South Africa.
Schaffer, B. and Whiley, A.W. (2002). Environmental physiology. pp. 135–160. *In*: Whiley, A.W., Schaffer, B. and Wolstenholme, B.N. (eds.). *The Avocado: Botany, Production and Uses*. CABI-Publishing, Homestead, USA.
Sedgley, M. and Grant, W.J.R. (1982). 'Effect of low temperatures during flowering on floral cycle and pollen tube growth in nine avocado cultivars'. *Scientia Horticulturae*, 18: 207–213.
Silva, S.R. (2015). Manejo da podridão radicular (Phytophthora cinnamomi Rands) na cultura do abacateiro. Tese (Livre Docência) – Universidade de São Paulo, Piracicaba, 2015. Access in: 03 jul 2022.
Sommaruga, R. and Eldridge, H.M. (2020). Avocado production: water footprint and socio-economic implications. *Eurochoices*, 20: 48–53.
Souza, A.P. and Frizzone, J.A. (2008). *Viabilidade econômica da irrigação no abacateiro*. pp. 81–112. *In*: Leonel, S. and Sampaio, A.C. (eds.). *Abacate: Aspectos Técnicos da Produção*. Cultura Acadêmica, São Paulo, Brasil.

Stefanoski, D.C., Santos, G.G., Marchão, R.L., Petter, F.A. and Pacheco, L.P. (2013). *Uso e manejo do solo e seus impactos sobre a qualidade física. Revista Brasileira de Engenharia Agrícola e Ambiental*, 17: 1301–1309.

Tango, J.S., Carvalho, C.R.L. and Soares, N.B. (2004). *Caracterização física e química de frutos de abacate visando a seu potencial para extração de óleo. Revista Brasileira de Fruticultura*, 26: 17–23.

Teixeira, C.G. (1991). Cultura. pp. 01–57. *In*: Teixeira, C.G., Bleinroth, E.W., Castro, J.V., Martin, Z., Tango, J.S., Turatti, J.M., Leite, R.A.S.S.F. and Garcia, A.E.B. (eds.). *Abacate: Cultura, Matéria Prima, Processamento e Aspectos Econômicos*. ITAL, Campinas, Brasil.

Teixeira, C.G., Bleinorth, E.W., Castro, J.V., Martin, Z.J., Tango, J.S., Turrati, J.M., Leite, R.S.S.F. and Garcia, A.E.B. (1992). *Abacate - Cultura, Matéria-prima, Processamento e Aspectos Econômicos*. 2nd. ed., ITAL, Campinas, Brasil, 250pp.

Vargas, L.M.T., Guzmán, A.L., Fernández, I.V., Santos, M.E.P. and Barradas, V.L. (2011). *Cambio climático en la zona aguacatera de Michoacán: Análisis de precipitación y temperatura a largo plazo. Revista Mexicana de Ciencias Agrícolas*, 2: 325–335.

Varillas, I.T., Velazco, C.A., Cuzquen, L.C. and Caiña, K.Q. (2013). *Variabilidad climatica: Percepciones e impacto en los cultivos de café, granadilla y palto en la subcuenca de Santa Teresa – Cusco, SENANHI*. Lima, Peru.

Whitmore, J.S. (1986). *The Climatic Suitability of South Africa for Production of Avocados*. NPWCAR project, CSIR, Pretoria.

Wolfe, H.S., Toy, L.R. and Stahl, A.L. (1942). *Avocado Production in Florida*. Agricultural Extension Service, Gainesville, USA.

Yusof, I.M., Buchanan, D.W. and Gerber J.F. (1969). The response of avocado and mango to soil temperature. *Journal of the American Society for Horticultural Science*, 94: 619–621.

9

Enhancing Climate Change Resilience in Guava (*Psidium* Species)

Muhammad Usman

1. Introduction

Long-term environmental changes including elevated temperatures, atmospheric CO_2 and altered precipitation patterns are referred as climate change (Korres et al., 2016). Climatic conditions are changing due to the increasing anthropogenic activities releasing greenhouse gases like CO_2, NO_2, CH_3 during the last century. There are predictions of net increase in mean annual temperatures and precipitation patterns (Lobell and Gourdi, 2012); however, the changing patterns may vary in their intensity in different parts of the world. Developing countries would be relatively more vulnerable to the impact of climate change by declining fruit production and rising hunger. Brazilian agriculture production is expected to lose 2–5 billion US$ by 2070 and guava is one of the major fruit crops of Brazil. The Mediterranean area stands amongst the most sensitive ones being affected by climate change (De Ollas et al., 2019). More frequent episodes of heat waves, drought, floods and other environmental disasters are expected. According to Intergovernmental Panel on Climate Change (IPCC), an increase in temperature of about 2–4°C is expected by 2050 (Pachauri et al., 2014).

Institute of Horticultural Sciences, University of Agriculture, Jail Road, Faisalabad 38040, Pakistan.
Email: m.usman@uaf.edu.pk

Pakistan stands as one of the most affected developing countries by the climate change and by the end of 21st century an increase of 3–5°C is expected. Changes in the precipitation intensity and patterns will have a strong impact on agricultural productivity. Summer rainfall over monsoon region has increased by 18–32%. Global warming is killing centuries-old green assets in Pakistan, Canada and USA (IGWT, 2000–2008; Khan et al., 2008; Rasul and Ahmad, 2012). Frequency of hot days and nights will rise and the projected temperature rise may be more than the global average temperature. Warming during winter will rise and summer season will be prolonged (Chaudhry, 2017).

Climate change has also reduced production of horticultural crops in India from 10–80% due to severe cold wave. Fruit crops will have to face major consequences of climatic changes. The adverse climatic conditions will affect crop productivity by influencing the overall plant growth patterns, phenology, extended plant growth season and frequent cold spells could lead to massive reductions in the productivity (Gordo and Sanz, 2010). All changes may not be detrimental, like rising CO_2 levels could enhance crop productivity as well (Kimball, 2016). In tropics, precipitation patterns are the major factors affecting phenological stages and yield in tree-fruit crops. Fruit producers are interested in adaptation strategies to the climatic changes to maintain production while the opportunity seekers are looking for introduction of tolerant cultivars and new crops. Climate change is also shifting the range of the plant habitat. Hence, emerging threats of climate change, opportunities for adaptations and fruit cultivation at new locations need to be addressed.

In perennial fruit crops, rising temperature will mainly affect the phenological states of the plant, including early bud induction during winter season, persistence of the developed flowers in spring, total yield and fruit quality during its development, fruit expansion and maturity phases (De Ollas et al., 2019). Collectively, water deficit, elevated CO_2 concentration and rising temperatures will have multiple effects on the fruit crop production and quality, including reduced stomatal conductance, transpiration up to 21% and enhanced water use efficiency (WUE), as reported by Medlyn et al. (2013) and Zandalinas et al. (2018). These events will also improve plant performance under water stress conditions. However, consistently rising temperatures will enhance evapotranspiration and reduce the net yield (Fares et al., 2017). Citrus and grapes showed more growth at higher CO_2 levels accompanied with increased temperature due to more WUE and higher photosynthesis rate indicating better acclimation capacity of these crops to the changing climate (Vu et al., 2002; Kizildeniz et al., 2018). Decrease in crop productivity, poor plant performance and reduced fruit quality will also adversely affect the grower's income and marketing competence. The fruit ripening, maturity time and harvesting span will be adversely affected (Fitchett et al., 2014).

1.1 Climate Change and Productivity in Guava

Guava (*Psidium guajava* L.), member of Family Myrtaceae, originated in the tropics of America and distributed widely in the tropical and sub-tropical regions (Pereira and Nachtigal, 2002; Pereira et al., 2017, 2019). Guava plant was domesticated about 2,000 years ago and its first commercial cultivation was reported in 1526 in the Caribbean islands. Guava is exceptionally enriched with ascorbic acid more than citrus. Global warming has reduced its plant vigor, fruit bearing, fruit size, juice content, fruit color, shelf life and has enhanced attack of insects and pests. Flower drop is common in mango and guava if flowering occurs at extremely low temperatures (Reddy, 2012; Reddy et al., 2017).

The rapidly changing climatic conditions suggest dire need to understand the climatic requirements of guava for an optimal plant growth, important phenological stages, impact of major climatic changes on plant growth and development and mitigation strategies, its sensitivity or tolerance to pollutants, abiotic stress tolerance and breeding tools for enhancing climatic resilience. Phenological models should be developed to estimate the impact of changing climatic conditions on plant development in different ecological regions of the world. It is important to understand its floral biology and its relationship with environmental components to devise mitigation strategies.

1.2 Floral Biology and Crop Regulation

Most of the cultivars of guava bloom two to three times a year under sub-tropical and tropical environmental conditions while in temperate regions, it blooms once a year. White colored, hermaphrodite flowers develop in two to three cymes in leaf axils and the middle flower has higher probability of producing fruit. Guava is mainly a self-pollinated crop which requires some external agents for more pollination due to stigmatic position stretched above the stamens. The intensity of cross-pollination may vary from 20–40%, depending upon the variety, availability of pollinizers (*Apis mellifera*) and climatic conditions (Medina, 1988). About 25–30 days are required from flower bud differentiation to calyx cracking stage which starts 20–24 hours before anthesis (Subramanyam and Iyer, 1993). Complete flower bud development, from 2 mm size to bloom, during winter season may take 45–50 days in *P. guajava* while in other species, it may take 35–45 days (Subramanyam et al., 1992). In most of the *Psidium* sp. pollen germination temperature is 30°C. Stigma is receptive a day after anthesis and remains so up to 30–35 hrs. Floral biology of different guava species is similar to each other except *P. cattleianum,* which show early floral bud maturation and fruit development. The fruit developing from the central bud grows faster as compared with the fruit developing from the adjacent buds. This important feature could be helpful in fruit thinning, germplasm characterization and genetic improvement. The fruit setting percentage in *P. guajava* is highly variable among varieties and may range from 30% (Pyriform) to 80% (apple color and Allahabad Safeda).

The floral buds always emerge on the newly developed vegetative sprouts, irrespective to the time of the year. Hence, bud formation, fruit setting and fruit development across seasons (summer and winter) and different years could be highly variable, depending upon the patterns of rainfall, irrigation, fertilization scheduling and other climatic factors, including day length and temperature (Usman et al., 2021). Vegetative growth could be promoted by defoliation induced by spray of 25% urea, pruning, nitrogen fertilizer application and irrigation. Irrigation and nitrogen application after flowering enhanced the yield. Moisture stress at fruiting increased fruit drop and splitting (Menzel and Paxton, 1986). Changes in temperature delay fruiting season while fruit setting and harvest were also affected (Reddy et al., 2017). There is need to have a deeper understanding of the mechanisms of flowering, fruiting, fruit expansion, maturity in relation to changing climatic conditions in the tropical and sub-tropical areas.

Guava is usually propagated by seed and 70–80% of the seedlings retain characteristics of the parent (Pereira et al., 2017). Seedlessness observed in guava (*P. guajava* L.) is attributed to self-incompatibility and chromosomal aberrations. A 'seedless' variety was found diploid (2n = 22) having low percentage of viable pollen grains showing meiotic abnormalities (Menzel and Paxton, 1986).

1.3 Germplasm Resources

A few out of 150 species of *Psidium* are widely cultivated while most of the species are native plants. Higher diversity of *P. guajava* is found in Central Brazil, North and South America and West Indies. There are more than 400 known varieties of guava available in different parts of the world. In guava, selection of superior strains followed by clonal propagation has been a very useful tool widely utilized by the amateur gardeners and researchers. Selections may also be made for greater shelf life and plants having more tolerance against biotic and abiotic stress factors (Usman et al., 2022). In most of the guava-growing regions of Asia, the nomenclature and description of varieties is confusing and same cultivar may be given different names at different places. Guava is being grown in Queensland since 1850 and most of the varieties were introduced from Hawaii and Asian regions. Most of the known selections have been named according to their shape (Gola, Surahi), size (Large Gola, Large Surahi), flesh color (Surkha) or skin color (Apple color, red skin) and place of origin (Allahabad Safeda, Sharqpuri Gola, Lucknow-49) in Pakistan and India (Pereira et al., 2017).

1.4 Climatic Requirements for Optimal Growth and Productivity

Most important factor affecting fruit marketing and profitability involves phenological changes including tree phenology, flowering and fruit development in response to the changing climate. Soil moisture also plays a direct role in the fruit growth and development. Guava requires full sunlight for optimal growth. Optimal temperatures are required for flower induction, fruit setting and fruit development. Guava can withstand a wide range of environmental conditions provided these are frost-free areas. It grows well in both humid and dry climates from sea level to 2100 m elevation, situated in 25–30° S having temperature range from 23–28°C and 4.5–9.5 soil pH (Verheij and Coronel, 1992; Reddy, 2012). Guava requires warm temperatures, vegetative growth starts at 15°C and peaks at 28–30°C (Arevalo-Marin et al., 2021). Guava can tolerate higher temperature up to 45–46°C (It needs at least three to six months of summer temperature above 16–18°C for blooming and fruiting). Both soil and air temperatures play a key role in determination of fruit setting. Temperature changes the endogenous hormonal levels required for the tree growth and development. Guava is well adapted to diverse rainfall conditions; however, it needs evenly distributed annual rainfall (100–200 mm). Excessive rains at fruit maturity lead to poor development of fruit color, fruit splitting and flavor loss. Cold winds restrict vegetative growth, flowering and delay fruit maturity while strong winds may lead to branch splitting.

Guava can survive flooding and continue plant growth in waterlogged soils for short periods. It is relatively more tolerant to drought, salinity and chilly season compared with other fruit crops of the tropics. It can produce throughout the year under tropical environmental conditions. Guava has a long harvesting span and is a potentially high-value cash crop. Summer and winter crops of guava have much variation in fruit size, quality and maturity due to change in temperature, sunlight and humidity (Hassan et al., 2018; Usman et al., 2021). Irrigation requirements of guava are less than citrus. Use of drip irrigation allows precise control of watering without significant loss in yield. Further, it facilitates management of fertilization and natural weeds. Guava is a shallow-rooted crop and its feeder roots are distributed from 20–25 cm deep. Dry period during flowering and fruit growth could be highly detrimental and lead to flower and fruit drop (Mishra et al., 2014). Extreme environmental phases are expected; hence, there is need to protect plants from abiotic stress factors.

The phenological studies in fruit crops are getting more important due to the rapidly changing climatic conditions and reliable long-term phenological datasets are lacking in guava, like other fruit crops. Guava is also heterozygous in nature and genotype × environment (G × E) interaction is much higher. Genotypic stability is important for consistent performance under variable environmental conditions.

2. Impact of Extreme Temperatures and Elevated Atmospheric Carbon Dioxide (CO_2) on Plant Growth, Productivity and Management

Changes in temperatures and elevated CO_2 in the tropical and sub-tropical areas have delayed the fruit season and fruit setting in guava and harvest has also been affected.

2.1 *Temperature*

2.1.1 Elevated Temperature, Flowering and Fruit Quality

The pollination frequency, fruit growth and color development are directly affected by the temperature fluctuations. Optimal air and soil temperatures are required for higher yields. Water availability is a major limiting factor in fruit growth, production and yield. Higher air temperatures at flowering time could increase pollen sterility and reduce fruit setting. However, high temperatures could be beneficial in the cold areas. Usually, warm climate affects fruit metabolism and lowers yield and fruit quality. Higher temperature increases total sugars and reduces acidity (TA) due to increased photosynthetic activity (Reddy, 2012). Ratio of total soluble solids (TSS) and TA are the key traits in determination of the fruit quality and maturity. Annual climatic variations also affect fruit sweetness and acidity more than total yield while crop management is less influential (Le Bourvellec et al., 2015). External fruit quality is affected in summer in tropical and sub-tropical areas while in winter season the fruit quality is better. Sweetness in summer crop is higher than winter crop and increases further with increase in the diurnal temperature differences (Usman et al., 2012). Fruit growing in areas having higher temperatures and moisture contents develop higher TSS (Reddy, 2012). Exceedingly higher temperature limits vegetative growth and fruit setting. Breeders have limited success in developing heat tolerant varieties due to complexity of inheritance and moderate heritability of heat and cold tolerance in fruit setting.

2.1.2 Elevated Temperature, Plant-Pollinator Interaction and Fruit Fly

Pollinators play an important role in maintaining higher productivity. Climate change is affecting pollinator populations and their activity in all major fruit crops (Costanza et al., 2014). About 85% of the fruit crops are pollinator dependent (Klein et al., 2007) and global warming has seriously affected the interaction of plants and pollinators (Morton and Rafferty, 2017). Guava is an essential source of pollen and nectar for the pollinators. Amount of pollination is critical for fruit yield and has been least studied in guava. Out of 13 Brazilian tree crops, guava was found greatly dependent upon insect-mediated pollination. Guava orchards are more likely to face 15–16% decrease in the insect pollinators in main guava producing regions and production may be greatly affected (Giannini et al., 2017). De-synchronization between blooming time, frequency and the natural pollinators could lead towards phenological mismatches, thus reducing the reproductive success (Kudo and Ida, 2013). Rising temperatures may also change the floral preferences of the pollinizing agents and they may shift to the cooler flowers for nectar collection (Shrestha et al., 2018). Thus, climate change may have direct effects on the plant-pollinator interaction and seriously affect pollination and fruit yield.

In guava, pest and disease incidences have increased due to changing climatic conditions. Fruit fly population is becoming alarming due to prevalence of hot and humid conditions (Hazarika, 2015; Mani and Suresh, 2018). The fruit fly population, diversity and sex ratio are also affected by environment and plantation (alternate host) surrounding the guava orchard. Hence, distant areas of the native forests shall be selected for guava plantations (Vargas et al., 2019). The fruit fly infestation decreases at higher temperatures and 25–27°C was found as suitable temperature for its survival (Reddy, 2012). Defoliation also reduced the harvesting span, cost and gave more economical control of fruit fly. Elevated temperatures could lead to higher transpiration losses and atmospheric moisture content favoring the pathogen and disease development.

2.1.3 Low Temperature, Peel Color Development and Plant Acclimation

In apple guava, the development of red color in the peel depends upon availability of cool nights at fruit maturity stage. An increase of 0.2–0.3°C temperature could significantly reduce the peel color development and increase of 0.5°C will drastically reduce the guava cultivation areas suitable for color development (Reddy et al., 2017). In colored guava, rise in temperature reduces skin color development, affecting the demand and marketing of the fruit (Rejab et al., 2008; Mani and Suresh, 2018).

In *P. guajava* the cambial activity increased with the rise in temperature above 15°C and is considered as temperature threshold for vegetative growth. It is also used as base temperature for calculating heat units required for pruning and fertilizing for a better harvest (Bittenbender and Kobayashi, 1990). Strawberry guava is relatively cold-resistant specie compared with common guava and can withstand frost without any severe injuries at minimum temperature threshold lower than common guava. The trees at higher elevation were tall and dense and this could be attributed to differences in soil, temperature and genotype. A strong correlation existed between flowering and shoot production; hence, poor yield could be related to poor shoot growth. Fruit setting and retention were higher with little fruit drop (Normand and Habib, 2001).

2.2 Carbon Dioxide (CO_2)

2.2.1 Plant Growth and Photosynthetic Efficiency

Carbon dioxide is responsible for 60% of the global warming. Rise in the atmospheric CO_2 and other greenhouse gases (GHG), including CH_4, N_2O and O_3 is raising the global temperature. Increasing CO_2 is also affecting the precipitation patterns and leading to reduced soil water availability (Keeling et al., 1995). Understanding the process of photosynthesis under rising CO_2 conditions could help in enhancing the efficiency of fruit production under changing climate. Prolonged exposure to elevated CO_2 could alter the photosynthetic biochemistry, physiology and frequency of stomata, plant morphology, anatomy, branching and biomass production. Gene expression of Ribulose-1,5-bisphosphate carboxylase-oxygenase (RuBisCO) was reduced while other photosynthetic genes showed a differential expression pattern after prolonged exposure to the elevated CO_2 (Moore et al., 1999). Photosynthesis is also affected by the elevated temperature which reduces RuBisCO activation and decreases solubility of CO_2 relative to O_2 leading to increased photo-respirational losses of CO_2. The elevated atmospheric CO_2 inhibits RuBisCO oxygenation and balances the adverse effects of rising temperature on net photosynthesis (Rehman et al., 2015).

Rise in atmospheric CO_2 changes the annual precipitation patterns, reduces stomatal conductance, increases WUE with a higher rate of assimilation and reduced water loss. Thus, elevated CO_2 delays onset of the adverse effects of drought stress, continues essential plant growth and enhances plant water conservation (Idso, 1998; Rehman et al., 2015). Photosynthesis can occur at high and low light intensities. It is enhanced by elevated CO_2 and higher temperature under limited light conditions (Drake et al., 1997). Elevated CO_2 also decreases nitrogen levels and increases Nitrogen Use Efficiency (NUE) by reallocation of proteins and accelerated leaf senescence.

2.2.2 Secondary Metabolites and Biotic Stress Tolerance

Changes in CO_2 concentration raises the annual temperature, altering physiological behavior of plants and secondary metabolite production. However, responses may vary with site and kind of plant (Korner, 2006; Wu et al., 2011). In guava (*P. guajava* L.), tannins and flavonoids are produced as major secondary metabolites (Mailoa et al., 2013; Yousaf et al., 2013). Guava showed alteration in phenolic production in the presence of atmospheric pollutants. The elevated CO_2 led to accumulation of tannins and starch in guava leaves with no significant increase in plant growth (de Rezende et al., 2015). Similar accumulation of extra carbon in leaves is well documented in other crops (Tausz et al., 2013). The higher accumulation of tannins in 'Pedro Sato' guava leaves increased protection against pathogens and herbivores and is useful for pharmacological extracts. The increase in C/N ratio due to elevated CO_2 may have decreased the demand of amino acids. The surplus carbon may be allocated to starch and tannin accumulation in leaves (de Rezende et al., 2015). Higher CO_2 led to more vegetative growth, dense canopies, lower plant decomposition, providing environment conducive for more microbial activities and diseases. Bactericide and fungicide activity may also vary with higher CO_2, temperature and moisture, leading to more frequent applications. There is need to evaluate the effects of elevated CO_2 for longer periods on fruit quantity and quality in guava. Guava has also been recommended for intercropping as a biological control for citrus greening disease.

2.3 *Stomatal Density and Responses to Abiotic Stress*

Stomata are the gateway for cellular gaseous exchange and stomatal conductance has strong correlation with photosynthesis in plant leaves (Kanemura et al., 2007). Researchers have reported responses of stomatal densities to different environmental factors, including elevated CO_2 (Woodward, 1987), drought stress (Galmes et al., 2007), salt stress (Zhao et al., 2006), heat stress (Beerling and Chaloner, 1993), changes in precipitation patterns (Yang et al., 2007) and planting density (Zhang et al., 2003). Intercellular CO_2 was found positively correlated to transpiration (E) and stomatal conductance (Gs) while Gs was also corelated to E in 22 Indian guava genotypes. However, no genotypic variation was reported for these traits (Shiva et al., 2017).

3. Sensitivity to Air Pollutants, Nutritional Contaminants and Potential as Bioindicator

In urban environments, increasing air pollution is a global concern (Cohen et al., 2017; WHO, 2016) and is a serious threat to the health of ecosystems. Identification of plants having tolerance to air pollutants, including particulate matter, SO_2, NO_2 and O_3 could help in mitigation of the issue (Ulmer et al., 2016; Manes et al., 2016). The plant responses to these pollutants (Gao et al., 2016) and the potential of guava against these contaminants need to be assessed.

3.1 *Ozone (O_3) Sensitivity, Stomatal Density and Role of Anthocyanins*

Ozone shields the plants from ultraviolet effects; however, at lower atmospheric levels, it is highly dangerous. Ozone, a secondary air pollutant, develops in the atmosphere by phytochemical reactions of volatile compounds and nitrogen oxide, leading to formation of a photochemical smog. It causes damage to the vegetation by inhibiting photosynthesis, increasing leaf injuries and production losses (Krupa et al., 2001). The O_3 level has increased fivefold (200 $\mu g/m^3$–1 h) due to rising anthropogenic activities (Derwent et al., 2002) and is higher during spring and summer seasons due to more light than in winter (CETESB-Companhia de Tecnologia de Saneamento Ambiental, São Paulo, 2013). In Pakistan, surface O_3 ranges from 3–46 ppb in major cities. Injuries may occur in the sensitive plant species after exposure to 30.58 ppb O_3 for eight hours (www.suparco.gov.pk).

Ozone-induced oxidative stress causes physiological and biochemical damage, alters photosynthesis, gas exchange, enzyme activities and DNA mutations. Plant exposure to higher concentrations of O_3 leads to chlorophyll degradation, necrosis and production of secondary metabolites (Heath, 2009; Sandre et al., 2014). Ozone intensifies ROS formation and affects the carbon allocation between shoot and root. Higher exposure may lead to plant death (Kangasjärvi et al., 2005). The visibility of O_3 symptoms shows variability which is attributed to the association of variable meteorological conditions at the time of application or uptake of O_3 (Pina and Moraes, 2007). Rainfall patterns, vapor pressure deficit (VPD) and amount of solar radiation influence the uptake of O_3 (Sandre et al., 2014).

Guava (*Psidium guajava*) cultivar 'Paluma' and Strawberry guava (*P. cattleianum*) are sensitive to O_3 and have been used as bioindicators, showing dark reddish, reddish-purple colored leaves and interveinal stippling in older leaves (Furlan et al., 2007; Pina and Moraes, 2007, 2010; Kateivas et al., 2018) as shown in Fig. 1.

The guava leaves show pigmentation with no cell death compared with other sensitive species, like *N. tabacum* and this could be attributed to its compact leaf arrangement. Further, this species has an efficient mechanism to regulate gas entrance in the leaves through substomatal chamber occlusion and to reduce transpirational water losses. The stomatal density in 'Paluma' plants exposed to accumulated O_3 showed a direct relationship with each other. Stomatal density increased with increasing O_3 in winter and was found lower in summer when O_3 was also reduced. Thus, anatomical changes in leaves of 'Paluma' when exposed to O_3 including compact nature of leaves, higher phenolic contents, occlusion of stomatal chamber and higher stomatal density enabled it to neutralize ROS and avoid cell death (Tresmondi and Alves, 2011).

Anthocyanins are efficient free radical scavenging molecules that act as nonenzymatic antioxidants and are produced when plants are exposed to abiotic stress, mitigating photo-inhibitory and photo-oxidative damages in leaves and conferring physiological advantages on plants (Neil and Gould, 2003). Moraes et al. (2004) fumigated guava (*P. guajava* L.) plants with O_3 and found it as most resistant to reduction in net photosynthesis, transpiration rate and stomatal conductance. Plants showed leaf injuries after five days

Fig. 1. Ozone injury in leaves of guava (*P. guajava* L.) cultivar 'Paluma' after 30 days of exposure in open top chambers show enhanced amount of anthocyanins. Figures show percentages of leaf-injury: (A), 5%; (B), 25%; (C), 50%; and (D), more than 75% (de Rezende and Furlan, 2009).

of exposure to O_3, indicating its potential as bioindicator of O_3 (Furlan et al., 2007; Pina and Moraes, 2007). The O_3 fumigation of guava trees induced dark reddish-colored stippling or dots which could be attributed to higher accumulation of anthocyanins and decreased carbon assimilation under saturating light conditions (A_{sat}) to neutralize ROS activity (Gould, 2004). These foliar symptoms of red stippling appear to protect leaves against O_3 injury (Tresmondi and Alves, 2011). Accumulation of phenolic compounds, including anthocyanin and tannins, was attributed to less intensive photo-inhibition in guava cultivar 'Paluma' (de Rezende and Furlan, 2009). Phenolic compounds including anthocyanins may neutralize the ROS (Soares and Machado, 2007). Anthocyanins are weak pollution indicators and show no relationship with O_3. Urban air pollutants, climatic conditions and biotic factors could be responsible for triggering the production of phenolics (Sandre et al., 2014). However, further investigation is desired to understand the O_3 uptake to check critical levels of O_3 uptake and leaf injuries using DO_3SE model (Assis et al., 2015).

3.2 *Industrial Contaminants and Leaf Functional Traits*

The potential of guava cultivar 'Paluma' was explored against atmospheric contaminants in industrial areas having oil refineries. In field areas, plant responses are not exclusively to a single pollutant, rather, it's a sum of environmental pollutants and climatic conditions. The alterations observed in 'Paluma' could be attributed to combined effect of O_3, NO_2 and other pollutants (da Silva et al., 2017). More leaves were produced to compensate for the leaf senescence induced by air pollutants, reduction in photosynthesis and maintenance of its growth rate (Amthor and McCree, 1990; Oksanen, 2001). The altered root shoot ratio in the plants was identified by photo-assimilates partitioning induced by pollutants (Szabo et al., 2003). The plantation of 'Paluma' and other sensitive guava cultivars in the urban areas of sub-tropical and tropical regions could be helpful as a potential bioindicator of the higher levels of O_3 and other industrial pollutants.

Leaf functional traits could be used to assess the response to air pollution, plant tolerance and are reliable markers against biotic and abiotic stress factors (Brandao et al., 2017). The assessment of leaf functional traits and tree characteristics was performed against industrial pollutants (particulate matter, SO_2, NO_2, O_3) in 13 different tropical plant species including fruit plants, like mango (*Mangifera indica* L.) and guava (*P. guajava* L.) in India. Proline acts as osmolyte, metal chelator, protects antioxidant enzymes and the photosynthetic apparatus (Szabados and Savouré, 2010). The increase in proline content was maximum in guava compared with other plant species. Relative water content (RWC) helps plants in regulation of osmotic potential and cellular turgidity and decreases with rise in the load of pollutants (Arena et al., 2014). Increase in RWC was maximum in guava when exposed to pollutants. Assessment of 15 different leaf functional traits revealed *P. guajava* as the second most tolerant species after *Caesalpinia sappan* having increased non-enzymatic antioxidants, pigments and RWC at higher load of the pollutants. *P. guajava* also showed maximum response against NO_2 accumulation by higher accumulation of polyphenols in the leaves. Among industrial pollutants, particulate matter (PM) was found as a major stress factor followed by O_3 (Mukherjee and Agarwal, 2018).

3.3 *Nutritional Contaminants and Heavy Metals*

Increasing industrialization, automobile fuel consumption, energy production systems, herbicides and fertilizer applications are major sources of anthropogenic nutrient deposition (Sawidis et al., 2011; Boian and Andrade, 2012). The deposition of pollutants including nitrogen, sulfur and iron on agricultural ecosystems up to toxic levels is an alarming health hazard, leading to reduced biomass production, alterations in the plant physiological, biochemical mechanisms and nutritional imbalances. These risks can be analyzed using bioindicator plants which help in prediction and quantification of the environmental disturbances (Abril et al., 2014; Bulbovas et al., 2015). Wild guava (*P. guajava* L.) is an efficient accumulator of nitrogen, fluorine, sulfur and a few heavy metals under tropical environmental conditions and cultivar 'Paluma' has been used as a bioindicator against toxic air pollutants (Moraes et al., 2002; Nakazato, 2014). Guava plants were found as the most suitable accumulator for biomonitoring of K, P, Ca and B depositions. Higher contamination risk was observed for K and B during the dry season (Bulbovas et al., 2015).

Heavy metals are another serious health hazard and guava orchards near big cities are usually irrigated with sewage water which is rich in heavy metals (Ozores-hampton et al., 2005). Fruit crops vary in uptake and tolerance towards heavy metals. Guava accumulated greater content of lead (Pb) and lower content of mercury (Hg) compared with mango and papaya (Ang and Ng, 2000; Hussain et al., 2021). Different plant genotypes showed differential behavior on heavy metal uptake (Khalid et al., 2017). Oil refineries are also a major source of atmospheric emissions including compounds like sulfur, carbon monoxide and emit particulates containing different heavy metals including nickel (Ni). Guava (*Psidium guajava* L.) cultivar was also found as a good accumulator of Ni after three months of exposure at different sites with atmospheric emissions from a nearby oil refinery and could be used for its biomonitoring (Perry et al., 2010). Further studies are suggested to investigate greater heavy metal tolerance in guava varieties for selection and multiplication.

4. Abiotic Stress Tolerance

4.1 Salt Stress

About 30% of the irrigated land is salt affected globally and 6.8 M ha are affected by salinity in Pakistan (Khan, 1998; Chaves et al., 2009). Under saline conditions, plant growth is affected by ion toxicity, osmotic stress, restricted uptake and translocation of nutrients leading to a disrupted ionic balance (Hameed and Ashraf, 2008). Under salinity-induced oxidative stress, plant height and number of leaves were reduced (Bhantana and Lazarovich, 2010). The NaCl dominates in salt-affected soils which may decrease potassium (K^+) concentration in plants (Kumar et al., 2008). Different fruit crops are sensitive to salinity, including mango (Zuazo et al., 2004), passion fruit (Cavalcante et al., 2005) and guava (Ebert et al., 2002; Cavalcante and Cavalcante, 2006) as shown in Fig. 2. However, guava is also considered as moderately salt tolerant (Syvertsen and Levy, 2005) with a salinity limit of 8–9 dS m^{-1} EC of the soil saturation extract and better tolerance to highly saline conditions is about > 16 dS m^{-1} (Thaipong and Boonprakob, 2019). Twenty genotypes of guava raised by stem cuttings were tested for salt tolerance up to 200 mM NaCl. Concentration of sodium and chloride ions was higher in leaves of sensitive genotypes compared

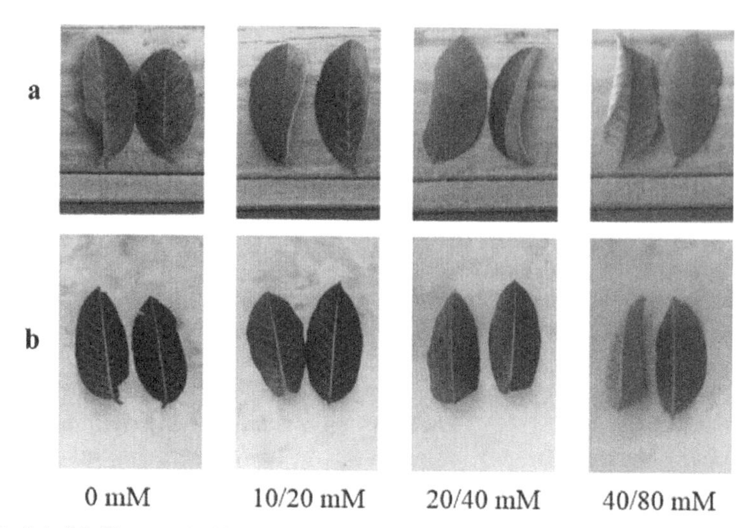

Fig. 2. Salt (NaCl) stress in Kenyan common guava (*P. guajava* L.) leaves: (A) upper side of the leaves, and (B) underside of the leaves. Symptoms were more pronounced at higher salt stress levels (40/80 mM) (Chiveu et al., 2020).

with the tolerant genotypes (KUHP38, KUHP12, Paen Seethong, Na Suan), indicating prevalence of a salt exclusion mechanism in these genotypes which need further exploration (Thaipong and Boonprakob, 2019). These genotypes could be used both as scion and rootstocks in saline areas. Increase in Cl^- concentration is the primary source of salt toxicity in guava (Moya et al., 2003).

Guava is more sensitive to salinity during seedling formation, seedlings died when irrigated with 4.5–6.0 dSm^{-1} saline water; hence it shall not be irrigated with saline water above 1.5 dSm^{-1}. Among cultivars, 'IPA B-38' and 'Surubim' were found relatively more salt-tolerant compared with 'Paluma' (Cavalcante et al., 2007). Somatic embryogenesis is a very useful tool for *in vitro* mass multiplication of the plant material and to evaluate abiotic stress tolerance. Somatic embryos were developed in guava from zygotic embryo on MS media containing 2,4-D and NaCl salt (0–200 mM). Callogenesis and embryogenesis were arrested at higher doses of salt. Embryos matured up to 100 mM NaCl. The germinated plants showed chlorosis at 100 mM NaCl. The chlorophyll content and carotenoids were reduced at higher salt concentrations. Leaf proline and glycine betaine content increased with increasing salt stress (Rai et al., 2010). Guava produces well in arid climates having soil salinity and low soil fertility and scarcity of fresh water. Shoot growth and stem girth of white flesh guava cultivar 'Round' were reduced at higher levels of NaCl (150–300 mM). Defoliation (%) increased with rising levels of NaCl and 50% defoliation was observed at 200 mM. Leaf length was reduced at

100 mM while leaf width expansion was restricted at 250 mM. Reduction in root shoot biomass was higher above 150 mM. In leaves, Na^+, K^+ and Cl^- ion concentration increased up to 150 mM NaCl and further uptake declined at higher salinity levels (Usman et al., 2015).

Cultivar Allahabad Safeda tolerated saline water irrigation having EC_{IW} 3–4 dS m^{-1} and prevented excessive accumulation of Na^+ and Cl^- in leaves by higher accumulation of proline and maintained higher levels of photosynthesis (Singh et al., 2018). Application of saline water reduced stomatal conductance, CO_2 assimilation transpiration, WUE, fruit number, fruit weight and size. Fruit weight was reduced as 8 gm per unit increase in EC of water at 0.3–3.5 dS m^{-1}. Addition of more N did not reduce the deleterious effects of salinity (Bezerra et al., 2018). The protease activity was directly proportional to the application of saline water and 40°C was optimal temperature for leaf protease activity (Ghalati et al., 2020). Exogenous foliar application of NaCl salt stress in guava seedlings increased catalase, polyphenol oxidase, carotenoids, proline and decreased peroxidase and chlorophyll (a, b and total). Salt stress changed the carotenoid content which quenches ROS and prevents lipid peroxidation (Loggini et al., 1999; Ben-Asher et al., 2006). However, putrescine-treated plants reduced catalase and peroxidase activities and increased the pigments. NaCl stress increased proline content in citrus and guava (Murkute et al., 2010; Sa et al., 2013). Proline and putrescine are connected through activity of diamine oxidase (DAO) and g-aminobutyric acid (GABA) metabolism (Aziz et al., 1998). The exogenous application of 250 and 500 ppm of putrescine (under 5 and 10 dS m^{-1} of NaCl, respectively) increased the antioxidant non-enzymatic compounds, including proline, which may provide better salt tolerance to guava seedlings (Ghalati et al., 2020). Summary of salt tolerance reports is presented in Table 1. These findings indicate that the potential of mild salt tolerance is available in guava and germplasm shall be screened for better tolerance followed by clonal propagation of the selected plant material for cultivation in moderately saline areas.

4.2 *Water Stress*

Response of different genotypes of the same species in a crop may show variable plant responses to abiotic stress factors, including drought stress (Lopes et al., 2015; Wani et al., 2015; Wehner et al., 2015). Guava is considered as relatively more drought tolerant than citrus and can survive water deficit conditions (Shaukat et al., 2015; Bokhari et al., 2018). In water-deficit studies, where 75% and 50% field capacity were maintained for 16 weeks in two-year-old guava seedlings, plants depicted more plant height, higher photosynthetic performance, chlorophyll content, WUE and reduced leaf drop in cv. Pyriform followed by rapid recovery compared with cv. 'Round' during summer (Fatima et al., 2018) and winter

Table 1. An overview of abiotic stress sensitivity and tolerance in guava cultivars.

Stress type	**Cultivars/plant age**	**Results**	**References**
Salt tolerance	Ganib (Red guava) and Pakistani (White guava)	Red guava was more salt tolerant (30 mM NaCl) compared with white guava	Ali-Dinar et al. (1999)
	IPA B-38, Surubim and Paluma	IPA B-38 and Surubim were more tolerant to saline water (1.5 dS m^{-1}) than Paluma	Cavalcante et al. (2007)
	Paluma (Seedlings 12 cm height)	Toxicity observed for saline water treatment (2 dS m^{-1})	Silva et al. (2008)
	Banarasi	Somatic embryo induction at 100 mM NaCl	Rai et al. (2010)
	Round/ one year	Significant reduction in root shoot biomass, shoot growth, stem girth and 50% leaf drop observed above 150 mM NaCl	Usman et al. (2015)
	Allahabad Safeda/ one year	Salinity tolerance (EC) threshold was about 1.5 dS m^{-1}	Singh et al. (2016)
	Crioula, Paluma and Ogawa (rootstocks)	Crioula was more tolerant to salinity (1.8 dS m^{-1})	Sa et al. (2016)
	Paluma	Saline ($<$ 0.3 dS m^{-1}) water irrigation hampered gas exchange and reduced fruit production	Bezerra et al. (2018)
	KUHP38 KUHP12 'Paen Seethong', 'Na Suan'	The genotype KUHP38 was highly tolerant while others were moderately tolerant to saline conditions (200 mM NaCl) among 20 genotypes.	Thaipong and Boonprakob (2019)
	Seedlings/ one year	Application of 250–500 ppm of putrescine (under 5–10 dS m^{-1} NaCl) increased protease activity and two proteases were reported	Ghalati et al. (2020)
Drought	Surahidar	Seedlings growth was arrested when irrigated after 10 days	Shaukat et al. (2015)
	Jen-Ju-Ba/ two years	Significant reduction in shoot growth and leaf CO_2 exchange rate (25%–85%)	Lin et al. (2015)
	Four guava landraces	Shoot formation tolerance up to 8% PEG	Youssef et al. (2016)
	Round and Pyriform/ two years	Plant growth and photosynthetic attributes were more affected in Round compared with Pyriform	Bokhari et al. (2018) Fatima et al. (2018)

Table 1 contd. ...

...Table 1 contd.

Stress type	Cultivars/plant age	Results	References
	Guava species	Guava species *P. friedrichsthalianum* and *P. polycarpum* were sensitive to cold compared with common guava, *P. guineense* and *P. cattleianum* var, lucidum which tolerated freezing temperature up to –5°C	Utsunomiya (1988)
Freezing tolerance and cold acclimation	Lucknow-49 and Ruby Supreme/ one year	Accumulation of leaf anthocyanins during cold acclimation, four proteins under low temp. and two proteins under drought stress. Ruby Supreme exhibited less injury	Hao et al. (2009)
Varietal trial for environmental suitability in hot arid zone	Allahabad Safeda, Lalit, Sweta and L-49/five years	[Summer Temp. (45–50°C), Wind velocity (20–30 km/hr), Winter Temp. (–0.5 to –1.5°C), low precipitation] Sweta and Lalit performed better	Singh et al. (2017)

seasons (Bokhari et al., 2018). *In vitro* methods offer great potential to screen germplasm for drought tolerance using mannitol and polyethylene glycol (PEG) in the nutrient media. Four landraces were used for micropropagation by nodal segments under water-deficit conditions using PEG (0–10%). Shoot induction and development were obtained up to 8% PEG in the Murashige and Skoog (1962) media (Youssef et al., 2016). The available germplasm shall be screened for drought tolerance potential under *in vitro* conditions and selected genotypes shall be tested for tolerance under greenhouse conditions for further confirmation of the inherent varietal potential.

4.3 *Cold Stress*

Guava (*P. guajava* L.) can withstand short periods of freezing temperatures up to –3°C to –5°C; however, it is frost sensitive. Guava species, *P. friedrichsthalianum* and *P. polycarpum*, are very sensitive to cold compared with common guava and can't withstand temperatures below –5°C while the other species, including *P. guineense* and *P. cattleianum* var, lucidum, tolerated freezing temperatures up to –5°C (Utsunomiya, 1988), indicating the potential of these species as parents in hybridization programs for cold tolerance. Mature trees recover after frost by producing suckers from the uninjured roots or wood. The mechanisms, like plant growth, reduction in leaf moisture content, accumulation of anthocyanins and higher turnover of stress-related proteins like dehydrins contribute to freezing-stress tolerance. Water stress increased cold tolerance in guava plants by higher proline accumulation. The timing of water stress was more critical in cold acclimation compared with length of stress (Utsunomiya, 1988). Two guava cultivars were compared with freeze tolerance and cold acclimation and cultivar 'Ruby x Supreme' showed less cold injury and more tolerance compared with 'Lucknow-49'. Four different proteins were produced in response to low temperature; two proteins were produced under drought stress while 17.4 kDA dehydrin accumulation was common under both types of stress conditions (Hao et al., 2009). Guava is also suitable for cultivation under moderately cold and drought conditions (Orwa et al., 2010; Simelton et al., 2015; Bokhari et al., 2018).

4.4 Rootstocks

Use of abiotic stress-tolerant rootstocks could help in inducing tolerance in scion cultivars. These species, *P. guineense* and *P. cattleianum* var, lucidum, could also be used as rootstocks for the cold tolerance (Utsunomiya, 1988). *P. cujavillis* is tolerant to drought and sodic soils and Chinese guava (*P. friedrichsthalianum*) could be used as dwarfing rootstock (Singh et al., 2009; Malhotra, 2017). These potential rootstocks shall be grafted by different stress-tolerant scion cultivars and their potential shall be explored under multiple stress conditions to select better germplasm for growers and researchers for further utilization in their respective domains.

5. Breeding Tools and Enhancing Climate Resilience

Development of climate-resilient guava varieties which are tolerant to rising temperature, elevated CO_2, air pollutants, water stress, salinity and frost tolerance through breeding and omics is essentially required.

5.1 Selection and Hybridization for the Development of Better Adapted Cultivars

Selection of better adapted cultivars could be a useful breeding tool for crop improvement under changing climatic conditions. The selected elite genotypes may be clonally propagated using soft wood cuttings and micropropagation (Usman et al., 2012). For better exploitation of the favorable traits, sexually reproduced progenies could be used for characterization and selection of elite accessions (Warschefesky et al., 2016). However, crop juvenility could be problematic in perennial crops. Guava starts bearing in one to two years and juvenile period is not as long as in citrus and other tree fruit crops (Ollitrault and Navarro, 2012). Germplasm resources shall be collected, conserved and screened for better adaptability to different biotic and abiotic stress factors. Based on this information, parents could be selected and crossed for the development of better adapted hybrids.

5.2 *Marker Assisted Selection (MAS), Genomic Selection (GS) and Sequencing Tools*

Advances in genomics and other omics tools have enhanced efficiency of the breeding and selection programs in perennial woody crops, particularly tropical fruits under changing environmental constraints (Scheben et al., 2016; Mathizhagan et al., 2021). DNA markers have been used to classify guava cultivars on the base of fruit shape (Usman et al., 2020). These tools could also help in underpinning the adaptation traits and the controlling genes in the germplasm (Brozynska et al., 2016). There has been limited genomic information available in guava. However, recently the genome of *P. guajava* cultivar 'New Age' has been sequenced using single molecule real-time (SMRT) sequencing (Feng et al., 2021). Another draft genome assembly of [*P. friedrichsthalianum* (O. Berg) Nied] is also available (Rojas-Gomez et al., 2021). Genetic markers linked to tolerance to changing climate shall be identified followed by quantitative trait loci (QTL) analysis and developing genetic associations of marker genes to the phenotypic traits. QTLs associated with guava fruit traits including its size, weight, number of seeds, ascorbic acid and plant height were found to be located among the linkage groups (11) related to the haploid genome (Padmaker et al., 2016). Linkage associations could also be developed in diverse germplasm or segregating populations using genome-wide association studies (GWAS), which could help in identification of monomorphic key genes in a hybrid population. The use of GWAS has been increased in perennial fruit crops, including guava (Farneti et al., 2017; Minamikawa et al., 2017). The transcriptomic analysis has revealed different pathways involved in fruit development, ethylene biosynthesis and few signal transduction genes were found upregulated. Forty families of different transcription factors were involved in the process of fruit development (Minamikawa et al., 2017). Whole genome sequencing (WGS) has revealed genes involved in ascorbic acid accumulation and fruit softening (Feng et al., 2021).

Genomic selection is based on the modulization of phenotypic values (high density marker information) over the whole genome (Desta and Ortiz, 2014). Combined analysis of multiple traits under different environmental conditions could enhance prediction accuracy and utilization of GS in plant improvement (De Ollas et al., 2019). Knowledge and information of genomics and genetic marker resources is required for QTL analysis, GS and GWAS. Next generation sequencing (NGS) is a promising method with reduced genomic representation in crops like guava. Genotyping by sequencing (GBS) is more promising for high-throughput genotyping of whole genomes, single nucleotide polymorphism (SNP) and are suitable for analysis of large segregating populations (Montero-Pau et al., 2017). These methods are more flexible than microarray and don't need pre-determined genetic polymorphisms. GS is more efficient than MAS to increase genetic gain of the complex traits. There is enormous potential in identification of the QTLs and genes of interest using the advanced sequencing and omics tools for crop improvement in guava.

5.3 *Polyploidization to Enhance Tolerance to Environmental Constraints*

During the evolutionary process, polyploidization has been a major driving force in enhancing adaptability to the environmental hazards and polyploids have higher heterozygosity compared with their diploid progenitors (Brochmann et al., 2004; Chen, 2010; Fatima et al., 2015). In woody plants, polyploidization has enhanced stress tolerance (Ruiz et al., 2016; Greer et al., 2018; Bokhari et al., 2019). In citrus, spontaneous polyploids were reported from apomictic seeds in most of the species (Usman et al., 2006, 2008; Fatima et al., 2010; Aleza et al., 2011). Guava lacks apomixes; however, natural polyploids have been reported occasionally. First triploid ($2n = 3x = 33$) seedless variety was verified by Kumar and Ranade (1952). Pollens of diploid guava ($2n = 2x = 22$) had higher germination rate compared with other species having a higher chromosome number (Pommer and Murakami, 2009). Polyploidy is less common in *P. guajava*; however, other species show ploidy levels as *P. polycarpum* ($2n = 2x = 22$), *P. cujavillus* and *P. guineense* ($2n = 4x = 44$), *P. friedrichsthalianum* ($2n = 6x = 66$) and *P. cattleianum* ($2n = 8x = 88$). Induction, selection and characterization of polyploids could be a suitable alternative tool in guava breeding (Bokhari et al., 2019).

Polyploids have several advantages over diploids towards adaptability to the adverse climatic conditions including dwarf trees, large and thick leaves, fewer and larger stomata, thick mesophyll cells and reduced net photosynthesis leading to dwarfism (Allario et al., 2011, 2012; Usman et al., 2012b; Padoan et al., 2013). Doubled diploid genotypes show better acclimation to abiotic stress tolerance with reduced respiration, maintenance of higher water content, enhanced radial hydraulic conductivities leading to increased tolerance to water deficit (Oliveira et al., 2017). The enhanced tolerance in tetraploids was linked to up-regulation of genes involved in osmolyte and hormonal biosynthesis and ROS detoxification (Tan et al., 2015). Polyploids are being developed and screened for morphological, physiological and genetic characterization using molecular markers in the elite Pakistani guava strains (Bokhari et al., 2019). Potential of abiotic stress tolerance in guava could be further enhanced using polyploidization which may also improve reproductive and yield-related traits.

5.4 Integration of Physio-biochemical and Molecular Data for Predictive Modeling

The ability of plants to survive in unfavorable environmental conditions is the outcome of complex interaction of diverse protective responses of genetic components with the environmental factors. Systems biology approaches study different biological systems utilizing mathematical modeling. Participation of different metabolic pathways in response to stress could be investigated and functions of different genes can be predicted using omics technologies (Rhee and Mutwil, 2014; Fukushima and Kusano, 2014). Combined utilization of transcriptomics and metabolomics data help to develop association of genes with metabolites which could be confirmed by protein-protein and DNA-protein interactions and using reverse genetic approaches (De Ollas et al., 2019). Systems biology resources are available for olives (http://www.bioinfo-cbs.org/ogdd/) and grapes (http://vespucci.colombos.fmach.it/) in fruit crops (Ben Ayed et al., 2016; Moretto et al., 2016); however, no such information is available for guava.

5.5 Role of miRNAs in Regulating Abiotic Stress Tolerance

The miRNAs are known to play an important role in regulating responses to abiotic stress factors in plants (Ragupathy et al., 2016). For instance, *Poncirus trifoliata* L. Raf a citrus rootstock known for its low temperature acclimation trait had five novel miRNAs which target the stress-responsive genes (Zhang et al., 2014). Similarly, 76 miRNA were altered in response to salt stress, water stress or both in the roots (Xie et al., 2017). Understanding of protein-protein interaction, their interaction with other biomolecules, protein sequence and post-translational modifications (PTMs) are essential for working under systems biology; however, such studies are lacking for guava.

5.6 Bioinformatic Database

Plastome of guava (*P. guajava* L.) was sequenced by Jo et al. (2016) and sequence length was found as 158, 841 bp, consisting of large single copy (87, 675 bp) and small single copy (18, 464 bp) separated by the inverted repeats (26, 351 bp). Reatini et al. (2018) also reported plastome sequences of two *Psidium* species, including *P. galapageium* and *Psidium* sp. from the Galápagos Islands. Guava (*P. guajava* var. Zhenzhu) genomic information has been provided by Hainan University and total sequence length was 386 Mbp (ID GenBank: 5984788). Recently the genome of *P. guajava* cultivar 'New Age' has been sequenced (Feng et al., 2021) and a draft genome assembly of [*P. friedrichsthalianum* (O. Berg) Nied] is also available (Rojas-Gomez et al., 2021). The available sequence and other molecular database will play a pivotal role in the development of biotechnological tools in guava. The strategy of matching available annotated ESTs from the genomic database can be used to study critical stress responses, including salt and drought stress (Gimeno et al., 2009; Allario et al., 2012). Guava cultivars, Round and Pyriform, were compared for drought tolerance and global gene expression analysis using microarray. Guava cultivar 'Pyriform' depicted greater drought stress acclimation due to increased activities of catalase (CAT) and peroxidase (POD) which was complimented by the upregulation of a higher number of related expressed sequence tags (ESTs) compared with cultivar 'Round' (Usman et al., 2022). Still little genomic information is available in guava compared to other major fruit crops and thus it needs to be addressed. Greater availability of genetic information will help researchers to utilize these databases for systems biology and other related omics for screening and development of resistant varieties against biotic and abiotic stress factors.

6. Mitigation and Crop Improvement Approaches

It is important to develop strategies to mitigate the climate change challenges by more efficient utilization of the natural resources and development of novel varieties having better tolerance to abiotic stress conditions. Prediction of impact of the changing climatic conditions on plant growth and phenology is also challenging. Also, it is difficult to study the impact of environmental factors on perennial tree crops, like guava, under controlled conditions. Better understanding of adaptability of the crop to different production systems in tropical and sub-tropical areas could be helpful in developing such models and mitigation strategies. Integrated efforts, effective strategy and good simulation models are required to mitigate the impact of climate change on guava plant growth and fruit production. Hence, there is need to develop innovative methodologies for such studies to formulate reliable strategies and effective predictions. Such studies will also help in identification of new varieties/strains for the new target zones of plantations. Following approaches can be used for better mitigation to climatic changes and for crop improvement:

1. Guava cultivation shall be shifted to areas having suitable environmental conditions to mitigate climate changes. Red-color guava shall be grown in areas having cool nights for better color development. Strawberry guava is more cold and frost tolerant than common guava and shall be grown in areas having low temperatures. It could also be used as a rootstock. Fruit growing in areas with more moisture and temperature have higher TSS, thus summer crop fruit is sweeter than fruit of winter crop.
2. Establishing phenological stages in stress-susceptible and resistant varieties using BBCH (Biologische Bundesanstalt, Bundessortenamt

und CHemische Industrie) scales for multiple seasons and developing their correlations with the environmental components will help in better understanding of plant responses to the changing climatic conditions.

3. Cultivars, like KUHP38, KUHP12, Paen Seethong, Na Suan, IPA B-38, Surubim, Allahabad Safeda are salt tolerant and can be planted in saline areas.
4. Cultivar Pyriform is more tolerant to water stress compared with Round (Usman et al., 2022). Water resources shall be more efficiently utilized for irrigation using modern tools like drip irrigation and sprinklers. Plant water use efficiency and water productivity (crop yield: water used) shall be enhanced by applying less water than evapotranspiration demand of the crop. Mild water deficit may reduce vegetative growth, transpiration, fruit quality and enhance average yield; however, it shall be carefully applied to get optimal yield (Fereres and Soriano, 2007; Fernandez et al., 2018). Supply of root-shoot assimilates and root activity need to be monitored under abiotic stress conditions.
5. Guava species, *P. guineense* and *P. cattleianum* var. lucidum, are more tolerant to freezing temperatures (–5°C). *P. cujavillis* is tolerant to drought and sodic soils while Chinese guava (*P. friedrichsthalianum*) could be used as dwarfing rootstock. Among *P. guajava* cultivars, Ruby × Supreme is more cold tolerant compared with 'Lucknow-49'.
6. Guava cv. 'Paluma' is more tolerant to O_3 and other industrial pollutants compared with tobacco and could be used as an effective biomonitoring plant in sub-topical and tropical regions. Wild guava is a good accumulator of N, S, Fl and other heavy metals and could be used for forest plantations. Higher proline accumulation in guava compared with other tree crops in response to industrial air pollutants indicates its inherent potential for abiotic stress tolerance. Guava (*P. guajava* L.) can be a recommended fruit crop in the industrial areas.
7. Plantation of guava in industrial areas as commercial orchards or forests could help in mitigation of air pollution. The plantation of 'Paluma' and other sensitive guava cultivars in the urban areas of sub-tropical and tropical regions could be helpful as a potential bioindicator to the higher levels of O_3 and other industrial pollutants. Use of cellular markers, including wart-like protrusions in cell wall, accumulation of plastoglobuli and swelling of mitochondrial and thylakoid membranes could be helpful in estimating the plant responses to oxidative pollutant, O_3. Accumulation of phenolics in the mesophyll tissue as an accelerated cell senescence could also be used as marker for oxidative damage in guava (Alves et al., 2016).
8. Long-term productivity of the renewable resources like biodiversity, forests, fresh water and soils must be ensured. Mulching helps in

soil conservation, improvement of soil microclimate and microbial activity. Use of plastic mulch increased fruit yield up to 25% compared to control (Rajatiya et al., 2018). Plantations of perennial horticultural crops which are at risk shall help in mitigation of the climate change by absorbing radiations.

9. Breeding programs for selection and development of drought, heat- and frost-tolerant cultivars shall be initiated. Genotypes having improved traits including Rubisco activity, stomatal density and size, genes of stress related proteins (aquaporins, chaperonins, dehydrins and enzymes regulating ROS detoxification), phytohormones, expression patterns in related genes and proteolysis shall be developed.
10. Integration of omics tools could be quite useful in identification of genes related to the development of phenotype under abiotic stress conditions. Advances in phenotyping and omics have enabled screening of tolerant varieties that can produce better yields with low water availability and under elevated temperatures. However, there is little information available about guava plant phenotypic and genetic responses to these stress factors. Better understanding of the crop ecophysiology, biochemical mechanisms and genetic pathways involved in abiotic stress tolerance and identification of related genes for future breeding applications is utmost requirement (Lawlor, 2013; De Ollas et al., 2019).
11. Identification of key enzymes and proteins, like dehydrins (LEA proteins) and other proteins involved in abiotic stress tolerance, and use of proteomics tools to identify additional players involved in stress tolerance is vital.
12. Antisense genes and RNAi technologies could be used for improvement of shelf life and reduction of post-harvest losses. Application of CRISPR/Cas9 genome editing tool could be used for mutation, activation, repression and epigenomic editing of the genes related to abiotic stress tolerance, including transcription factors like DREB1, WRKY, MAPK, etc.
13. Guava could also be used for intercropping as a biological control against citrus greening.
14. Use of infra-red cameras for thermography (non-invasive analysis of aerial plant parts) for high throughput analysis of abiotic stress tolerant (heat, drought, frost) genotypes is recommended (Moller et al., 2007; Jones et al., 2009).
15. Development of close collaborations of horticulturists, physiologists and geneticists is important for comprehensive stress-tolerance studies and developing climate resilient superior cultivars.

7. Conclusion and Way Forward

Global fruit production has been seriously affected by drastic climatic changes and perennial crops like guava are subjected to such changes throughout their life cycle. Guava has been relatively more tolerant to abiotic stress, including drought, salt, elevated CO_2 and environmental pollutants like O_3. Selection of superior strains, developing hybrids and transgenics which could perform better and produce more under rising CO_2 levels accompanied with episodes of higher temperature and drought stress is desired. Selection and breeding for the development of more climate-resilient varieties and their clonal multiplication could be a primary way out. Extensive studies shall be conducted under contrasting climatic conditions for multiple seasons to develop reliable associations of phenotypic traits to abiotic stress and environmental components. Further, systems biology approaches shall be preferred to study abiotic stress-tolerance traits due to their complex nature involving interaction of multiple groups of genes. Integration of the omics technologies could relate the phenotypic, biochemical, physiological and genetic datasets (gene expression) for the development of association of the relevant candidate genes and proteins for further genetic improvement programs. The outcomes of basic research will be helpful to develop mechanisms for improved farming, develop exogenous applications for better tolerance and develop transgenics for biotic and abiotic stress resistance. Collectively, these strategies will enable to mitigate the rapidly changing climatic conditions and enhance productivity of guava under stressful environment.

References

Abril, G.A., Wannaz, E.D., Mateos, A.C. and Pignata, M.L. (2014). Biomonitoring of airborne particulate matter emitted from a cement plant and comparison with dispersion modeling results. *Atmospheric Environment*, 82: 154–163.

Aleza, P., Froelicher, Y., Schwarz, S., Agusti, M., Hernandez, M., Juarez, J., Luro, F., Morillon, R., Navarro, L. and Ollitrault, P. (2011). Tetraploidization events by chromosome doubling of nucellar cells are frequent in apo mictic citrus and are dependent on genotype and environment. *Annals of Botany*, 108(1): 37–50.

Ali-Dinar, H.M., Ebert, G. and Ludders, P. (1999). Growth, chlorophyll content, photosynthesis and water relations in guava (*Psidium guajava* L) under salinity and different nitrogen supply. *Gartenbauwissenschaft*, 64(2): 84.

Allario, T., Brumos, J., Colmenero-Flores, J., Tadeo, F., Froelicher, Y., Talon, M., Navarro, L., Ollitrault, P. and Morillon, R. (2011). Large changes in anatomy and physiology between diploid Rangpur lime (*Citrus limonia*) and its autotetraploid are not associated with large changes in leaf gene expression. *Journal of Experimental Botany*, 62(8): 2507–2519.

Allario, T., Brumos, J., Colmenero-Flores, J.M., Iglesias, D.J., Pina, J.A., Navarro, L., Talon, M., Ollitrault, P. and Morillon, R. (2012). Tetraploid Rangpur lime rootstock increases drought tolerance via enhanced constitutive root abscisic acid production. *Plant Cell and Environment*, 36(4): 856–868.

Alves, E.S., Moura, B.B., Pedroso, A.N.V., Tresmondi, F. and Machado, S.R. (2016). Cellular markers indicative of ozone stress on bioindicator plants growing in a tropical environment. *Ecological Indicators*, 67: 417–424.

Amthor, J.S. and McCree, K. (1990). Carbon balance of stressed plants: a conceptual model for integrating research results. pp. 1–15. *In*: Alscher, R.G. and Cumming, J.R. (eds.). *Stress Responses in Plants: Adaptation and Acclimation Mechanisms*. Wiley-Liss, New York, USA.

Ang, L.H. and Ng, L.T. (2000). Trace element concentration in Mango (*Mangifera indica* L.), seedless guava (*Psidium guajava* L.) and papaya (*Carica papaya* L.) grown on agricultural and ex-mining lands of Bidor, Perak. *Pertanika Journal of Tropical Agricultural Science*, 23: 15–22.

Arena, C., De Maio, A., De Nicola, F., Santorufo, L., Vitale, L. and Maisto, G. (2014). Assessment of eco-physiological performance of *Quercus ilex* L. leaves in urban area by an integrated approach. *Water, Air and Soil Pollution*, 225–1824.

Arévalo-Marín, E., Casas, A., Landrum, L., Shock, M.P., Alvarado-Sizzo, H., Ruiz-Sanchez, E. and Clement, C.R. (2021). The taming of *Psidium guajava*: natural and cultural history of a neotropical fruit. *Frontiers in Plant Science*, p.2138.

Assis, P.I., Alonso, R., Meirelles, S.T. and Moraes, R.M. (2015). DO_3SE model applicability and O_3 flux performance compared to AOT40 for an O_3-sensitive tropical tree species (*Psidium guajava* L. 'Paluma'). *Environmental Science and Pollution Research*, 22(14): 10873–10881.

Aziz, A., Martin-Tanguy, J. and Larher, F. (1998). Stress-induced changes in polyamine and tyramine levels can regulate proline accumulation in tomato leaf discs treated with sodium chloride. *Plant Physiology*, 104: 195–202.

Beerling, D.J. and Chaloner, W.G. (1993). The impact of atmospheric CO_2 and temperature change on stomatal density: Observations from Quercusrobur Lammad leaves. *Annals of Botany*, 71: 231–235.

Ben-Asher, J., Tsuyuki, I., Bravdo, B.A. and Sagih, M. (2006). Irrigation of grapevines with saline water. I. Leaf area index, stomatal conductance, transpiration and photosynthesis. *Agricultural Water Management*, 83(1-2): 13–21.

Ben Ayed, R., Ben Hassen, H., Ennouri, K., Ben Marzoug, R. and Rebai, A. (2016). *OGDD (Olive Genetic Diversity Database): A microsatellite Markers' Genotypes Database of Worldwide Olive Trees for Cultivar Identification and Virgin Olive Oil Traceability*. Database 2016: bav090. Doi: 10.1093/database/bav090.

Bezerra, I.L., Gheyi, H.R., Nobre, R.G., Barbosa, J.L., de Fátima, R.T., Elias, J.J., de Pádua Souza, L. and de Azevedo, F.L. (2018). Physiological alterations and production of guava under water salinity and nitrogen fertilizer application. *Semina: Ciências Agrárias*, 39(5): 1945–1956.

Bhantana, P. and Lazarovitch, N. (2010). Evapotranspiration, crop coefficient and growth of two young pomegranate (*Punica granatum* L.) varieties under salt stress. *Agricultural Water Management*, 97(5): 715–722.

Bittenbender, H.C. and Kobayashi, K.D. (1990). Predicting the harvest of cycled `Beaumont' guava. *Acta Horticulturae*, 269: 197–204.

Boian, C. and Andrade, M.F. (2012). Characterization of ozone transport among Metropolitan regions. *Revista Brasileira de Meteorologia*, 27(2): 229–242.

Bokhari, S.A.M., Fatima, B., Usman, M. and Rashid, B. (2018). Morpho-physiological responses of guava cultivars to water deficit. p. 45. *In*: Abbasi, N.A., Ahmad, T., Qureshi, A.A. and Hasan, S.Z.U. (eds.). *The Proceedings of the International Horticulture Conference*. 25–27 April 2018, Pir Meher Ali Shah Arid Agriculture University, Rawalpindi, Pakistan.

Bokhari, S.A.M., Usman, M., Jehangir, M., Arif, R., Fatima, B. and Naseer, M. (2019). Potential of mutation breeding in guava crop improvement. pp. 191–201. *In*: Hafiz, I.A., Ahmad, T., Qureshi, A.A. and Hasan S.Z. (eds.). *Proceedings of the International Horticultural Conference*. 25–27 April, 2018, PMAS Arid University, Rawalpindi, Pakistan.

Brandão, S.E., Bulbovas, P., Lima, M.E.L. and Domingos, A. (2017). Biochemical leaf traits as indicators of tolerance potential in tree species from the Brazilian Atlantic Forest against oxidative environmental stressors. *Sci. Total Environ.*, 575: 406–417, 10.1016/j. scitotenv.2016.10.006.

Brochmann, C., Brysting, A., Alsos, I., Borgen, L., Grundt, H., Scheen, A. and Elven, R. (2004). Polyploidy in arctic plants. *Biological Journal of the Linnean Society*, 82: 521–536.

Brozynska, M., Furtado, A. and Henry, R.J. (2016). Genomics of crop wild relatives: Expanding the gene pool for crop improvement. *Plant Biotechnology Journal*, 14(4): 1070–1085.

Bulbovas, P., Camargo, C.Z. and Domingos, M. (2015). Ryegrass cv. Lema and guava cv. Paluma biomonitoring suitability for estimating nutritional contamination risks under seasonal climate in southeastern Brazil. *Ecotoxicology and Environmental Safety*, 118: 149–157.

Cavalcante, Í.H.L., Cavalcante, L.F., Hu, Y. and Beckmann-Cavalcante, M.Z. (2007). Water salinity and initial development of four guava (*Psidium guajava* L.) cultivars in north-eastern Brazil. *Journal of Fruit and Ornamental Plant Research*, 15: 71–80.

Cavalcante, L.F. and Cavalcante, I.H.L. (2006). *Uso da água salina na agricultura*. pp. 1–12. *In*: Cavalcante, L.F. and Lima, E.M. (eds.). *Algumas frutíferas tropicais e a salinidade*. FUNEP, Jaboticabal, Brazil.

CETESB, *Companhia de Tecnologia de Saneamento Ambiental*, São Paulo. (2013). *Relatório de qualidade do ar no estado de São Paulo*, 2012. CETESB, São Paulo.

Chaudhry, Q.Z. (2017). *Climate Change Profile of Pakistan*. Asian Development Bank, Philippines.

Chaves, M.M., Flexas, J. and Pinheiro, C. (2009). Photosynthesis under drought and salt stress: Regulation mechanisms from whole plant to cell. *Annals of Botany*, 103(4): 551–560.

Chen, Z.J. (2010). Molecular mechanisms of polyploidy and hybrid vigor. *Trends in Plant Science*, 15(2): 57–-71.

Chiveu, J., Ubbenjans, U., Kehlenbeck, K., Pawelzik, E. and Naumann, M. (2020). Partitioning of dry matter and minerals in Kenyan common guava under salt stress: Implications for selection of adapted accessions for saline soils. *Forests, Trees and Livelihoods*, 29(2): 99–118.

Cohen, A.J., Brauer, M., Burnett, R., Anderson, H.R., Frostad, J., Estep, K. et al. (2017). Estimates and 25-year trends of the global burden of disease attributable to ambient air pollution: An analysis of data from the global burden of diseases study, 2015. *The Lancet*, 389(10082): 1907–1918.

Costanza, R., de Groot, R., Sutton, P., Ploeg, S., Anderson, S.J., Kubiszewski, I., Farbere, S. and Turner, R.K. (2014). Changes in the global value of ecosystem services. *Global Environmental Change*, 26: 152–158.

Da Silva, E.M., Nobre, R.G., de Pádua Souza, L., Pinheiro, F.W.A., de Lima, G.S., Gheyi, H.R. and de Sá Almeida, L.L. (2017). Physiology of 'Paluma' guava under irrigation with saline water and nitrogen fertilization. *Semina: Ciências Agrárias*, 38(2): 623–634.

De Ollas, C., Morillón, R., Fotopoulos, V., Puértolas, J., Ollitrault, P., Gómez-Cadenas, A. and Arbona, V. (2019). Facing climate change: Biotechnology of iconic Mediterranean woody crops. *Frontiers in Plant Science*, 10.

de Rezende, F.M. and Furlan, C.M. (2009). Anthocyanins and tannins in ozone-fumigated guava trees. *Chemosphere*, 76(10): 1445–1450.

de Rezende, F.M., de Souza, A.P., Buckeridge, M.S. and Furlan, C.M. (2015). Is guava phenolic metabolism influenced by elevated atmospheric CO_2? *Environmental Pollution*, 196: 483–488.

Derwent, R., Collins, W., Johnson, C. and Stevenson, D. (2002). Global ozone concentrations and regional air quality. *Environmental Science and Technology*, 36: 379A–382A.

Desta, Z.A. and Ortiz, R. (2014). Genomic selection: Genome-wide prediction in plant improvement. *Trends in Plant Science*, 19(9): 592–601.

Drake, B.G., Gonzalez-Meler, M.A. and Long, S.P. (1997). More efficient plants: A consequence of rising atmospheric CO_2? *Annual Review of Plant Physiology and Plant Molecular Biology*, 48: 609–639

Ebert, G., Eberle, J., Ali-Dinar, H. and Ludders, P. (2002). Ameliorating effects of Ca $(NO_3)_2$ on growth, mineral uptake and photosynthesis of NaCl-stressed guava seedlings (*Psidium guajava* L.). *Scientia Horticulturae*, 93(2): 125–135.

Fares, A., Bayabil, H.K., Zekri, M., Mattos, D.J. and Awal, R. (2017). Potential climate change impacts on citrus water requirement across major producing areas in the world. *Journal of Water and Climate Change*, 8(4): 576–592.

Farneti, B., Guardo, M., Di Khomenko, I., Cappellin, L., Biasioli, F., Velasco, R. and Costa, F. (2017). Genome-wide association study unravels the genetic control of the apple volatilome and its interplay with fruit texture. *Journal of Experimental Botany*, 68(7): 1467–1478.

Fatima, B., Usman, M., Khan, I.A., Khan, M.S. and Khan, M.M. (2010). Exploring citrus cultivars for underdeveloped and shriveled seeds: A valuable resource for spontaneous polyploidy. *Pakistan Journal of Botany*, 42(1): 189–200.

Fatima, B., Usman, M., Khan, M.S., Khan, I.A. and Khan, M.M. (2015). Identification of citrus polyploids using chromosome counts, morphological and SSR markers. *Pakistan Journal of Agricultural Sciences*, 52(1): 107–114.

Fatima, B., Batool, F., Usman, M. and Bokhari, S.A.M. (2018). Screening of guava cultivars for tolerance under water deficit. pp. 131. *In*: *The Proceedings of the 2nd International Conference on Plant Sciences*. 5–7 December 2018, Government College University, Lahore, Pakistan.

Feng, C., Feng, C., Lin, X., Liu, S., Li, Y. and Kang, M.A. (2021). Chromosome-level genome assembly provides insights into ascorbic acid accumulation and fruit softening in guava (*Psidium guajava*). *Horticulture Research*, 19: 717–730.

Fereres, E. and Soriano, M.A. (2007). Deficit irrigation for reducing agricultural water use. *Journal of Experimental Botany*, 58(2): 147–159.

Fernandez, E., Baird, G., Farías, D., Oyanedel, E., Olaeta, J.A., Brown, P., Zwieniecki, M., Tixier, A. and Saa, S. (2018). Fruit load in almond spurs define starch and total soluble carbohydrate concentration and therefore their survival and bloom probabilities in the next season. *Scientia Horticulturae*, 237: 269–276.

Fitchett, J.M., Grab, S.W., Thompson, D.I. and Roshan, G. (2014). Spatiotemporal variation in phenological response of citrus to climate change in Iran: 1960–2010. *Agricultural and Forest Meteorology*, 198-199: 285–293.

Fukushima, A. and Kusano, M. (2014). A network perspective on nitrogen metabolism from model to crop plants using integrated 'omics' approaches. *Journal of Experimental Botany*, 65(19): 5619–5630.

Furlan, C.M., Moraes, R.M., Bulbovas, P., Domingos, M., Salatino, A. and Sanz, M.J. (2007). *Psidium guajava* 'Paluma', as a new bio-indicator of ozone in the tropics. *Environmental Pollution*, 147(3): 691–695.

Galmes, J., Flexas, J., Save, R. and Medrano, H. (2007). Water relations and stomatal characteristics of Mediterranean plants with different growth forms and leaf habits: Responses to water stress and recovery. *Plant and Soil*, 290(1-2): 139–155.

Gao, F., Calatayud, V., García-Breijo, F., Reig-Armiñana, J. and Feng, Z. (2016). Effects of elevated ozone on physiological, anatomical and ultrastructural characteristics of four common urban tree species in China. *Ecological Indicators*, 67: 367–379.

Ghalati, R.E., Shamili, M. and Homaei, A. (2020). Effect of putrescine on biochemical and physiological characteristics of guava (*Psidium guajava* L.) seedlings under salt stress. *Scientia Horticulturae*, 261: 108961.

Giannini, T.C., Costa, W.F., Cordeiro, G.D., Imperatriz-Fonseca, V.L., Saraiva, A.M., Biesmeijer, J. and Garibaldi, L.A. (2017). Projected climate change threatens pollinators and crop production in Brazil. *PLoS ONE*, 12(8): e0182274.

Gimeno, J., Gadea, J., Forment, J., Pérez-Valle, J., Santiago, J., Martínez-Godoy, M. A., Yenush, L., Bellés, J.M., Brumós, J., Colmenero-Flores, J.M, Talón, M. and Serrano, R. (2009). Shared and novel molecular responses of mandarin to drought. *Plant Molecular Biology*, 70(4): 403–420.

Gordo, O. and Sanz, J.J. (2010). Impact of climate change on plant phenology in Mediterranean ecosystems. *Global Change Biology*, 16: 1082–1106.

Gould, K.S. (2004). Nature's Swiss army knife: The diverse protective roles of anthocyanins in leaves. *Journal of Biomedicine and Biotechnology*, 5: 314–320.

Greer, B.T., Still, C., Cullinan, G.L., Brooks, J.R. and Meinzer, F.C. (2018). Polyploidy influences plant-environment interactions in quaking aspen (*Populus tremuloides* Michx.). *Tree Physiology*, 38(4): 630–640.

Hameed, M. and Ashraf, M. (2008). Physiological and biochemical adaptations of Cynodon dactylon (L.) Pers. from the Salt Range (Pakistan) to salinity stress. *Flora-Morphology, Distribution, Functional Ecology of Plants*, 203(8): 683–694.

Hao, W., Arora, R., Yadav, A.K. and Joshee, N. (2009). Freezing tolerance and cold acclimation in guava (*Psidium guajava* L.). *HortScience*, 44(5): 1258–1266.

Hassan, S.M.M., Lata, P., Singh, S.K. and Yadav, D.K. (2018). Seasonal manifestations on the productivity of guava (*Psidium guajava* L.) at Tekari Orchard, Gaya, Bihar. *Journal of Global Biosciences*, 7(7): 5504–5512.

Hazarika, T.K. (2015). Climate change and indian horticulture: opportunities for adaptation and mitigation strategies. pp. 227–232. *In*: Sati, V.P. (ed.). *Climate Change and Socio-Ecological Transformation*. TTP Publishers.

Heath, R.L., Lefohn, A.S. and Musselman, R.C. (2009). Temporal processes that contribute to nonlinearity in vegetation responses to ozone exposure and dose. *Atmospheric Environment*, 43(18): 2919–2928.

Hussain, S., Hassan, A., Arshad, P. and Anjum, M.A. (2021). *Different Sources of Irrigation Water Affect Heavy Metals Accumulation in Soils and Subsequently on Physiological Determinants and Physico-chemical Properties of Guava Fruits*. https://doi.org/10.21203/rs.3.rs-498294/v1.

Idso, S.B. (1998). Three phases of plant response to atmospheric CO_2 enrichment. *Plant Physiology*, 87(1): 5–7.

IGWT(a). 2008. Population and climate change. www.globalissues.org. Accessed 17 April 2008.

IGWT(b). 2008. The axis of evil shortages: Food, water, fuel. www,livescience.com. 28 March 2008.

IGWT(c). 2008. Global warming, climate, air pollution. www.population-awareness.net. 26-1-2008.

IGWT(d). 2008. Warming affects trees, streams in West. News.yahoo.com. 29 March 2008.

IGWT(e). 2007. Rising temperatures threaten tree cover. thenews.com.pk.

IGWT(f). 2007. Global warming and Pakistan. www.dawn.com. 26 March 2007.

IGWT(g). 2007. Global warming effects on trees studied. 2007. Earthtimes.com. 25 September 2007.

IGWT(h). 2007. Sierra trees dying faster as climate warms. Msnbc.com. 2007.

IGWT(i). 2007. Global warming to stunt growth of forest trees. News.mongabay.com. 12 August 2007.

IGWT(j). 2007. World's largest field experiment examines long term effects of global change on forest, dynamics. 2007. Lacking water, trees turn to carbon dioxide. www.haaretz.com 2008.

IGWT(k). 2007. Mortality of plants could increase by 40% if land temperature increases 4°C. 2007.

IGWT(l). 2007. At the poles, melting occurring at alarming rate, Washington Post Oct 2007.

IGWT(m). 2005. Growing more forests in USA could contribute to global warming. LLNL.2005.

IGWT(n). 2005. Can we burry global warming? Scientific America. 2005.
IGWT(o). 2004. Cloud forest plants may not survive climate change. 2004.
IGWT(p). 2000. Impact of a climate warming scenario on tree growth. Sciencedirect.com. 27 May 2000.
IGWT(q). 2000. Potential impacts of climatic change on natural forests and forestry in Brazilian Amazonia. ScienceDirect com.20 April 2000.
IGWT(r). 2000. Quiver tree. www. global-greenhouse-warming. Com.
Jo, S., Kim, H.W., Kim, Y.K., Cheon, S.H. and Kim, K.J. (2016). Complete plastome sequence of *Psidium guajava* L. (Myrtaceae). *Mitochondrial DNA Part B*, 1(1): 612–614.
Jones, H.G., Serraj, R., Loveys, B.R., Xiong, L., Wheaton, A. and Price, A.H. (2009). Thermal infrared imaging of crop canopies for the remote diagnosis and quantification of plant responses to water stress in the field. *Functional Plant Biology*, 36(11): 978–989.
Kanemura, T., Homma, K., Ohsumi, A., Shiraiwa, T. and Horie, T. (2007). Evaluation of genotypic variation in leaf photosynthetic rate and its associated factors by using rice diversity research set of germplasm. *Photosynthesis Research*, 94(1): 23–30.
Kangasjärvi, J., Jaspers, P. and Kollist, H. (2005). Signalling and cell death in ozone exposed plants. *Plant Cell and Environment*, 28(8): 1021–1036.
Kateivas, K.S.B., Cairo, P.A.R., Neves, P.H.S., Ribeiro, R.S.S., Leitão, C.A.E. and Machado, L.M. (2018). Morphophysiological changes in *Psidium cattleianum* caused by air emissions from a mining industry in Brumado, Bahia, Brazil. *Revista Brasileira de Ciências Agrárias*, 13(4): 1–8.
Keeling, C.D., Whorf, T.P., Wahlen, M. and Van der Plicht, J. (1995). Interannual extremes in the rate of rise of atmospheric carbon dioxide since 1980. *Nature*, 375: 660–670.
Khalid, S., Shahid, M., Niazi, N.K., Murtaza, B., Bibi, I. and Dumat, C. (2017). A comparison of technologies for remediation of heavy metal contaminated soils. *Journal of Geochemical Exploration*, 182: 247–268.
Khan, G.S. (1998). Soil salinity and sodicity status in Pakistan, *Soil Survey of Pakistan*, Lahore, Pakistan, 12: 59–60.
Khan, N., Abas, N. and Mariun, N. (2008). Impact of global warming on trees. pp. 1–12. *In*: *The Proceedings of the Global Conference on Global Warming*. 6–10 July 2008, Istanbul, Turkey.
Kimball, B.A. (2016). Crop responses to elevated CO_2 and interactions with H_2O, N, and temperature. *Current Opinion in Plant Biology*, 31: 36–43.
Kizildeniz, T., Pascual, I., Irigoyen, J.J. and Morales, F. (2018). Using fruit-bearing cuttings of grapevine and temperature gradient greenhouses to evaluate effects of climate change (elevated CO_2 and temperature, and water deficit) on the cv. red and white Tempranillo. Yield and must quality in three consecutive growing seasons (2013–2015). *Agricultural Water Management*, 202: 299–310.
Klein, A.-M., Vaissiere, B.E., Cane, J.H., Steffan-Dewenter, I., Cunningham, S.A., Kremen, C. and Tscharntke, T. (2007). Importance of pollinators in changing landscapes for world crops. *Proceedings of the Royal Society of London*, 274: 303–313.
Korner, C. (2006). Plant CO_2 responses: an issue of definition, time and resource supply. *New Phytologist*, 172(3): 393–411.
Korres, N.E., Norsworthy, J.K., Tehranchian, P., Gitsopoulos, T.K., Loka, D.A., Oosterhuis, D.M., Gealy, D.R., Moss, S.R., Burgos, N.R., Miller, M.R. and Palhano, M. (2016). Cultivars to face climate change effects on crops and weeds: A review. *Agronomy for Sustainable Development*, 36(1): 12.
Krupa, S., McGrath, M.T., Anderson, A.H., Chevone, B.I., Pell, E.L. and Zilinskas, B.A. (2001). Ambient ozone and plant health. *Plant Disease*, 85(1): 4–12.
Kudo, G. and Ida, T. (2013). Early onset of spring increases the phenological mismatch between plants and pollinators. *Ecology*, 94(10): 2311-–2320.
Kumar, L.S.S. and Ranade, S.G. (1952). Autotriploidy in guava (*Psidium guajava*, Linn.). *Current Science*, 21: 75–76.

Kumar, N., Pamidimarri, S., Kaur, M., Boricha, G. and Reddy, M. (2008). Effects of NaCl on growth, ion accumulation, protein, proline contents and antioxidant enzymes activity in callus cultures of *Jatropha curcas*. *Biologia*, 63(3): 378–382.

Lawlor, D.W. (2013). Genetic engineering to improve plant performance under drought: Physiological evaluation of achievements, limitations, and possibilities. *Journal of Experimental Botany*, 64(1): 83–108.

Le Bourvellec, C., Bureau, S., Renard, C.M.G.C., Plenet, D., Gautier, H., Touloumet, L., Girard, T. and Simon, S. (2015). Cultivar and year rather than agricultural practices affect primary and secondary metabolites in apple fruit. *PLoS ONE*, 10(11): e0141916. Doi:10.1371/journal.pone.0141916.

Lin, T.W., Pan, M.W., Lee, Y.C., Hsieh, H.Y., Chu, Y.C. and Chang, J.C. (2015). Assessing the physiological responses and shoot growth of potted 'Jen-Ju-Ba' guava (*Psidium guajava* L.) plants to drought. *Acta Horticulturae* 1166: 173–182.

Lobell, D.B. and Gourdji, S.M. (2012). The influence of climate change on global crop productivity. *Plant Physiology*, 160: 1686–1697.

Loggini, B., Scartazza, A., Brugonli, E. and Navari-Izzo, F. (1999). Antioxidative defense system, pigment composition, and photosynthetic efficiency in two wheat cultivars subjected to drought. *Plant Physiology*, 119: 1091–1099.

Lopes, M.S., El-Basyoni, I., Baenziger, P.S., Singh, S., Royo, C., Ozbek, K., Aktas, H., Ozer, E., Ozdemir, F., Manickavelu, A., Ban, T. and Vikram P. (2015). Exploiting genetic diversity from landraces in wheat breeding for adaptation to climate change. *Journal of Experimental Botany*, 66(12): 3477–3486.

Mailoa, M.N., Mahendradatta, M., Laga, A. and Djide, N. (2013). Tannin extract of guava leaves (*Psidium guajava* L) variation with concentration organic solvents. *International Journal of Scientific and Technology Research*, 2(9): 106–110.

Malhotra, S.K. (2017). Horticultural crops and climate change: A review. *Indian Journal of Agricultural Sciences*, 87(1): 12–22.

Manes, F., Marando, F., Capotorti, G., Blasi, C., Salvatori, E., Fusaro, L., Ciancarella, L., Mircea, M., Marchetti, M., Chirici, G. and Munafò, M. (2016). Regulating ecosystem services of forests in ten Italian metropolitan cities: Air quality improvement by PM10 and ozone removal. *Ecological Indicators*, 67: 425–440.

Mani, A. and Suresh, C.P. (2018). Climate Resilient Fruit Crops—Possible Solution to Ensure Nutritional Security in Changing Climate Scenario, *Training Manual*, pp. 51.

Mathizhagan, M., Chidambara, B., Hunashikatti, L.R. and Ravishankar, K.V. (2021). Genomic approaches for improvement of tropical fruits: fruit quality, shelf life and nutrient content. *Genes*, 12(12): 1881.

Medina, J.C. (1988). Goiaba I - Cultura. *In:* ITAL (Campinas, SP). *Goiaba: Cultura, matéria-prima, processamento e aspectos econômicos. Rev Ampl Campinas*, 2: 1–120.

Medlyn, B.E., Duursma, R.A., De Kauwe, M.G. and Prentice, I.C. (2013). The optimal stomatal response to atmospheric CO_2 concentration: Alternative solutions, alternative interpretations. *Agricultural and Forest Meteorology*, 182-183: 200–203.

Menzel, C.M. and Paxton, B.F. (1986). The pattern of growth, flowering and fruiting of guava varieties in subtropical Queensland. *Australian Journal of Experimental Agriculture*, 26(1): 123–128.

Minamikawa, M.F., Nonaka, K., Kaminuma, E., Kajiya-Kanegae, H., Onogi, A., Goto, S., Yoshioka, T., Imai, A., Hamada, H., Hayashi, T., Matsumoto, S., Katayose, Y., Toyoda, A., Fujiyama, A., Nakamura, Y., Shimizu, T. and Iwata, H. (2017). Genome-wide association study and genomic prediction in citrus: potential of genomics-assisted breeding for fruit quality traits. *Scientific Reports*, 7: 4721.

Mishra, D., Shukla, S.K., Ravishankar, H. and Adak, T. (2014). Impact of weather on phenology of guava in Uttar Pradesh: A cursory analysis. *Agricultural Sciences*, 6(1): 74–75.

Moller, M., Alcanatis, V., Cohen, Y., Meron, M., Tsipris, J., Naor, A., Ostrovsky, V., Sprintsin, M. and Cohen, S. (2007). Use of thermal and visible imagery for estimating crop water status of irrigated grapevine. *Journal of Experimental Botany*, 58(4): 827–838.

Montero-Pau, J., Blanca, J., Esteras, C., Martínez-Pérez, E.M., Gómez, P., Monforte, A.J., Cañizares, J. and Picó, B. (2017). An SNP-based saturated genetic map and QTL analysis of fruit-related traits in zucchini using genotyping-by-sequencing. *BMC Genomics*, 18(1): 94.

Moore, B.D., Cheng, S.H., Sims, D. and Seemann, J.R. (1999). The biochemical and molecular basis for photosynthetic acclimation to elevated atmospheric CO_2. *Plant, Cell and Environment*, 22(6): 567–582.

Moraes, R.M., Klumpp, A., Furlan, C.M., Klumpp, G., Domingos, M., Rinaldi, M.C.S. and Modesto, I.F. (2002). Tropical fruit trees as bioindicators of industrial air pollution in southeast Brazil. *Environment International*, 28(5): 367–74.

Moraes, R.M., Furlan, C.M., Bulbovas, P., Domingos, M., Meirelles, S.T., Salatino, A., Delitti, W.B.C. and Sanz, M.J. (2004). Photosynthetic responses of tropical trees to short-term exposure to ozone. *Photosynthetica*, 42: 291–293.

Moretto, M., Sonego, P., Pilati, S., Malacarne, G., Costantini, L., Grzeskowiak, L., Giorgia, B., Stella, G.M., Claudio, M. and Kristof, E. (2016). VESPUCCI: exploring patterns of gene expression in grapevine. *Frontiers in Plant Science*, 7: 633. Doi: 10.3389/fpls.2016.00633.

Morton, E.M. and Rafferty, N.E. (2017). Plant-pollinator interactions under climate change: The use of spatial and temporal transplants. *Applications in Plant Science*, 5(6): 1600133.

Moya, J.L., Gómez-Cadenas, A., Primo-Millo, E. and Talon, M. (2003). Chloride absorption in salt-sensitive Carrizo citrange and salt-tolerant Cleopatra mandarin citrus rootstocks is linked to water use. *Journal of Experimental Botany*, 54(383): 825–833.

Mukherjee, A. and Agrawal, M. (2018). Use of GLM approach to assess the responses of tropical trees to urban air pollution in relation to leaf functional traits and tree characteristics. *Ecotoxicology and Environmental Safety*, 152: 42–54.

Murashige, T. and Skoog, F. (1962). A revised medium for rapid growth and bioassays with tobacco tissue cultures. *Physiologia Plantarum*, 15(3): 473–497.

Murkute, A.A., Satyawati, S.H. and Singh, S.K. (2010). Biochemical alterations in foliar tissues of citrus genotypes screened in vitro for salinity tolerance. *Journal of Plant Biochemistry and Biotechnology*, 19(2): 203–208.

Nakazato, R.K. (2014). *Caracterização de riscos à Floresta Atlântica associados à contaminação atmosférica por elementos tóxicos, no entorno de uma refinaria de petróleo, em Cubatão/São Paulo, com plantas acumuladoras*. M.S. thesis, *Instituto de Botânica da Secretaria de Estado do Meio Ambiente*.

Neil, S.O. and Gould, K.S. (2003). Anthocyanins in leaves: Light attenuators or antioxidants? *Functional Plant Biology*, 30(8): 865–873.

Normand, F. and Habib, R. (2001). Phenology of strawberry guava (*Psidium cattleianum*) in Reunion Island. *The Journal of Horticultural Science and Biotechnology*, 76(5): 540–545.

Oksanen, E.J. (2001). Increasing tropospheric ozone level reduced birch (*Betula pendula*) dry mass within a five-year period. *Water, Air and Soil Pollution*, 30: 947–952.

Oliveira, T.M., Yahmed, J.B., Dutra, J., Maserti, B.E., Talon, M., Navarro, L., Ollitraut, P., Gesteira, A.D.S. and Morillon, R. (2017). Better tolerance to water deficit in doubled diploid 'Carrizo citrange' compared to diploid seedlings is associated with more limited water consumption. *Acta Physiologiae Plantarum*, 39: 204.

Ollitrault, P. and Navarro, L. (2012). Citrus. pp. 623–662. *In*: Badenes, M. (ed.). *Fruit Breeding*. Springer, London, UK.

Orwa, C., Mutua, A., Kindt, R., Jamnadass, R. and Simons, A. (2010). *Agroforestry Database: A Tree Reference and Selection Guide Version 4.0*. World Agroforestry Centre, Nairobi.

Ozores-hampton, M., Stansly, P.A. and Obreza, T.A. (2005). Heavy metal accumulation in a sandy soil and in pepper fruit following long-term application of organic amendments. *Compost Science and Utilization*, 13: 60–64.

Pachauri, R.K., Allen, M.R., Barros, V.R., Broome, J., Cramer, W., Christ, R., Church, J.A., Clarke, L., Dahe, Q., Dasgupta, P. and Dubash, N.K. (2014). Climate change, 2014: Synthesis report, *Contribution of Working Groups I, II and III to the Fifth Assessment Report of the Intergovernmental Panel on Climate Change, IPCC*, 151 pp.

Padmaker, B., Kanupriya, C., Latha, P.M., Vasugi, C., Dinesh, M.R., Sailaja, D. and Aswath, C. (2016). Enrichment of genetic linkage maps and mapping qts. specific to seed strength—hardness/softness—in guava (*Psidium guajava* L.). *Scientia Horticulturae*, 11: 13–20.

Padoan, D., Mossad, A., Chiancone, B., GermanÃ, M.A. and Khan, P.S.S.V. (2013). Ploidy levels in citrus clementine affects leaf morphology, stomatal density and water content. *Theoretical and Experimental Plant Physiology*, 25: 283–290.

Pereira, F.M. and Nachtigal, J.C. (2002). Goiabeira. pp. 267–289. *In*: Bruckner, C.H. (ed.). *Melhoramento de Fruteiras Tropicais*. Vicosa, UFV.

Pereira, F.M.P., Usman, M., Mayer, N.A., Nachtigal, J.C., Maphanga, O.R. and Willemse, S. (2017). Advances in guava propagation. *Revista Brasileira de Fruticultura*, 39(4): e-358.

Pereira, P.A.P. and de Souza, V.R. (2019). Influence of gelling agent concentration on the characteristics of functional sugar-free guava preserves. *Emirates Journal of Food and Agriculture*, 501–510.

Perry, C.T., Divan Jr, A.M., Rodriguez, M.T.R. and Atz, V.L. (2010). *Psidium guajava* as a bioaccumulator of nickel around an oil refinery, southern Brazil. *Ecotoxicology and Environmental Safety*, 73(4): 647–654.

Pina, J.M. and Moraes, R.M. (2007). Ozone-induced foliar injury in saplings of *Psidium guajava* 'Paluma' in São Paulo, Brazil. *Chemosphere*, 66(7): 1310–1314.

Pina, J.M. and Moraes, R.M. (2010). Gas exchange, antioxidants and foliar injuries in saplings of a tropical woody species exposed to ozone. *Ecotoxicology and Environmental Safety*, 73(4): 685–691.

Pommer, C.V. and Murakami, K.R. (2009). Breeding guava (*Psidium guajava* L.). pp. 83–120. *In*: *Breeding Plantation Tree Crops: Tropical Species*. Springer, New York, USA.

Ragupathy, R., Ravichandran, S., Mahdi, S.R. and Huang, D. (2016). Deep sequencing of wheat sRNA transcriptome reveals distinct temporal expression pattern of miRNAs in response to heat, light and UV. *Scientific Reports*, 6: 39373.

Rai, M.K., Jaiswal, V.S. and Jaiswal, U. (2010). Regeneration of plantlets of guava (*Psidium guajava* L.) from somatic embryos developed under salt-stress condition. *Acta Physiologiae Plantarum*, 32(6): 1055–1062.

Rajatiya, J., Varu, D.K., Gohil, P., Solanki, M., Halepotara, F., Gohil, M., Mishra, P. and Solanki, R. (2018). Climate change: impact, mitigation and adaptation in fruit crops. *International Journal of Pure and Applied Biosciences*, 6(1): 1161–1169.

Rasul, G., Mahmood, A., Sadiq, A. and Khan, S. (2012). Vulnerability of the Indus delta to climate change in Pakistan. *Pakistan Journal of Meteorology*, 8: 89–107.

Reatini, B., Torres, M.L., Valdebenito, H. and Vision, T. (2018). Complete plastome sequences of two Psidium species from the Galápagos Islands. *F1000 Research*, 7: 1361.

Reddy, A.G.K., Kumar, J.S., Maruthi, V., Venkatasubbaiah, K. and Rao, C.S. (2017). Fruit production under climate changing scenario in India: A review. *Environment and Ecology*, 35(2B): 1010–1017.

Reddy, B.M.C. (2012). Physiological basis of growth and fruit yield characteristics of tropical and sub-tropical fruits to temperature. *Tropical Fruit Tree Species and Climate Change*, 45.

Rehman, M.U., Rather, G.H., Gull, Y., Mir, M.R., Mir, M.M., Waida, U.I. and Hakeem, K.R. (2015). Effect of climate change on horticultural crops. *In*: Hakeem, K. (ed.). *Crop Production and Global Environmental Issues*. Springer, Cham.

Rejab, M., Teck, C.S., Zain, K.M. and Muhamad, M. (2008). Mangosteen. pp. 155–174. *In*: Kwok, C.Y., Lian, T.S. and Jamaluddin, S.H. (eds.). *Breeding Horticultural Crops*. MARDI, Malaysia.

Rhee, S.Y. and Mutwil, M. (2014). Towards revealing the functions of all genes in plants. *Trends in Plant Science*, 19(4): 212–221.

Rojas-Gómez, M., Jiménez-Madrigal, J.P., Montero-Vargas, M., Loaiza-Montoya, R., Chavarría, M., Meneses, E. and Fuchs, E.J. (2021). A draft genome assembly of 'Cas' (*Psidium friedrichsthalianum* (O. Berg) Nied.): An indigenous crop of Costa Rica untapped. *Genetic Resources and Crop Evolution*, 1–9.

Ruiz, M., Quiñones, A., Martínez-Alcántara, B., Aleza, P., Morillon, R., Navarro, L., Primo-Millo, E. and Martínez-Cuenca, M.R. (2016). Effects of salinity on diploid (2x) and doubled diploid (4x) citrus macrophylla genotypes. *Scientia Horticulturae*, 207: 33–40.

Sá, F.V.S., Brito, M.E.B., Melo, A.S., Antônio Neto, P., Fernandes, P.D. and Ferreira, I.B. (2013). *Produção de mudas de mamoeiro irrigadas com água salina. Revista Brasileira Engenharia Agrícola Ambiental*, 17(10): 1047–1054.

Sandre, A.A., Pina, J.M., Moraes, R.M. and Furlan, C.M. (2014). Anthocyanins and tannins: is the urban air pollution an elicitor factor? *Brazilian Journal of Botany*, 37(1): 9–18.

Sawidis, T., Breusteb, J., Mitrovic, M., Pavlovic, P. and Tsigaridas, K. (2011). Trees as bioindicator of heavy metal pollution in three European cities. *Environmental Pollution*, 159(12): 3560–3570.

Scheben, A., Yuan, Y. and Edwards, D. (2016). Advances in genomics for adapting crops to climate change. *Current Plant Biology*, 6: 2–10.

Shaukat, F., Balal, R.M., Shahid, M.A., Akhtar, G., Akram, A., Khan, M.W. and Zubair, M. (2015). Identification of drought induced drastic effects on Guava (*Psidium guajava*) at seedling stage. *International Journal of Chemical and Biochemical Sciences*, 8: 15–27.

Shiva, B., Nagaraja, A., Srivastava, M. and Goswami, A.K. (2017). Determination of correlation between stomatal density and gas exchange traits in guava. *International Journal of Current Microbiology and Applied Sciences*, 6(9): 1328–1334.

Shrestha, M., Garcia, J.E., Bukovac, Z., Dorin, A. and Dyer, A.G. (2018). Pollination in a new climate: assessing the potential influence of flower temperature variation on insect pollinator behavior. *PLoS ONE*, 13(8): e0200549. Doi: 10.1371/journal.pone.0200549.

Silva, A.B.F.da, Fernandes, P.D., Gheyi, H.R. and Blanco, F.F. (2008). Growth and yield of guava irrigated with saline water and addition of farmyard manure. *Revista Brasileira de Ciências Agrárias*, 3(4): 354–359.

Simelton, E., Dam, B.V. and Catacutan, D. (2015). Trees and agroforestry for coping with extreme weather events: experiences from northern and central Vietnam. *Agroforestry Systems*, 89(6): 1065–1082.

Singh, A., Kumar, A., Yadav, R.K., Dutta, A. and Sharma, D.K. (2016). Growth and Mineral Nutrition in Salt Stressed Guava (*Psidium guajava* L.) cv. Allahabad Safeda. *Journal of AgriSearch*, 3(1): 21–25.

Singh, A., Kumar, A.S.H.W.A.N.I., Datta, A.S.H.I.M. and Yadav, R.K. (2018). Evaluation of guava (*Psidium guajava*) and bael (*Aegle marmelos*) under shallow saline water-table conditions. *Indian Journal of Agricultural Sciences*, 88(5): 720–5.

Singh, H.P., Shukla, S. and Malhotra, S.K. (2009). Ensuring quality planting material in horticulture crops. pp 469–84, *In*: *9th Agricultural Science Congress*. 22–24 June 2009, Srinagar, India.

Singh, S.K., Malhotra, S.K., Bhargava, R., Singh, R.S. and Shukla A.K. (2017). Morphological and physiological characterization of guava (*Psidium guajava*) under hot-arid zone of Rajasthan. *Indian Journal of Agricultural Sciences*, 87(4): 491–5.

Soares, A.M.S. and Machado, O.L.T. (2007). *Defesa de plantas: Sinalização química e espécies reativas de oxigênio. Revista Trópica*, 1(1): 9–19.

Subramanyam, M.D., Dinesh, M.R. and Braganza, M. (1992). Varietal evaluation and floral biology studies in the genus Psidium. *Acta Horticulturae*, 321: 211–219.

Subramanyam, M.D. and Iyer, C.P.A. (1993). Improvement of guava. pp. 337–347. *In*: Chadha, K.L. and Pareek, O.P. (eds.). *Advances in Horticulture: Fruit Crops*. vol. 1, Malhotra Publishing House, New Delhi, India.

Syvertsen, J. and Levy, Y. (2005). Salinity interactions with other abiotic and biotic stresses in citrus. *HortTechnology*, 15(1): 100–103.

Szabados, L. and Savouré, A. (2010). Proline: A multifunctional amino acid. *Trends in Plant Science*, 15(2): 89–97.

Szabo, A., Domingos, M., Rinaldi, M.C.S. and Delitti, W.B.C. (2003). *Acumulo foliarde enxofre e suas relacoes com alteracoes no crescimento de plantas jovens de Tibouchina pulchra Cogn. (Melastomataceae) expostas nas proximidades do po´ lo industrial Cubatao*, SP. *Revista Brasileira De Botanica*, 26(3): 379–390.

Tan, F.-Q., Tu, H., Liang, W.-J., Long, J.-M., Wu, X.-M., Zhang, H.-Y. and Guo, W.W. (2015). Comparative metabolic and transcriptional analysis of a doubled diploid and its diploid citrus rootstock (*C. junos* cv. *Ziyang xiangcheng*) suggests its potential value for stress resistance improvement. *BMC Plant Biology*, 15: 89.

Tausz, M., Tausz-Posch, S., Norton, R.M., Fitzgerald, G.J., Nicolas, M.E. and Seneweera, S. (2013). Understanding crop physiology to select breeding targets and improve crop management under increasing atmospheric CO_2 concentrations. *Environmental and Experimental Botany*, 88: 71–80.

Thaipong, K. and Boonprakob, U. (2019). Salt tolerance evaluation in guava germplasm. *International Journal of Agricultural Technology*, 15(5): 791–796.

Tresmondi, F. and Alves, E.S. (2011). Structural changes in *Psidium guajava* 'Paluma' leaves exposed to tropospheric ozone. *Acta Botanica Brasilica*, 25(3): 542–548.

Ulmer, J.M., Wolf, K.L., Backman, D.R., Tretheway, R.L., Blain, C.J., O'Neil-Dunne, J.P. and Frank, L.D. (2016). Multiple health benefits of urban tree canopy: the mounting evidence for a green prescription. *Health and Place*, 42: 54–62.

Usman, M., Saeed, T., Khan, M.M. and Fatima, B. (2006). Occurrence of spontaneous polyploids in citrus. *Horticulture Science*, 33(3): 124–129.

Usman, M., Fatima, B., Gillani, K.A., Khan, M.S. and Khan, M.M. (2008). Exploitation of potential target tissues to develop polyploids in citrus. *Pakistan Journal of Botany*, 40(4): 1755–1766.

Usman, M., Fatima, B., Usman, M., Samad, W.A. and Bakhsh, K. (2012). Embryo culture to enhance efficiency of colchicine induced polyploidization in grapefruit. *Pakistan Journal of Botany*, 44: 399–405.

Usman, M., Amin, T., Rashid, H., Anwar ul Haq, H., Sabir, M. and Fatima, B. (2015). Growth response of guava (cv. White Fleshed Round) under salt stress. pp. 130. *In: The Proceedings of the International Conference on Soil Sustainability for Food Security*. 15–17 November 2015, Institute of Soil and Environmental Sciences, University of Agriculture, Faisalabad, Pakistan.

Usman, M., Zaman, Q., Fatima, B., Rana, I.A. and Awan, F.S. (2020). Morpho-chemical diversity and RAPD fingerprinting in white flesh guava cultivars. *Journal of Animal and Plant Sciences*, 30: 398–409.

Usman, M., Shah, M.H., Fatima, B. and Ahmad, Z. (2021). Seasonal and genotypic correlations of guava and implications for breeding. *Pakistan Journal of Agricultural Sciences*, 58(3): 937–944.

Usman, M., Bokhari, S.A.M., Fatima, B., Rashid, B., Nadeem, F., Sarwar, M.B., Nawaz-ul-Rehman, M.S., Shahid, M. and Ayub, C.M. (2022). Drought stress mitigating morphological, physiological, biochemical, and molecular responses of Guava (*Psidium guajava* L.) cultivars. *Frontiers in Plant Science*, 13: 878616. doi: 10.3389/fpls.

Utsunomiya, N. and Shigenaga, S. (1988). Effect of NaCl on the growth and salt accumulation in young plants of giant granadilla (*Passiflora quadrangularis* L.) and yellow passion fruit (*Passiflora edulis* Sims forma flavicarpa Degener). *Japanese Journal of Tropical Agriculture*, 32(3): 147–153.

Vargas, K.C., Abot, A.R., Acosta, V.R., Telles, M.H.F., do Nascimento, J.N. and Garcia, F.R.M. (2019). Does the surrounding vegetation influence the fruit fly assemblage in guava orchards? *Bioscience Journal*, 35(4): 1245–1255.

Verheij, E.W.M. and Coronel, R.E. (1992). *Edible Fruits and Nuts; 2: Plant Resources of South East Asia.* Prosea Foundation, Bogor, Indonesia, 128–131 pp.

Vu, J.C.V., Newman, Y.C., Allen, Jr, L.H.A., Gallo-Meagher, M. and Zhang M.-Q. (2002). Photosynthetic acclimation of young sweet orange trees to elevated growth CO_2 and temperature. *Journal of Plant Physiology*, 159(2): 147–157.

Wani, S.H., Sah, S.K., Sagi, L. and Solymosi, K. (2015). Transplastomic plants for innovations in agriculture: A review. *Agronomy for Sustainable Development*, 35(4): 1391–1430.

Warschefesky, E.J., Klein, L.L., Frank, M.H., Chitwood, D.H., Londo, J.P., von Wettberg, E.J.B. and Miller, A.J. (2016). Rootstocks: diversity, domestication, and impacts on shoot phenotypes. *Trends in Plant Science*, 21(5): 418–437.

Wehner, G.G., Balko, C.C., Enders, M.M., Humbeck, K.K. and Ordon, F.F. (2015). Identification of genomic regions involved in tolerance to drought stress and drought stress-induced leaf senescence in juvenile barley. *BMC Plant Biology*, 15: 125.

WHO, World Health Organization. (2016). *Ambient Air Pollution: A Global Assessment of Exposure and Burden of Disease.* World Health Organization, 121 pp.

Woodward, F.I. (1987). Stomatal numbers are sensitive to increases in CO_2 from pre-industrial levels. *Nature*, 327(6123): 617–618.

Wu, G., Chen, F.J., Ge, F. and Xiao, N.W. (2011). Impacts of elevated CO_2 on expression of plant defensive compounds in Bt-transgenic cotton in response to infestation by cotton bollworm. *Agricultural and Forest Entomology*, 13(1): 77–82.

Xie, R., Zhang, J., Ma, Y., Pan, X., Dong, C., Pang, S., Shaolan, H., Deng, L., Yi, S., Zheng, Y. and Lv, Q. (2017). Combined analysis of mRNA and miRNA identifies dehydration and salinity responsive key molecular players in citrus roots. *Scientific Reports*, 7: 42094.

Yang, L., Han, M., Zhou, G. and Li, J. (2007). The changes in water-use efficiency and stoma density of *Leymus chinensis* along northeast China transect. *Acta Ecologica Sinica*, 27(1): 16–23.

Yousaf, M., Zahoor, A.F., Ali, K.G., Anjum, A., Shahzadi, H., Mubeen, Z., Awais, M. and Purveen, S. (2013). Qualitative determination of allelochemicals in *Psidium guajava* L. leaves for their effect against wheat and canary grass. *Pakistan Journal of Weed Science Research*, 19(3): 239–256.

Youssef, M., Mohamed, E.A., El-Sayed, E.N., and Abouzaid, E. (2016). Establishment of an *in vitro* evaluation method for drought tolerance in guava. *Research Journal of Applied Biotechnology (RJAB)*, 5(1): 32–38.

Zandalinas, S.I., Mittler, R., Balfagon, D., Arbona, V. and Gomez-Cadenas, A. (2018). Plant adaptations to the combination of drought and high temperatures. *Physiologia Plantarum*, 162(1): 2–12.

Zhang, X.N., Li, X. and Liu, J.H. (2014). Identification of Conserved and Novel Cold-Responsive microRNAs in trifoliate orange (*Poncirus trifoliata* (L.) Raf.) using high-throughput sequencing. *Plant Molecular Biology Reporter*, 32(2): 328–341.

Zhang, X.Y., Wang, H.M., Hou, Z.D. and Wang, G.X. (2003).Stomatal density and distributions of spring wheat leaves under different planting densities and soil moisture levels. *Chinese Journal of Plant Ecology*, 27(1): 133–136.

Zhao, S., Chen, W., Ma, D. and Zhao, F. (2006). Influence of different salt level on stomatal character in rice leaves. *Reclaiming and Rice Cultivation*, 6: 26–29.

Zuazo, V.H.D., Martínez-Raya, A., Ruiz, J.A. and Tarifa, D.F. (2004). Impact of salinity on macro- and micronutrient uptake in mango (*Mangifera indica* L. cv. Osten). *Spanish Journal of Agricultural Research*, 2(1): 121–133.

10

Development of the Jujube Industry under Changing Climates

Xingang Li,[1,*] *Zhong Zhang*[1] and *Xiaopeng Li*[2]

1. Introduction

Jujube, *Ziziphus jujuba* Mill., also known as the Chinese jujube or Chinese date, is an ancient fruit tree native to China, where it has been cultivated for more than 3,000 years. Jujube was domesticated from its wild relative, the acid jujube (*Z. jujuba* var. *spinosa* Hu.) or the wild jujube (Qu and Wang, 1993; Chen and Schirarend, 2007; Li, 2015; Huang et al., 2016). Based on their different usages, jujube varieties are traditionally prepared dry, fresh, or candied. The developmental phases of jujube fruit can generally be divided into enlarging, white mature, crisp mature and fully mature stages, based on the number of days after flowering and peel color. Fruits of dry varieties are harvested during the fully mature stage and processed into dried fruit (red dates). Varieties that are served fresh are usually harvested during the crisp-mature stage for direct sale in the market or for short-term storage. Finally, varieties for candying are harvested and processed at the white-mature stage. In addition, some varieties can be processed into a candy-like product, known purple-crystal

[1] College of Forestry, Research Centre for Jujube Engineering and Technology of State Forestry and Grassland Administration, Key Comprehensive Laboratory of Forestry of Shaanxi Province, Northwest A & F University, Yangling 712100, China.

[2] College of Agricultural and Forestry Economics and Management, Lanzhou University of Finance and Economics, Lanzhou 730020, China.

* Corresponding author: xingangle@nwsuaf.edu.cn

jujube (Zijingzao in Chinese), if harvested during the crisp-mature stages (Qu and Wang, 1993; Li, 2015).

Traditional jujube-producing regions are distributed along the middle and lower reaches of the Yellow River in northern China which includes Shandong, Hebei, Shanxi, Shaanxi and Henan provinces (Fig. 1). The dominant climatic factors of these regions are listed in Table 1. These are important because both the annual factors and several specific seasonal factors are important during the flowering and fruiting phases and can determine fruit growth and yield. These jujube-producing regions have annual climatic factors that include a mean annual temperature of 10–14°C, annual rainfall of 400–650 mm, cumulative temperature higher than 10°C of > 3000°C, frost-free season of 170–230 d, mean temperature in July of 23–27°C and latitude of 35–39°C. These factors must be sustained to maintain environmental suitability for jujube cultivation. Meanwhile, other factors are specific to the flowering and fruiting stages, such as a temperature of 22–27°C, relative humidity of 40–80%, rainfall during flowering of < 65 mm, fruitlet growth of > 200 mm, and crisp ripening of < 45 mm. These factors have indispensable impacts on the quality and yields of both dry and fresh jujube varieties (Table 2).

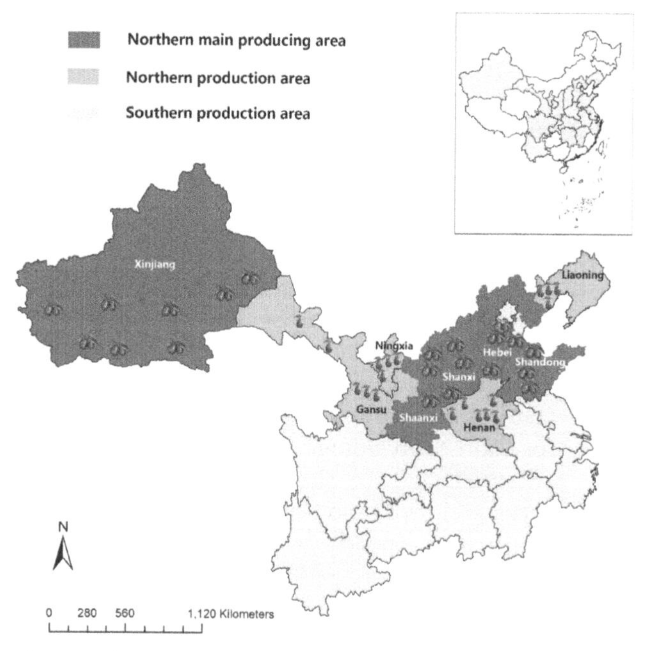

Fig. 1. Distribution of jujube-producing regions in China. Jujube is distributed in more than 20 provinces of China. Traditional cultivation is mainly concentrated in the regions of northern China (*Source*: Xingang Li; fig. constructed by Xiaopeng Li).

Table 1. Climate factors in jujube producing regions. The annual sunshine in Xinjiang was the sum of daylight hours from June to October.

Production region	Major Climate Factors						
	Mean annual temp. (°C)	Annual rainfall (mm)	Cumulative temp. ≥ 10°C	Annual sunshine (h)	Frost-free season (d)	Temperature in July (°C)	Latitude range
Ideal region	10.0–14.0	450–680	> 3,000	2300–3400	170–230	23.0–27.0	35–39°
Jinsixiaozao	11.5–14.2	610–650	3600–4700	2600–3000	180–210	26.5–29.0	37°30′–38°
Zanhuangdazao	12.4–13.3	520–680	3900–4500	2700	180–200	25.8–26.6	37°30′–39°
Taigu, Shanxi	9.0–10.9	400–500	3200–3700	2500–3200	170–200	25.0–27.0	37°–37°60′
Zhongyangmuzao	9.6–11.1	430–580	3000–3200	2400–3400	165–180	22.0–28.0	36–39°
East Guanzhong	10.0–13.5	510–660	3800–4500	2200–2400	210–240	25.0–27.5	34°20′–36°
South Xinjiang	11.0–14.0	< 200	4000–4800	1258–1580	210–250	25.1–30.0	36–44°
Xiaokouzao, Gansu	9.6–11.8	230–300	3200–3300	2700–2725	140–180	22.0–26.0	36°18′–37°
Chaoyang, Liaoning	8.3–8.9	400–600	3400–3700	2830–2860	140–150	22.4–25.0	40°55′–42°

Table 2. Climate factor measures during jujube flowering and fruiting.

Stage	Flowering			Fruiting	
	Temperature (°C)	Relative humidity (%)	Rainfall (mm)	Fruitlet-growth rainfall (mm)	Crisp-mature rainfall (mm)
Measures	22–27	40–80	< 65	> 200	< 45

2. Impacts of Climate Change on Jujube Production

2.1 Dry Jujube

Traditional dry jujube producing regions are distributed along the middle and lower reaches of the Yellow River in northern China, including Shandong, Hebei, Shanxi, Shaanxi and Henan provinces (Qu and Wang, 1993). Before 2000, these regions produced nearly 80% of the national yield of dried fruit. Among them, Shandong and Hebei accounted for almost 60% of the national jujube yield. The cv. Jinsixiaozao and Dongzao are the most popular dry and fresh jujube varieties, respectively. They are produced mostly at Shandong, Hebei and Tianjin in China. Furthermore, cv. Jinsixiaozao is an important export commodity from China.

With the changing climate, there have been rapid increases in rainfall levels during the crisp-mature stages, leading to significantly decreased yields (> 30%) over the past 20 years. Rainfall during mature phases can result in serious fruit cracking, in some cases causing yield losses as high as 70%. To date, the largest jujube yield losses in Shanxi and Shaanxi producing regions have reached 75% and were associated with continuously low temperatures and rainfall in 2013 and 2018. Therefore, traditional farmers in these jujube producing regions are facing great challenges due to climate change. In Hebei and Shandong, there were record yield losses of 59.95% and 76.77% respectively in 2006, due to fruit cracking. Jujube yields in these regions have been faltering in recent decades. The decreasing yields can be partly explained by the rapid development of new production regions (i.e., Xinjiang Autonomous Region). However, climate change is a more direct threat to the jujube industry in traditional jujube farming regions. In many cases, farmers have lost confidence and given up growing due to the unsustainable and continuous losses.

Shanxi and Shaanxi basin of the Yellow River Canyon are known as some of the original jujube producing regions (Qu and Wang, 1993). The cv. Zhongyangmuzao is a landrace here with a growth area of 213,300 ha in 2000. The growth area rapidly increased to 333,300 ha by 2010 due to the high economic value. However, since then, the growth area has continuously decreased, dropping to 200,000 ha by 2019 due to increasing rainfall. Historically, northern Shaanxi was generally semi-arid with a mean annual rainfall ranging between 410–500 mm (1961–1990; Fig. 2). Furthermore, the rainfall levels have usually been lower than 45 mm during the crisp-mature stages (20 September to 10 October; Table 3) before 2000. However, the climate has been constantly changing since 1998 under the implementation of 'Prohibition of Grazing and Return the Grain Plots to Forestry' by the Chinese government. Along with the increase in forest coverage, the mean temperature has decreased and rainfall has increased in the Loess Plateau regions. The low temperature and high rainfall during flowering stages have delayed jujube fruit setting and the low cumulative temperature has reduced fruit qualities. In 2015–2017, the low temperatures (13~15°C) that occurred between June 18th and 21st caused fruitlets to drop. Moreover, the frequency of fruit cracking caused by excessive rainfall has increased approximately three times per decade. These changing climates have limited the development of Zhongyangmuzao. Local jujube growth regions have contracted back to areas extending only 5~10 km from the banks of the Yellow River.

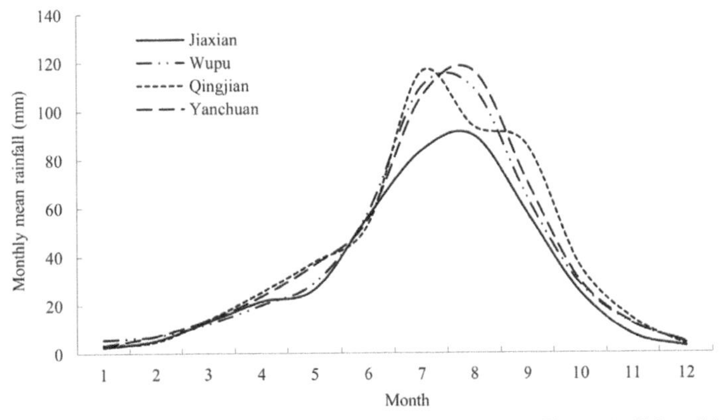

Fig. 2. Distribution of monthly mean rainfall (mm) in northern Shaanxi, China (1961–1990) (*Source*: Xingang Li; fig. constructed by Xiaopeng Li).

Table 3. Rainfall (mm) during the growth periods of jujube in northern Shaanxi, China.

Stage	Dormancy (Nov.–Mar.)	Germination (Apr.–May)	Flowering (June)	Fruitlet growth (Jul.–mid Sep.)	Crisp mature (late Sep.–early Oct.)	Defoliation (late Oct.)	Annual
Jiaxian	31.0	57.0	56.0	219.2	20.8	17.0	410.0
Qingjian	39.7	63.2	53.0	280.3	21.5	29.4	487.1
Yanchuan	40.6	60.3	55.2	272.6	27.5	23.0	478.8

2.2 *Fresh Jujube*

Fresh jujube makes up a large proportion of the total jujube yield and economic value. Fresh jujube fruits are best tasting when crisp with sweet flavor. Jujube is also known as an excellent source of vitamin C and full of other nutrients (Huang et al., 2016). Until now, more than 264 fresh jujube varieties have been documented in China. However, historically, fresh jujube has not been cultivated on a very large scale; with the most important reason being its short storage/shelf life (Qu and Wang, 1993; Li, 2015). Until 1980s, the cv. Linyilizao was the most popular fresh variety to be cultivated in the traditional growing regions of China. The development of advanced storage techniques also facilitated the development of the fresh jujube industry. By 1995, the cv. Dongzao had rapidly grown in popularity at Shandong, Hebei and Tianjin. Currently, it is one of the leading fresh cultivars in China, thanks to its delicious taste and crisp texture. However, fruit cracking also presents an extreme threat to the cultivation of cv. Dongzao. For example, in Weinan, Shaanxi Province, excessive rainfall during the crisp-mature stages led to fruit-cracking losses during 2001–2003.

3. Influence of Climate Changes on Jujube Cultivation

3.1 *Dry Jujube: From East to West*

Increases in rainfall levels have also led to great losses in the dry jujube industry. This has led to a revolution 'from east to west', meaning that jujube production has begun to move from the moist to semi-arid/arid regions (annual rainfall < 200 mm, mature-stage rainfall < 15 mm) in China over the last 20 years. In arid/semi-arid western China, especially in southern Xinjiang Uygur Autonomous Region, several dry varieties (i.e., cv. Huizao and Junzao) have been successfully introduced from traditional producing regions (Henan, Shandong and Shanxi). They have exhibited good adaptability to local conditions and as such have rapidly increased the production area in southern (including Aksu, Kashi, Hetian and Bazhou) and eastern (including Hami and Turpan) Xinjiang.

The wild jujube (*Z. jujuba* var. *spinosa* Hu) is generally used as rootstock. Seeds of wild jujube are collected from traditional production areas and were grown for one year in preparation for grafting the next year. This rapid propagation method has greatly enhanced the development of jujube industry in Xinjiang (Zhou et al., 2015b). Moreover, the wild jujube can be specially adapted to desert-oasis conditions with abundant sunshine and heat, and relatively low rainfall (< 10 mm) during the fruit-ripening stage. This effectively circumvents the problem of fruit cracking. Moreover, the fruit grown in Xinjiang is of a higher quality than the fruit grown in traditional production regions, having much higher accumulated sugars (Shi et al., 2013; Song et al., 2010; Zhang et al., 2011; Zhou et al., 2015a). Furthermore, the yield of dried fruit from traditional production regions of eastern China ranged from 4500 to 6000 kg·ha^{-1}, and occasionally even less during years with bad conditions. However, the yield of jujube in Xinjiang has always ranged from 9000 to 12,000 kg·ha^{-1}. Therefore, Xinjiang has become the most important region for producing dried jujube. In addition, some desert-oases located in Gansu Province and Ningxia Autonomous Region have also proven suitable for jujube cultivation, so the local jujube industries have also gradually grown in recent years (Liu, 2008; Li, 2015).

3.2 *Fresh Jujube: Protected Cultivation*

The growth area of cv. Dongzao reached 0.13 million ha in 2019, thus accounting for nearly 65% of all fresh jujube cultivation area. However, the fruit has a low resistance to cracking. Researchers from Northwest A & F University in China have worked to develop effective strategies for reducing cracking losses. Protected cultivation has been successfully applied to Dongzao cultivation in Dali, Shaanxi Province. With protection provided by warm helio-greenhouses or cold sheds (Fig. 3), jujube fruit cracking has been significantly reduced. Local jujube growth area has also rapidly increased from 0.2 ha in 2007 to 26,667 ha in 2019 (Gao et al., 2015).

Protected cultivation aims to reduce fruit cracking losses. However, the new cultivation strategy has also enhanced fruit quality and prolonged the fruit harvesting time and market-cycling period (Kang et al., 2019). In Dali, Dongzao grown in warm helio-greenhouses ripened from late May to late June, which was four months earlier than those grown without protection. Meanwhile, Dongzao cultivated in cold sheds generally ripened from early July to late September. Each helio-greenhouse is about 900 m^2 (9 × 100 m) and can contain 450 spindle-shaped trees planted rows spaced 1 × 2 m apart. Net profits were approximately 565.06–706.31 USD per

Fig. 3. 'Dongzao' jujube grown in protected cultivation facilities (helio-greenhouse) in Dali, Shaanxi Province, China (*Source*: Xingang Li; photo by Xingang Li).

helio-greenhouse, with 2000 kg yields for the whole fruit-harvesting period which was sustained over about 40 d. Cold sheds were designed with widths of 8–12 m and lengths of 80–120 m with trees grown in row spacing of 1.7 × 3.3 m or 2.0 × 3.0 m with an open-center shape achieved via trimming. The mean yield of Dongzao grown in cold sheds is approximately 22,500 kg·ha^{-1} and the profit ranges from 21,189.6–42,379.2 USD·ha^{-1}.

3.3 Candied Jujube: New Strategies for Production

Traditionally, candied jujube fruit are popular in southern China. Fruits are harvested at the white-mature stage for processing, which means that losses caused by rainfall are avoided (Li, 2015). However, in northern China, the relatively new purple-crystal jujube has been developed. For this, fruit of cv. Zhongyanmuzao are harvested at the late-crisp mature period and further processed until they have a moisture content of 13% and can be stored for a long time. Recent annual yields of purple-crystal jujube have reached 20,000 tons in Shanxi and Shaanxi Provinces. Moreover, cv. Zhongyangtuanzao, has also been processed as candied jujube in these regions. The changing climate will affect the production patterns of both traditionally candied jujube and purple-crystal jujube in these regions and new strategies for cultivars or products may be needed to enhance and sustain the development of the local jujube industry.

4. Effect of Climate Change on Jujube Cultivation Outside of China

4.1 South Korea

South Korea has a history of jujube cultivation that dates back to more than 1,000 years (Qu and Wang, 1993). In the 1990s, the cv. Huizao was first introduced in South Korea. Since then, many new cultivars have been developed for both dry and fresh jujube production, including the cv. Yuechu, Hongyan, Fuzhong, etc. Climate change has also influenced cultivation of jujube in South Korea due to increased rainfall during the flowering and mature fruit periods (Li, 2015). Protected cultivation patterns have been used to grow cv. Fuzhong in Cheong Ju, South Korea since 2000 (Fig. 4). By 2018, the total jujube growing area had reached 2,744 ha, of which about 80% utilized protected cultivation approaches. These approaches also integrate the supplementation of water and fertilizer to maximize the yield. During the flowering and mature periods, sensors will automatically control the facilities in response to measured rainfall levels. These structures play the dual role of protecting against excessive rainfall as well as against high temperature burns during the mature-fruit stages.

Fig. 4. Automatic rain roof facilities in Korea (*Source*: Xingang Li; photo by Xingang Li).

4.2 Iran

Jujube plants were spread to ancient Persia along the Silk Road over 2,000 years ago (Qu and Wang, 1993). Nowadays Iran has a jujube growing area of 1,350 ha, which is mainly concentrated in the South Khorasan Province with an annual yield of 3,000 tons produced in arid-dessert conditions (Liu et al., 2016). The fruit produced in Iran is like semi-wild type, having small fruit sizes and a relatively low consumption rate. But the fruit performs well against cracking and returns a stable yield. These fruits are generally used to produce dried products. In this situation, the type of fruit products are determined by the limitations of the arid conditions. In recent years, however, many big fruit varieties have been introduced in Iran with the advent of protected cultivation modes (Liu et al., 2016). The spread of these cultivars will force the local jujube industry to address the increases in rainfall and other climate change-associated shifts.

4.3 Other Countries

Jujube has been spread to many places around the world, including Western Australia, Israel and America (California and New Mexico) in the last 20 years. The jujube tree has a wide tolerance range and is highly adaptable to high temperatures and arid-desert conditions. These conditions benefit the growth of fresh jujube cultivars, such as cv. Lizao, Langzao, Kongfusucui, etc. Due to the adoption of wide-row and sparse-planting patterns, fruits are of good quality and a sustainable yield has emerged. Artificial irrigation and fertilizer integration measures enacted to meet normal growth requirements have also enhanced the fruit quality and ensured consistent market quality fruit.

5. Natural Adaptation of Jujube Genetic Resources

More than 700 jujube varieties have been developed in China, including dry, fresh, candied and multi-use varieties. These varieties mostly originated from the Shanxi, Shaanxi, Hebei, Henan and Shandong Provinces. The traditional production regions are responsible for producing 83.29% of the total number of jujube varieties (Qu and Wang, 1993; Li et al., 2013). Among them, 207 are registered by the Committee of Forest Tree Validation in China and include 87 dry, 114 fresh and six candied varieties (Li, 2015). In addition, sources of 630 varieties have been stored in the National Jujube Resource Germplasm at the Shanxi Institute of Pomology, Taigu, Shanxi, China (Li et al., 2013; Guo et al., 2020).

Jujube producing regions are divided into northern and southern subregions according to the 15°C isothermal line and annual rainfall of 650 mm, coinciding with the boundaries formed by the Qinling Mountains and the Huaihe River (Qu and Wang, 1993). Southern production regions have slightly higher temperatures, humidity and rainfall levels, which are not conducive to high-quality fruit products. In contrast, northern production regions have mean annual temperatures lower than 15°C with rainfall levels lower than 650 mm. This is why more than 90% of the total yield, growth area and varieties are concentrated in this region (*China Forestry Statistical Yearbook 2006–2020*), which produces both dry and fresh fruit that have excellent flavor and higher accumulated dry matter and sugar contents.

Recent phylogenetic studies have further divided cultivars from the northern producing regions into two geographical groups: varieties coming from the Northern China Plain (located to the east of the Taihang Mountains) and those from the Loess Plateau (Huang et al., 2016; Li, 2015; Zhang et al., 2015). The cv. Jinsixiaozao and Dongzao are native to the Northern China Plain group, which is near the Bohai Gulf and China's eastern coast. Dongzao is highly adaptable to different conditions and has been successfully introduced in both moist and arid areas of China. However, Jinsixiaozao is poorly adapted to the arid desert conditions of Xinjiang and other places in western China, but performs well when cultivated in southern China. Examples such as cv. Jinsi No. 4 have shown broad success on being grown in Yunnan Province. As for the Loess Plateau, the annual rainfall ranges between 390–600 mm, the mean annual temperature ranges between 9.6–13.5°C and the relative humidity is generally low, but there are relatively large differences in temperature between day and night. The cv. Zhongyangmuzao, Junzao, Hupingzao, Huizao, Lingbaodazao and Dalishuizao are the native landraces (Li, 2015). Many of the varieties that come from this region can be easily adapted to the arid desert conditions of Xinjiang, Gansu and Ningxia.

6. Mitigation Strategies against Biotic and Abiotic Stressors

6.1 *Strategies against Abiotic Stressors*

Strategies to combat climate change are urgently required to ensure the continued development of the jujube industry. Over the past 20 years, Xinjiang has become the largest jujube producing region in China, with an estimated yield of 5.6247 million tons from a growing area of 473,000 ha in 2017. In the near future, the cultivation of jujube in Xinjiang will continue to expand due to its ideal growth conditions. According to estimates, the growing area should be adjusted to 400,000 ha to account for the wide-row and sparse-planting (plant spacing of 1.5 × 4 m) patterns. These patterns will limit production to a moderate yield, but will further enhance fruit quality and economic benefits.

As for the fruit quality, the climate in Xinjiang is suitable for jujube production, with relatively high temperatures and strong sunlight. However, the temperature can exceed 35°C in July and August, which will injure the fruit peel. The mean number of days per year with extreme high temperatures in Kashi, Ruoqiang and Shanshan Counties have reached 3.75, 41.15 and 50.00 d, respectively, with maximum temperatures of 38.5–45.4°C. Long-term exposure to high temperatures (> 35°C) will reduce the fruit setting rate, cause fruit peel injures and promote the formation of disease during the maturation period (Li, 2015). Therefore, a shelter forest system would benefit jujube production by reducing losses induced by extreme high temperatures.

Climate change may also result in freezing injuries, owing to sudden drops in temperature due to early winters or long-term low temperatures. The mean temperatures in southern and eastern Xinjiang in winter are generally around –10.67 ± 0.56 and –4.24 ± 0.55°C, respectively, with minimum temperatures dropping below –20°C. In Hami, the minimum temperature has dropped as low as –28.9°C and the number of days with mean temperatures below –20°C per year is around 8.65 d (Zhang et al., 2009; Wang et al., 2011; Li, 2015). Continuous observations suggested that adult jujube trees can survive freezing injuries from temperatures of –24°C for up to seven days or –28°C for no longer than 24 h. However, young trees can only survive freezing injuries from temperatures of –23°C, and for no longer than 3 d (Zhou et al., 2008). Therefore, the occurrence and duration of extreme low temperatures are major limiting factors for jujube production in Xinjiang (Li, 2015).

6.2 *Strategies against Biotic Stressors*

Climate change has altered management strategies for preventing pests and diseases in jujube planting orchards, especially regarding *Lygocoris lucorum* Meyer-Dür, *Scythropus yasumatsui* Kono et Mpriomoto, *Phakopsora*

Fig. 5. Control of *Scythropus yasumatsui* using sticky insect glue (*Source*: Xingang Li; photo by Xingang Li).

ziziphi-vulgaris Dietel, jujube witches' broom, black spot disease and fruit shrink disease. However, chemical and physical control methods remain the most efficient and effective approaches. For example, jujube rust disease can be controlled by spraying antiseptics (e.g., triadimefon) from late June to early July, and both *L. lucorum* and *S. yasumatsui* infestations can be prevented by using sticky insect glue (Fig. 5).

Jujube witches' broom is generally induced by *Candidatus Phytoplasma ziziphin* (Zhou et al., 2021). It is an infamous disease commonly known as the cancer of the jujube tree. This disease is highly destructive due to its high infection rate and difficulty in controlling. It has spread rapidly in deserted planting orchards in recent decades in traditional jujube producing regions. Leafhopper (i.e., *Hishimonoides chinensis*) is one of the main vectors of jujube witches' broom disease. Cross-infection via pruning tools may also enhance the spread. Even now, there is still no efficient manner to protect and cure infected trees. To stop the spread, farmers must burn all infected plants and even the wild jujube trees located near planting orchards. Promisingly, leafhoppers can be controlled during growing periods by eliminating their alternate hosts, such as Chinese arbor-vitae (*Platycladus orientalis*) and *Pinus* spp. (Liu, 2010; Zhang et al., 2017).

In Xinjiang, the major insect pests are *Tetranychus cinnabarinus* (red spider mite), *Eriophyes annaltuevals* Keife., *Dasineura* spp. and *Lygus lucorum*. In addition, fruit shrinking and blackhead disease can be induced by excessive rainfall during fruit ripening. The dry-variety Junzao is particularly susceptible to these diseases.

7. Conclusions and Prospects

Climate change has led to de-allocation of the jujube industry as increasing rainfall levels have caused serious losses in traditional producing regions. A revolution 'from east to west' has been rapidly progressing in recent decades. By 2017, the Xinjiang Uygur Autonomous Region had become the largest jujube producing region in China, accounting for 48.91% of the national yield. Meanwhile, the relative growing area of the traditional producing regions decreased by 56.67%.

As the industry adapts to the changing climate, cultivation patterns using wide-row and low-density planting will continue to grow as the leading model for dry jujube cultivation in Xinjiang. Furthermore, in traditional producing regions of northern China, the industry will continue to be optimized by adjusting the quantity and quality of dried products. As for the fresh jujube industry, protected cultivation in warm helio-greenhouses or cold sheds has proven to be effective in reducing losses caused by increased rainfall, while also synchronously extending the market-cycling period and improving fruit qualities. Currently, protected cultivation accounts for 23.33% of the total growing area of fresh jujube and it will continue to increase with further development.

In the near future, the cultivation of cv. Dongzao and Qiyuexian will continue to spread in fresh jujube production. The total cultivation area of Xinjiang is now 5000 ha and there are still additional development opportunities for the fresh jujube industry. The only disadvantage of expanding jujube cultivation into this region is the distance from the east-central Chinese markets as this increases the cost of transportation and selling.

References

Chen, Y.L. and Schirarend, C. (2007). Rhamnaceae. *In*: Wu, Z.Y., Raven, P.H. and Hong D.Y. (eds.). *Flora of China*, 12: 119–123, Science Press & Beijing & Missouri Botanical Garden Press, St. Louis.

Gao, W.H., Zhou, A.Y. and Zhao, J.M. (2015). *Cultivation Techniques of Fresh-eaten Jujube under Protected Culture*. Jindun Publishing House, Beijing, China.

Guo, M.X., Zhang, Z.R., Li, S.P., Lian, Q., Fu, P.C., He, Y. L., Qiao, J.X., Xu, K.K., Liu, L.P., Wu, M.Y., Du, Z.R., Li, S.N., Wang, Y.K., Tian, S., Zhao, J., Feng, X., Li, R.Q., Jiang, W.K. and Zhao, X.S. (2020). Genomic analyses of diverse wild and cultivated accessions provide insights into the evolutionary history of jujube. *Plant Biotechnology Journal*, 19(3): 517–531.

Huang, J., Zhang, C.M., Zhao, X., Fei, Z.J., Wan, K.K., Zhang, Z., Pang, X.M., Yin, X., Bai, Y., Sun, X.Q., Gao, L.Z., Li, R.Q., Zhang, J.B. and Li, X.G. (2016). The jujube genome provides insights into genome evolution and the domestication of sweetness/acidity taste in fruit trees. *PLoS Genetics*, 12(12): e1006433.

Kang, C.X., Su, J.J., Zhang, Z., Shi, Q.Q. and Li, X.G. (2019). *Study on the Fruit Quality of 'Dongzao' Jujube under Protected Cultivation*, 197(3): 67–71.

Li, D.K., Niu, X.W. and Tian, J.B. (2013). *Atlas of Chinese Jujube Variety Resources*. China Agriculture Press, Beijing, China.

Li, X.G. (2015). *Chinese Jujube Industry*. China Forestry Publishing House, Beijing, China.

Liu, D.H., Ye, X.Q. and Jiang, Y.M. (2016). *Chinese Dates: A Traditional Functional Food*. CRC Press, Taylor & Francis Group.

Liu, M.J. (2008). *Report on Development of Chinese Jujube Industry*. China Forestry Publishing House, Beijing, China.

Liu, M.J., Zhao, J. and Zhou, J.Y. (2010). *Jujube Witches' Broom Disease*. China Agriculture Press, Beijing, China.
Qu, Z.Z. and Wang, Y.H. (1993). *Chinese Fruit Tree Records: Chinese Jujube*. China Forestry Publishing House, Beijing, China.
Shi, Y.J., Wu, Z.B., Hadier, Y., Yu, T., Song, H.F. and Zhuoremu, T. (2013). The suitable harvest time of 'Junzao' cultivated in different ecological area of the Tarim Basin. *Journal of Xinjiang Agricultural Science*, 50(8): 1475–1481.
Song, H.F., Hadier, Y., Shi, Y.J., Zhang, P. and Luo, Q.H. (2010). Correlation between fruit nutrition and soil nutrition for 'Junzao' cultivated in the Tarim Basin, Xinjiang. *Journal of Fruit Science*, 27(4): 626–630.
State Forestry Administration: *China Forestry Statistical Yearbook 2006–2017*, China Forestry Publishing House, Beijing, China.
Wang, C.Z., Gao, J.C., Li, X.G., Gao, W.H., Xin, W.J. and Wu, P. (2011). Study on cold hardiness of major jujube cultivars in northwestern China. *Journal of Fruit Science*, 28(5): 898–902.
Zhang, C.M., Yin, X., Li, X.G., Huang, J., Wang, C.Z. and Lian, C.L. (2015). Genetic diversity and population structure of sour jujube. *Ziziphus acidojujuba, Tree Genetics & Genomes*, 11(1): 809.
Zhang, P., Shi, Y.J., Song, H.F., Zhuoremu, T. and Wu, Z.B. (2011). Variation of nutrition in 'Huizao' fruit cultivated in south Xinjiang. *Journal of Fruit Science*, 28(1): 77–81.
Zhang, S.Y., Zhang, Z., Zhang, C.M., Li, H. and Li, X.G. (2017). Transcriptomic analysis on pathogenesis of witches' broom disease in *Ziziphus jujuba. Acta Horticulturae Sinica*, 44(7): 1287–1298.
Zhang, Y.X., Liu, H.R., Li, M.C. and Aiyiti. (2009). Types of freezing injury to jujube and influence of climate factors in Aksu area. *Desert and Oasis Meteorology*, 3(6): 43–46.
Zhou, L., Yang, W.Z., Wang, C.Z. and Li, X.G. (2015a). Eugenics area of jujube cultivated in Xinjiang. *Journal of Fruit Science*, 32(3): 1017–1021.
Zhou, L., Wang, C.Z. and Li, X.G. (2015b). Cultivation model of jujube in Xinjiang. *Journal of Northwest Forestry University*, 30(2): 139–143.
Zhou, L.R., Li, J., Qu, H.L. and Yu, S.S. (2008). Study on the rational cultivation distribution of jujube in Aksu region. *Journal of Tarim University*, 21(2): 7–14.
Zhou, J.Y., Ma, F.L., Yao, Y., Deng, M.S., Chen, M.T., Zhang, S.Q., Li, Y.M., Yang, J., Zhang, N., Huang, J.Q., Sun, Q.B. and Sun, J. (2021). Jujube witches' broom phytoplasma effectors SJP1 and SJP2 induce lateral bud outgrowth by repressing the ZjBRC1-controlled auxin efflux channel. *Plant, Cell & Environment*, 44(10): 3257–3272.

11

Climate Resilience in Indian Jujube (*Ziziphus mauritiana* Lam.) and Other Jujube Species

Muhammad Usman,[1,*] *Safdar Ali Mirza*[2] and *Bilquees Fatima*[1]

1. Introduction

Jujube species are the most eminent members of *Rhamnaceae* (Buckthorn) family having 57 genera and 900 species (Richardson et al., 2000; Hauenschild et al., 2016). Jujube is mainly found in tropical and subtropical areas. Jujube is native to China and has been cultivated since 7,000 years ago (Qu et al., 1993). Afterwards, it was widely dispersed in Europe, Australia and East-Asian continents. In nearly 50 countries including Europe, Asia, Africa, Oceania and the Americas, Jujube has been introduced (Liu et al., 2014). Indian jujube (*Ziziphus mauritiana* Lam.) is frequently known as Indian plum, *ber*, Indian cherry, or desert apple (Kaaria, 1998) and was domesticated in the Indo-Pak subcontinent about 11,000 years ago (Gupta, 2004). In the Indian sub-continent, even 400-year-old trees are still existing (Bal et al., 2013). China has a major share in global exports of Chinese jujube (*Ziziphus jujuba* Mill.) fruits (Gao et al., 2011) and currently generates 99% of the world's jujube production. In the Chinese domestic market, this fruit is primarily consumed as fresh fruit. Pakistan is producing 22.16 thousand tons of fresh fruit of Indian

[1] Institute of Horticultural Sciences, University of Agriculture, Jail Road, Faisalabad 38040, Pakistan.
[2] Department of Botany, Government College University, Lahore, Pakistan.
Emails: safdaralimirza@gcu.edu.pk; dr.fatimausman@uaf.edu.pk
* Corresponding author: m.usman@uaf.edu.pk

jujube from an area of 4.3 thousand hectares per annum with potentially lower yield of 5.15 tons/ha. Since 2013, there has been a decline in both area (897 hectares) under cultivation and production (3,142 tons) per annum (www.faostat.org).

Jujube species, including *Z. mauritiana*, are known to have a diverse range of horticultural traits which highlight their immense potential in crop improvement and ecological restoration of the rain-fed and arid regions. It also has a great contribution towards economic development of arid regions and is an ideal cash crop in this zone where common fruits cannot be grown.

Vitamin C, flavonoids, phenolics, triterpenic acids and polysaccharides are the major biologically active components. The fruit is enriched with high ascorbic acid (vitamin C) contents, even higher than citrus and Kiwi fruits, sugars (25–30%) twice as compared to other common fruits and greater than sugar beet and sugarcane (Cui et al., 2008). Jujube fruit has very smooth and bright surface with sweet and crispy flesh, just like some pome fruits (apple and pear). The fruit contains plenty of nutritious constituents, cyclic nucleotides, phenolic content, flavonoids (Pawlowska et al., 2009), amino acids (Choi et al., 2011), cerebrosides, triterpenic acid (Gao et al., 2011), proline content, polysaccharides (Huang et al., 2008) and minerals including iron, zinc, potassium (Li et al., 2007) and cyclic adenosine monophosphate (cAMP), 100 to 1000 times greater than that of apple (Liu et al., 2020). Furthermore, jujubes are closely related to Rosacea, most well-known fruit-producing family consisting of many deciduous fruits, including apple, pear, peach and strawberry. Consequently, jujube could provide great basis for genes of interest for genetic improvement of fruit crops and further understanding of the diversified jujube genome (Liu et al., 2014).

Jujube is an excellent source of nutritional and nutraceutical compounds with active biological components. Various parts of jujube can also be used to treat different diseases. The remedies of bark boil are perfect for dysentery and diarrhea diagnosis. Leaves are antifungal and decrease obesity. The aqueous leaf part is applied on the outside to alleviate burning problems whereas roots are used to treat dysentery (Mahajan and Chopda, 2009). Indian jujube is used as fodder for animals, fuel wood for the household and pruned wood could be used as biomass for fencing (Singh et al., 2020). Jujube trees could be used for sand fixation in coastal dunes (Dagar et al., 2001), can provide permanent soil cover and maintain soil structure due to its deep root system (Arndt et al., 2001).

1.1 *Biology, History and Domestication*

The jujube plant may be a shrub or tree with soft and hard wood of 6–9 m, depending on the place of cultivation (Yao, 2013). Jujube species are distributed between latitude 18° S to 51° N, longitude of 76° to 124° and grow well above sea level (up to 2800 m). Jujube has extensively spread to the distant areas due to its higher adaptability, ease in cultivation, xerophytic attributes, early bearing habit, quick bud differentiation, about 50–60 day-long blooming period, long life cycle (1000 reproductive cycles), remarkable endurance under harsh environmental conditions, higher resistance to abiotic stresses (salinity and drought), least input requirements and higher nutraceutical benefits (Bhatt et al., 2008). Moreover, it is the best example of a drought-hardy plant, which is a dominant part of the Indo-Pak desert's natural vegetation. The tree is semi-deciduous, having a thick canopy. The leaves are smooth and shiny, ovate or oval shaped with 2.5–5.5 cm length and 2–4 cm width. Flowers are scented, light greenish, tiny, ranging from 4–8 mm in diameter. Jujube flower has five complete whorls of sepals, petals, anthers and one ovary containing two ovules. Flowers may appear single or in cluster from each leaf axil. The flower cluster (cyme) has nearly 13 flowers, depending upon the cultivar and location of flower on the branchlet. Most of the *Ziziphus* species are self-incompatible; however, *Z. jujuba* and *Z. mucronata* species may develop self-pollinated fruit that usually aborts later (Zeistman and Bohta, 1992; Weekley and Race, 2001; Asatryan and Tel-Zur, 2013). *Ziziphus* flowers also exhibit protandrous dichogamy, anthers dehisce well before the receptivity of stigma and promote cross-pollination (Tel-Zur and Schneider, 2009). Sequences coding for S-locus have been identified in *Z. celata* suggesting presence of gametophytic self-incompatibility system 'GSI' (Edwards et al., 2012). Two types of flowers are reported in *Z. mauritiana*, *Z. spina-christi* and *Z. mucronata*. In the morning and afternoon, A and B type flowers were found in male phase, respectively (Pareek et al., 2007; Tel-Zur and Schneider, 2009), in *Z. mauritiana* cultivars, Seb and Umran (Shukla et al., 2006). The receptive life of the flower is short (two to three days); hence many flowers remain un-pollinated and drop (Pareek et al., 2007). This non-synchronization behavior in male and female flowering phases seemed to be higher temperature (up to 45°C) related in *Z. mauritiana*, cultivated in the semi-arid and arid regions (Asatryan and Tel-Zur, 2013). Pollen viability in *Z. mauritiana* and other species ranged from 10–91% in different cultivars (Tel-Zur and Schneider, 2009) while it was reported to remain low 25–58% due to prevalence of extreme temperature conditions and the growth of pollen tube ceased in all cultivars of *Z. mauritiana* (Asatryan and Tel-Zur, 2013).

The fruit is a drupe, having one to two seeds with a pit (stone) in the center. The fruit comes from the nectary disk and its ovary. Fruit size ranges from thumb size to golf-ball size which is truly dependent upon the variety. There is large diversity in fruit shape, like oblong, obovate, oval, oblate, round, apple-like or sometimes abnormally shaped (Yao,

2013). The pulp is sweet, acidic, greenish, yellow or even reddish in color (Mahajan and Chopda, 2009). Seedless fruit development has been reported in *Z. sativa* and *Z. mauritiana* due to gametic self-incompatibility system (Tian and Ma, 1987). *Ziziphus* species can also produce fruit due to self-pollination; however, it usually drops off prematurely and has no viable seeds (Godt et al., 1997). In *Z. mauritiana* cultivars, including Umran and Q-29, self-pollination was observed; however, there was massive flower drop and fruit drop in early stages and fruit set remained low with non-viable seeds. In B5/4, Kaithali and Gola cultivars, flowers dropped two to three days after pollination and no fertilization was observed. The bark gets darker longitudinally and is grayish brown or reddish in color. The plant is usually spiny. Branches of tree are compact, white pubescent at young stage and are inclined to be zigzag. Branches spread upright, becoming light brown to gray. Fruit-giving branches of the tree are mostly not deciduous in nature (Mahajan and Chopda, 2009).

Jujube species have been cultivated in China, Pakistan and India since a long time and are well recognized species in these areas compared with other related species in rest of the world (Cherry, 1985). Many species grow wild, including *Z. xylocarpa*, *Z. rotundifolia* (rootstock use) and *Z. ocenoplia* (Jackson et al., 2011). Other species, such as *Z. jujuba* Mill and *Z. mauritiana* Lam., have been domesticated and are being widely cultivated (Memon et al., 2012). However, jujube species had caught little attention for research and development (Azam-Ali et al., 2001) until the recent past when these species were properly investigated under different research projects. Currently, it is one of the major cultivated species, prominent dried fruit and major source of income of about 20 million farmers in China. During the last 2000 years, its cultivation has spread to the neighboring States, including Japan and Korea. It is being commercially produced in China, Iran, South Korea, Israel, Italy, United States and Australia.

1.2 Climatic Requirements for Optimal Growth and Productivity

Jujube blooms and bears fruit in the same year of grafting and planting under high density plantation (HDP) system, but produces the regular yield after three to five years of orchard establishment. It has no low chilling requirements for early flowering, hence, it is appropriate for protected cultivation. Flowering and fruit setting occur in a series of steps. First fruit setting starts from April when flowers bloom and take almost 100 days for maturity with relatively larger sized fruit. Second generation bloom and fruit occur in June with medium-sized fruit and takes 60–75 days for fruit maturity. The third generation of fruit setting is witnessed in July and takes 40–50 days to attain maturity and fruit remain smaller in size (Kohandel et al., 2011). Jujube trees rarely miss a crop due to their late bearing habit when frost threat is over and there is a prolonged blooming season. Like many other tree fruit crops, jujube shows leaf emergence four to six weeks after blooming and continues blooming for two months and more (Yao, 2013).

Jujube plants perform well under wide soil and temperature conditions. Soil pH, suitable for jujube cultivation, ranges from 5.5 to 8.5. During summer season, jujube can withstand higher temperature up to 50°C and during winters, it tolerates low temperatures up to –30°C. The jujube trees are well adapted to extreme drought conditions due to its major cultivation in the rain-fed regions and produce an adequate yield (Sulusoglu et al., 2014). The trees can survive well with 200 mm annual precipitation; however, more rainfall and additional irrigation is required for enhanced productivity and better fruit quality (Yao, 2013).

Jujube requires less water and fertilizer; hence, the costs for crop management and cultivation are considerably less compared with other tree fruit crops (Liu et al., 2020). Indian jujube (*Z. mauritiana*), due to a vast range of climatic adaptability, is well-known for its desertification and for preventing soil erosion. Moreover, it is also tolerant to flooding, salinity and drought conditions (Maruza et al., 2017). Its cultivation can reduce land degradation and can grow well on the poor and degraded lands (Pareek, 2001).

1.3 *Climate Change Influence on Crop Productivity*

Climate change is mainly caused by natural factors, calamities or most importantly, due to interference of human anthropogenic factors. This ultimately led to drastic changes in the current environment, including remarkable increase in temperature (global warming), changing precipitation patterns leading to non-availability of water for crop survival, drought conditions, changing growing seasons and seasonal patterns. Agriculture sector is the most vulnerable to climate change due to its higher dependence on climate and weather conditions. Most of the people involved in agriculture are poor compared with the urban residents. Consistent warming trends with higher frequency and intensely extreme weather conditions have been observed in Asia and the Pacific region in recent decades. Fruit crops are a valuable source to retain nutritional security. Climate change events, like rise in temperature, has worst impact on crop productivity. Many fruit crops are facing extinction, leading to lack of food availability and terribly affecting the food chain. Jujube is a relatively climate-resilient fruit crop and is categorized under climate smart crops. It is a hardy fruit which gives better return in terms of food, fodder and fuel wood to the resource-lacking farmers. However, at higher temperatures up to 45°C under arid conditions, *Z. mauritiana* has shown separate male and female phases in flowers while pollen viability reduced to 25–58% (Tel-Zur and Schneider, 2009; Asatryan and

Tel-Zur, 2013). Fruit drop has been reported in *Ziziphus* species due to frost or higher temperatures (Pareek, 2001). The higher temperatures and extreme environmental conditions might have evolved gametophytic self-incompatibility (GSI) and synchronous dichogamy which prevented inbreeding in *Z. jujuba, Z. mauritiana* and other species (Asatryan and Tel-Zur, 2013). Further studies are suggested to investigate the evolution of both these inbreeding-preventing mechanisms in *Ziziphus* species and to develop correlations with the key environmental factors.

1.4 Germplasm Resources

Genetic variability in jujube species is desired for collection and preservation of elite plant stocks for future use. Characterization of jujube germplasm has been neglected and was a major hindrance in its domestication (Soltani et al., 2006). Jujube species depict great genetic diversity due to cross-pollination which highlights the significance of its selection and screening like other fruit crops (Pommer, 2012). Moreover, diversity assessment studies offer new scientific evidence for ambiguous jujube taxonomy in addition to utilization of its germplasm for academic exchange (Liu et al., 2008). *Ziziphus mauritiana* is the most important fruit crop in Pakistan, India and southern regions of China with diversified germplasm above 330 accessions stated in India (Awasthi et al., 2009; Sivalingam et al., 2013) whereas above 100 accessions are present in Pakistan (Fig. 1, Table 1). Among Indian varieties, Gola has round shape fruit and matures early; Umran is late variety and has better shelf life. These varieties are mainly grown in the Haryana state (Singh et al., 2017). Many new varieties and accessions have been selected in China as well.

After a long period (1950s to 1980s) of evaluation and selection, the first germplasm was introduced in China. After the nationwide germplasm findings, National Chinese Jujube Repository was established in Taigu (Shanxin province), China. A total of 930 jujube cultivars, 90% of total jujube genotypes worldwide, have been evaluated. Another germplasm repository was established in Cangxian (Hebei province), China. About 640 accessions and variants of local cultivars, such as Wuhezao, 'Dongzao' and Jinsixiaozao were preserved. Furthermore, Hebei Agricultural University collected and conserved 200 accessions of sour jujube germplasm and other species, including *Z. mauritiana* Lam., *Z. nummularia Burm* and *Z. spina-christi* Wild as given in Table 2 (Liu et al., 2020). In India, the *ber* cultivars Ponda, Narendra Ber Sel-2 and Banarsi Karaka are tolerant while the cultivars Gola and Narendra Ber Sel-1 are semi-tolerant to salinity (Verma et al., 2018a). Important breeding material of *Ziziphus* species, including several triploids, mixoploids, male sterile lines, seedless and disease resistant germplasm, is presented (Table 3).

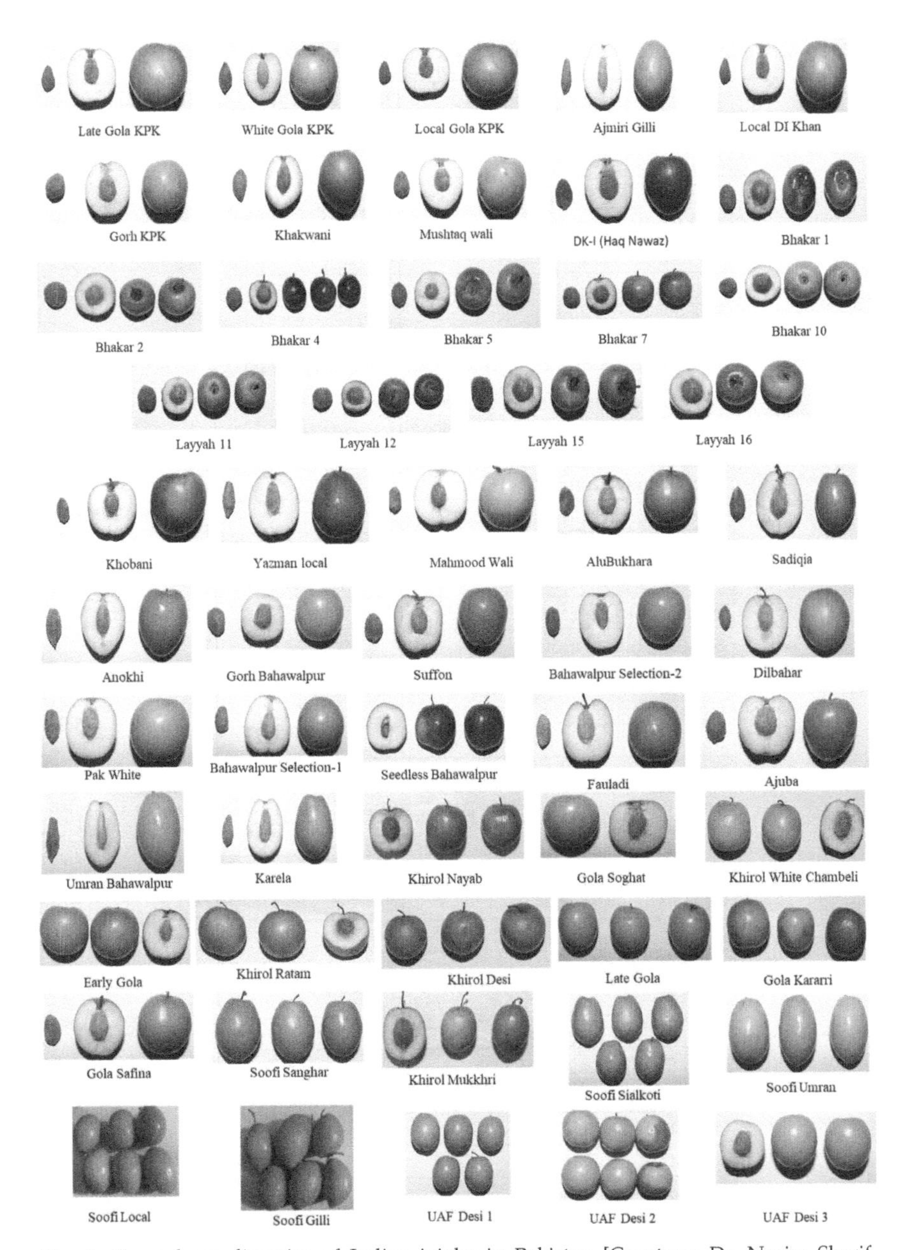

Fig. 1. Germplasm diversity of Indian jujube in Pakistan [Courtesy: Dr. Nasim Sharif, Horticulture Research Station, Sahiwal, Pakistan] (Source: Sharif et al., 2019).

Table 1. Germplasm collections of jujube accessions in Pakistan, India and Bangladesh.

Country	Name of Variety	Collection site	References
Pakistan	Suffon	Bahawalpur	Sharif et al. (2019)
	Yazman Local	"	
	Sadiqia	"	
	Pak White	"	
	Mehmood Wali	"	
	Fauladi	"	
	Karela	"	
	Alu Bukhara	"	
	Anokhi	"	
	Dilbahar	"	
	Umran Bahawalpur	"	
	Gorh Bahawalpur	"	
	Khobani	"	
	Ajuba	"	
	Bahawalpur Selection-1	"	
	Bahawalpur Selection-2	"	
	Seedless Bahawalpur	"	
	Akasha	"	Anjum et al. (2018)
	Umran-13	"	
	Karella	"	
	Gorh	"	
	Early Gola	Tandojam	Sharif et al. (2019)
	Late Gola	"	
	Gola Kararri	"	
	Gola Safina	"	
	Gola Soghat	"	
	Soofi Sanghar	"	
	Soofi Local	"	
	Soofi Sialkoti	"	
	Soofi Gilli	"	
	Soofi Umran	"	
	Khirol Mukkhri	"	
	Khirol White Chambeli	"	
	Khirol Desi	"	
	Khirol Ratam	"	
	Khirol Nayab	"	
	UAF-1	Faisalabad	
	UAF-2	"	
	UAF-3	"	

Table 1 contd. ...

...Table 1 contd.

Country	Name of Variety	Collection site	References
	Late Gola KPK	Dera Ismail Khan	
	Khakwani	"	
	Mushtaq Wali	"	
	Local DI Khan	"	
	DK-1 (Haq Nawaz)	"	
	Amiri Gilli	"	
	White Gola	"	
	Local Gola	"	
	Gorh KPK	"	
	Bhakar 1 to Bhakar 10	Bhakar	
	Layyah 11 to Layyah 16	Layyah	
India	BC-1	Rajasthan	Singh et al. (2017)
	Seb	"	
	Tikadi	"	
	CAZRI-Gola	"	
	Katha	"	
	Maharwali	"	
	S x K hybrid	"	
	Umran	Maharashtra	
	Chhuhara	"	
	Illaichi	Punjab	
	Dandan	"	
	Z-G-3	"	
	Bagwadi	"	
	Gola	Haryana	
	Sua	"	
	Reshmi	"	
	Mundia	"	
	Banarasi Karaka	Uttar Pradesh	
	Aliganj	"	
	Banarasi Pebandi	"	
Bangladesh	Dumki-1 to 21	Dumki	Bhattacharjee et al. (2019)
	Dumki-22 to 27	"	
	Dumki-28 to 32	"	
	Dumki-33 to 37	Rajasthan	

Table 2. Jujube germplasm literature/database/online resources.

Books/Descriptors/Online Resources	**Information**	**Citation/Weblinks**
China Fruit Tree Records –Jujube	700 Jujube cultivated varieties and 30 wild sour genotypes	Wang et al. (2019), http://www.ziziphus.net/zzzy
Germplasm Resources of Chinese Jujube	1033 accessions of Jujube including sour Jujube and Indian Jujube	Li et al. (2013)
The Illustrated Germplasm Resources of Jujube	250 genotypes 57 (table Jujubes) 80 (dried Jujubes) 82 (dried and fresh) 17 (processed prod.) 14 (ornamental)	
Descriptors and Data Standards for Jujube Germplasm	-	Li et al. (2013)
Test Guidelines for Distinctness, Uniformity and Stability-jujube	-	Li (2006)
Technical Regulators for the Identification of Jujube Cultivars – SSR Marker Method	-	
Weblinks		
Internet Information System for Perennial and Asexual Crop Germplasm Resources	Datasets observed at National Jujube Repository in Taigu, Shanxi, China	http://www.ziyuanpu.net.cn/
Chinese Crop Germplasm Resources Information System—jujube	Germplasm checking and analysis	http://www.cgris.net/query/croplist.php
Internet Information System for Jujube Germplasm	Information collected from 700 accessions of local area germplasm	http://www.ziziphus.net/zzzy

Table 3. Elite germplasm of jujube species with distinct features.

Germplasm Type	**Genotypes**	**Citation**
Triploids	Zanhuangdazao Jinglinyihaozao Pingguozao, Shanxitedage Zhenhuluzao, Hengshuibianzhizao	Yan (1984) Liu et al. (2013) Liu et al. (2018)
Mixoploids	Dongzao 2	Liu et al. (2016)
Male-sterile germplasm	JMS1, JMS2	Yan et al. (2009)
Seedless Germplasm	Wuhexiaozao	Liu (2006)
Resistant germplasm against WBD	Xingguang	Liu et al. (2010)

2. Impact of Climate Change on Crop Management

Climate change has a pronounced impact on plant growth and yield management. Impact of changing climate on crop management has other aspects also like increased concentrations of CO_2, fluctuations in speed of wind, animals and their caused diseases and related other indirect variations resulting from interactions between crops and agrological bionetworks. All these have substantial effects on this fruit crop's harvest. It disrupts the availability of food, reduces food accessibility and affects the quality of food. The predictable rise in temperatures, fluctuation in rainfall patterns, variations in extreme weather conditions and decreased availability of water ultimately lead to reduced agronomic yield along with disturbance in crop yield delivery and increased prices. Indian Jujube plants have natural adaptability to extreme temperature, drought and waterlogging. Meghwal et al. (2018) reported that the western Rajasthan in India has harsh environmental conditions in addition to low and unpredictable rainfall. This region is also facing regular drought conditions that lead to crop failures. Indian Jujube is known as king of the arid region fruits for its better adaptation to biotic and abiotic stress tolerance (Nizamani et al., 2015). It is the most suitable plant to be grown in such regions as it has capability to adjust well in the unified agricultural system. The recent decade has observed rapid climate changes specifically after monsoon season in this region. This change is also attributed to higher temperatures and reduced relative humidity. The two-year weather data after every monsoon season have discovered the changing climatic conditions that caused decreased fruit set and eventually reduced the overall fruit yield of the plant. Another notable observation in changing climatic conditions was response of varieties, such as Gola and CAZRI Gola, which sustained to produce more yield. It provides nutritious content to the masses (Haldhar et al., 2016, 2018). Under varying environmental conditions in tropical, sub-tropical and arid climates of the world, the cultivation and survival of this perennial, hardy, drought-tolerant and multipurpose fruit crop imparts resilience to the threatened bionetwork. Jujube species have become popular and more important in semi-arid and arid marginal lands due to their remarkable tolerance and adaptability to water-deficit conditions as well as salty, barren soil and deserve to be comprehended as a super fruit for future use because of its unique features and benefits (Liu et al., 2020). Jujube species are also suitable for cultivation in the temperate and subtropical areas where other common fruit crops are less productive and has significantly contributed to reduce poverty and reforestation in China (Liu et al., 2014; Song et al., 2019).

2.1 Extreme Temperature

Temperature is an important abiotic stress factor imparting dramatic effects on production of crops in general. Rate of plant development, plant color, stem elongation, number of flowers and flower size can all be manipulated by altering temperatures (Erwin, 1991). More climatic variability has been witnessed in the recent past, particularly in post-monsoon season due to rising temperature and declining relative humidity. The post-monsoon weather data (2015–2017) have revealed a change in temperature and humidity, leading to decreased fruit set and reduced yield. The increased harsh temperature affects the growth of this fruit crop. Extreme temperatures are accompanied by floods and famines and eventually reduce the harvests as reported in South India where about 3–6°C temperature increased (Lonergan, 1998). The fifth report of IPCC has predicted an increase of 2–4.8°C temperature till 2080 (Chaturvedi et al., 2012). In contrast, similar situation was observed in USA during 2010–2012 in the region's corn-belt that led to premature budding owing to warmer winter season that ultimately caused loss of net worth $220 M (USGCRP, 2014).

Indian jujube is one of such fruit crops that survived well under varying conditions (Meghwal et al., 2018). Indian Jujube can tolerate a wide range of temperature (5–49°C) and requires less rainfall (Hocking, 1993). The plant can grow in areas where low temperature ranges from 7°C to 13°C while the highest temperature is around 50°C (Kaaria, 1998). This advantage allows the farmers to grow crops in newly warm regions when these were previously grown only in existing warm areas. On the contrary, the increasing warmer temperature exceeding the optimal temperature of specific crop results in yield decline. In these regions, crops like *Z. mauritiana* are favorably cultivated as they produce large quantities of non-wood products (Sadio and Negreros-Castillo, 2006). Crop management is also crucial to extremely increasing temperatures in their reproductive and maturation stages (Hatfield et al., 2011). The non-synchronization in flowering patterns as male and female flowering phases and reduced pollen viability (up to 25–58%) at higher temperature up to 45°C seemed to be a key factor in reducing yield of *Z. mauritiana* (Tel-Zur and Schneider, 2009; Asatryan and Tel-Zur, 2013). Nevertheless, the responses of these crops to a variety of temperature ranges are different in each species. Increasing temperatures during the crop growing season has a consequence of forcing the plants to consume extra energy for respiration; for an increase of 1°C in average temperature in a region, about 5–10% yield of the crop species decreases (Lobell and Field, 2007; Hatfield et al., 2011). An example to such an effect was shown in Indian jujube that exhibited higher respiration rate in addition to ethylene production (Jat et al., 2013). However, the shelf life of the fruit was observed to be enhanced at lower temperatures. The increasing average temperatures also tend the plants to complete their life cycle faster (Hatfield et al., 2011). Lack of enough time for reproduction causes rapid reproductive failures along with lowering of crop yield (Craufurd and Wheeler, 2009). Another noteworthy fact is that the process of photosynthesis present in C_3 plants is more sensitive than C_4 crops to increasing temperatures (Lipiec et al., 2013). However, *Z. mauritiana* bears two crops every year which is evidence of its higher tolerance against temperature changes. Early Gola varieties like CAZRI and 'Gola' maintained greater yield under variable climatic conditions. Climate change is a global issue which urgently calls for adaptive measures, like microclimate modification and rescheduled pruning to minimize the adverse effects of climatic irregularities (Meghwal, 2018).

2.2 *Elevated Atmospheric Carbon Dioxide*

Elevated CO_2 reduced the nutritional quality of alfalfa and soybean. On the other hand, there are some crop plants that have the tendency to enhance the levels of CO_2 within their leaves prior to start of photosynthesis, e.g., *Z. mauritiana*. Owing to this tendency, the enhanced levels of CO_2 in the air do not offer any aid to these plants in normal circumstances. However, in extreme humid environment, plants showed reduced transpiration losses and little effects on the overall yield (Simpson, 2017). In such crucial conditions, the best strategy used by local farmers is cultivation of *Z. mauritiana* that plays its role in mitigating the changing climate owing to its tree component as well as amassing CO_2 in its biomass (Toppo and Raj, 2018).

3. Sensitivity to Air Pollutants and Nutritional Contaminants

Crops are generally sensitive to several air pollutants, including oxidants, sulfates, fluorides and particulate matter. Out of these, the most crucial to crops is ozone which too is the main air pollutant. Ozone is the main apprehension for global food security along with nitrogen oxides and unstable organic compounds (Griffiths, 2003), but all the crops behave in a different manner to ozone. Some crops are more sensitive towards ozone than others. There are some most sensitive crops (wheat and soybean); some others are moderately sensitive (potato, rice and maize). At the same time, Indian jujube and barley (Kaur et al., 2017) are ozone-resistant crops. They have high air pollution-tolerance indexes. Additionally, the particulate matter, like dust or carbon deposited on crop plants, hinders the processes of photosynthesis and respiration within their leaves. Furthermore, gathering of alkaline dust within the soil enhances pH of soil to a certain level that is unfavorable for crops to grow. However, *Z. mauritiana* is found to be tolerant to air pollutants (Dagar et al., 2016). In a previous study, the epidermal characters of *Z. mauritiana* showed reduction in stomatal index and trichrome length but an increase in trichome frequency. Significant epidermal changes were observed in species at all the polluted sites owing to air pollution (Rao and Dubey, 1992). The study was based on differential accumulation of metals in relation to metal content of the deposited dust and the modified epidermal traits that showed a significant scavenging potential in the species (Rao and Dubey, 1992).

Heavy metals, like copper, nickel, zinc, cobalt, copper or iron are essentially required for plant growth. These become contaminants when absorbed in surplus within soil (Khan et al., 2018; Narendrula-Kotha et al., 2019). Recently, researchers have observed an upsurge in nutrient contamination of soils that is leading to loss in productivity of crop (Proshad et al., 2018). The jujube crop yield is exposed to metal toxicity owing to inappropriate irrigation approaches and unnecessary quantities of synthetic fertilizers and nutrients (Pasala et al., 2017). Some other heavy metals, such as arsenate, lead or cadmium in trace amounts harm the overall productivity of crops when they are absorbed from the soil (Khalid et al., 2018). The accumulation of these nutritious contaminants in crop plant reduces productivity and is a considerable risk for animal and human health (Couto et al., 2018). However, recent studies have shown that *Z. mauritiana* possesses characteristics to reduce heavy metals from wastewater, using bark of the plant (Sudhir et al., 2020).

4. Abiotic Stress Tolerance

Increase in abiotic stress due to climate change has escalated the dryness and desertification in different regions of the world. There are several factors that cause abiotic stress and limit crop yield (Canter, 2018; Zörb et al., 2019). They distress the crop plants in terms of growth and net yield (Vaughan et al., 2018; Waqas et al., 2017; Zafar et al., 2018). Most of cultivated fertile agricultural fields are likely to have one of the major abiotic stress factors which cause yield losses in major food crops (dos Reis et al., 2012; Mantri et al., 2012). In this decade, there were huge shifts of climate change, due to which it was important to identify those plant crops that can survive low-water environments, help in reducing the local desertification and mainly mitigate the changing climate, including the greenhouse emissions. Genus *Ziziphus* has a tremendously tolerant nature towards abiotic stress. A recent study in Sudan confirmed the characteristics of *Z. spina-christi* as a thermophilic, highly heat- and drought-tolerant plant (Zait and Schwartz, 2018). Recently its trees have become more abundant in the Mediterranean region which suggests its tolerance to changing climate. The results have resulted in the selection of *Z. spina-christi* as an attractive tree for afforestation in these regions (Zait and Schwartz, 2018). In sodic lands, 10-year-old trees of *Z. mauritiana* and *Psidium guajava* produced 12–25 Mg ha fruit per annum that indicates their potential as candidate crops for agroforestry (Dagar and Yadev, 2017). Heavy metal stress also plays its role in stunting the plant growth. However, recent studies have confirmed the role of *Z. mauritiana* in reducing heavy metals (mainly cadmium) using bark of the plant (Sudhir et al., 2020).

4.1 Effect of Drought Stress on Yield

Availability of water is the main driver of the ecosystem processes. According to Farooq et al. (2009), the impact of drought stress on nutrient supply through mineralization depends on the severity and duration of drying and rewetting cycles. This is attributed to various factors, such as soil type affecting water potential, evapotranspiration demand, frequency and occurrence of rainfall. Hence, drought stress influences change in the plant nitrogen and phosphorous cycles (Sardans and Penuelas, 2012). Concentrations of nitrogen (N) and phosphorous (P) in plants are basic indicators of limitation of N and P elements (Aerts and Chapin, 2000; Gusewell, 2004).

Drought stress retards growth of the plant because of reduction in N and P uptake, re-distribution and transport (Rouphael et al., 2012). In most of the studies, plants have shown decreased N and P uptake with a decline in the soil moisture (Sardans and Penuelas, 2012). Reduction in soil moisture and drought stress reduces nutrient diffusion and mass flow in soil (Lambers et al., 2008) and decrease in nutrient uptake by reducing nutrient supply through mineralization (Sanaullah et al., 2012). Drought stress could be removed by rewetting of soil frequently which results in improved mineralization, accredited to release of nutrients from dead organisms that have accumulated during the drying phase (Austin et al., 2004).

Drought stress is the most alarming factor with the speedy change in global climate, thus affecting the plant growth and plant yield adversely (Vurukonda et al., 2016; Chandra et al., 2018). Jujube species are sclerophyllous shrubs or trees having a great degree of resistance against drought stress. Several studies have examined their morphological and physiological adaptations against water deficit (Maraghni et al., 2011; Kulkarni et al., 2010). The capability of jujube species to thrive under drought conditions has been characterized due to a blend of tolerance and avoidance mechanisms, including stomatal control and osmotic adjustment (Maraghni et al., 2011). During water-deficit state, generally the plant started to respond against biochemical adaptations which increased the synthesis of organic solutes, including sugar alcohol, soluble sugars, betaines and proline (Serraj and Sinclair, 2002). These organic solutes cause cellular osmotic modification, membrane integrity protection, ROS detoxification and enzyme stabilization (Attipali et al., 2004).

In water-deficit state, plant triggers continuous reactive oxygen species (ROS) production in mitochondria, peroxisomes and chloroplast (Apel and Hirt, 2004). Moreover, excessive production of ROS in plant cells has the tendency to damage the cellular components including protein, lipids, DNA and ultimately destroy the photosynthetic pigments (Mittler, 2002). Plant cells have modified their enzymatic [Guaiacol peroxidase (POD), Superoxide dismutase (SOD), catalase (CAT)] and non-enzymatic (β-carotenes, ascorbate, atocopherol, glutathione) defense system to scavenge ROS (Sofo et al., 2008). Under drought stress the behavior of many shrubs showed that antioxidant enzymes activity directly correlates with their resistance or tolerance to abiotic stress (Dichio et al., 2003; Sofo et al., 2004; Guo et al., 2006). Though jujube is considered as a resilient crop but over the time, drought stress has become a more severe threat in arid regions which limit growth of jujube plant and reduce its productivity. Hence, pruning (Chen et al., 2016) plant cover and gene breeding (Ji et al., 2019) are the approaches which are mostly used to alleviate the drought stress in jujube but these are time consuming methods. Zhang et al. (2020) reported that rhizosphere microbial populations transfer in drought-tolerant plants assist their host plants to adapt their drought stress conditions.

Genetic studies of jujube gene families revealed their role in stress and defense responses. About 854 gene families have been functionally marked for their response to stress. Jujube-specific genes are enriched with the arginine metabolism pathway that plays a significant role in perception and adaptation of plants in response to the climatic fluctuations. During osmotic stress conditions, some genes showed higher expression at fruit development stages which strongly supported the jujube's potential towards drought resistance. Some pathogenesis proteins, including chitinases, were activated against biotic and abiotic stresses in plants. Jujube is the most eminent fruit crop of the rain-fed regions and performs well under severe drought conditions. Branching physiology and hairy leaves of jujube help it to minimize water loss; hence, it can withstand long dry spells and has the marvelous tendency to endure drought conditions (Mani and Suresh, 2018). Indian jujube is resistant to extreme drought conditions during the growing season, low winters during dormancy and saline water (Jain and Dass, 1988; Maruza et al., 2017). Cruz et al. (2012) demonstrated that *Z. jujuba* can cope with extreme water deficit while maintaining leaf turgor that provides good gas exchange and leaf productivity. However, in low cascade regions of Pennsylvania and Washington state, the jujube plants bear fruit. Fruiting was also noticed in a few cultivars at northern Florida where Indian jujube was found to be more frost-sensitive; however, its fruit was of inferior quality. Northwest China has arid climate and well-known Chinese jujube cultivars. Annual precipitation is normally less than 200 mm in arid, 200–450 mm in semi-arid and 450–650 mm in the sub-humid regions (Qu and Wang, 1993). In cold areas, deserts and mountains, jujubes can grow well across diverse climates (Rushforth, 1999).

4.2 *Salinity*

Salinity is considered as the most important abiotic stress that drastically affects the plant growth. The salt tolerance can mitigate adverse effects of salinity by inducing changes in photosynthesis (Ebert, 2000), stomatal behavior (Pradeep and Jumbhale, 2000), chlorophyll content (Pandey et al., 1991) and accumulation of metabolites (Hooda et al., 1990). The most crucial part of plant is its cell membrane that is affected by salt stress injuries (Levitt, 1980). Jujube can be cultivated with little care even in marginal and saline soils. Research conducted on two species of jujube, i.e., *Z. rotundifolia* and *Z. nummularia,* indicated that the exposure to saline conditions decreased the rates of photosynthesis and transpiration (Agarwal et al., 2013). In addition, the membrane injury and proline content increased. Sodium content was the highest in roots and leaves of *Z. rotundifolia* and *Z. nummularia,* respectively. The potassium content was higher in leaves of *Z. rotundifolia* and both species were found tolerant to salinity.

Fruit trees including *ber* (*Z. mauritiana*) could be successfully cultivated in salt-affected black soils and it could tolerate sodic vertisols up to ESP 25–60. *Z. mauritiana* has also been found suitable for cultivation in soils with ECe up to 10–20 dSm^{-1} and could produce good biomass (Dagar and Yadev, 2017). In jujube cultivars, higher salinity level had negative effect on plant survival, its establishment and chlorophyll content whereas protein hydrolysis stimulated higher accumulation of free amino acids in leaves (Verma et al., 2018a). High levels of salts in soil restricted plant growth. Plant height, stem diameter, number of leaves, dry matter, mineral composition of leaves and yield were reduced above 12 dSm^{-1} ECe conductivity. The foliar content of Na^+ and Cl^- increased at higher levels of salinity, while N, P, K, Ca and Mg contents were reduced. Specific injury because of higher accumulation of chloride and sodium ions in plant foliage appeared to be the main impact of salinity. Jujube cultivars, Banarsi Karaka, Narendra Ber Sel-2 and Ponda, were considered as salt-tolerant cultivars and Narendra Ber Sel-1 and Gola were classified as semi-tolerant cultivars. The identification of higher and medium salt-tolerant cultivars may help in commercial production of jujube in high salinity regions (Verma et al., 2018b). An overview of the salt-tolerance studies and salient findings are presented in Table 4.

During osmotic stress, some genes were highly expressed at fruit development stages strongly supporting the potential of jujube species

Table 4. Salt stress tolerance in jujube species.

Jujube Species/ Varieties	Chemicals/ treatments	Results	Conclusion/ Suggestions	References
Z. mauritiana	Salinity levels (EC 0–16 dSm^{-1})	Higher salinity level decreased plant growth and leaf mineral composition. Na^+ and Cl^- content of leaves increased and caused leaf injury.	Tolerant varieties (Banarsi Karaka, Narendra Ber Sel.-2 and Ponda)	Verma et al. (2018a)
Banarsi Karaka Narendra Ber Sel. (1, 2, & 3), Ponda and Gola	Salts ($CaCl_2$, $MgCl_2$, NaCl and Na_2SO_4) at different ratio (1:1:1:2)	Salinity stress had negative impact on plant establishment, survival and chlorophyll content across varieties. Free amino acids increased in leaves.	Semi-tolerant varieties (Narendra Ber Sel-1 and Gola)	Verma et al. (2018b)
Z. mauritiana *Z. rotundifolia* *Z. nummularia*	Saline irrigation (EC 0–20 dSm^{-1}) by mixing $NaCl_2$, $NaSO_4$ and $MgCl_2$ (3:1 of chlorides and sulphates)	Photosynthetic activity decreased at higher salt stress in *Z. nummularia*. Proline anabolic enzyme, pyrolline-5-carboxylate reductase, pyrolline-5-carboxylate synthase, ornithine, sodium contents were more in root zone of *Z. rotundifolia*.	*Z. rotundifolia* is more salt tolerant due to better physical attributes and osmolytic defense mechanism.	Bagdi and Bagri (2016)
Z. mauratiana (Delhi white, Suffan, Karela, Mehmud wali)	Normal water, brackish water (11 dsm^{-1}) and 50% brackish + 50% normal water	Brackish water least affected fruit production and quality.	Delhi White is better for cultivation in marginal lands.	Sherani et al. (2017)
Z. nummularia *Z. rotundifolia*	Seedlings grown under salt stress (100 mM and 200 mM)	High salinity promoted the MDA, H_2O_2 and proline contents and inhibited CAT, PER and SOD activities in both species and high antioxidants were maintained in *Z. rotundifolia*. At low salinity, both species showed enhanced levels of GSH and ACS.	*Z. rotundifolia* performed better at higher salinity level compared with *Z. nummularia*.	Sankhla et al. (2008)
Z. mauritiana (Gola and Umran)	NaCl (EC 0–16 dS m^{-1})	Photosynthetic activity was lower in Umran. Leaf relative water content, chlorophyll (Chl) and membrane stability decreased under stress. Enzyme activity and K^+/Na^+ ratio was higher in Gola.	Salt tolerance was higher in cultivar Gola.	Agrawal et al. (2013)

towards salt tolerance. Proteins like chitinases are activated against abiotic stresses in plants. Jujube shoots also showed higher expression of chitinases-regulating tolerance towards higher salinity.

5. Breeding and Climate Resilience

Genetic diversity profiling provides the basis for establishment of crop evolutionary trends and provision of basic knowledge for crop improvement programs (Awasthi and More, 2008). Greater genetic diversity is required in the germplasm collections for more stringent selection of the desired individuals having traits of interest. During 1950's to 1980's, 930 jujube varieties (90% of global jujube genotypes) were evaluated against biotic and abiotic stresses and other fruit quality attributes (Wang et al., 2019). Though many germplasm accessions are available, however, there are a few improved cultivars compared with other major tree-fruit crops. Most of the available cultivars are farmer-selected landraces and traditional varieties (Li et al., 2015; Liu et al., 2016).

Jujube breeding is still in its infancy and there are various knowledge gaps existing in jujube breeding methods and understanding its genetics. Like many other fruit crops, including mango, flower size in jujube is very small (5–7 mm in diameter), hence emasculation and pollination procedures are quite difficult. Further, rate of flower drop, embryo abortion are quite high, whereas fruit setting frequency is low (1%). Recovery of hybrids through conventional breeding techniques is quite low (0.01%) mainly because of limited knowledge about reproductive biology of the jujube germplasm, particularly regarding male sterility and self-incompatibility patterns (Wang et al., 2016). Asatryan and Tel-Zur (2013) reported existence of gametophytic self-incompatibility system in three *Ziziphus* species, including *Z. mauritiana*. Out of 179 jujube cultivars, 87.8% were found to be self-fruitful (Liu and Wang, 2009) whereas out of 56 cultivars studied for their floral biology and seed development in New Mexico, only seven varieties were found to be self-compatible and self-fruitful (Yao et al., 2013).

Utilization of genomic approaches has enhanced our understanding of the molecular mechanism of self-compatibility in jujube to enhance fruit yield. The genetic mechanism of gametophytic self-incompatibility systems was found to be similar, based on re-sequencing results of 31 accessions of wild (*Z. jujuba* var. spinosa) and cultivated jujubes (Asatryan and Tel-Zur, 2013; Huang et al., 2016). Haplotype sharing in the parents resulted in development of seedless jujube kernels (Huang et al., 2016) and the incompatibility system could be overcome by crossing two different haplotypes.

Male sterility phenomenon has been reported in five Chinese jujube cultivars which could help to overcome the cross-hybridization barrier by eliminating flower emasculation and reducing embryo abortion (Liu et al., 2018). Liu et al. (2014) developed a controlled crossing method by using male sterility and planting parents together under a net covering during flowering season. These plants were naturally pollinated by honeybees and hybrid jujube populations were obtained for further molecular screening (Qi et al., 2009; Cui, 2011). Great frequency of the mislabeled germplasm (48% out of 947 accessions) was reported by Xu et al. (2016) in the Chinese jujube germplasm. A similar report has been documented by Song (2019) using SNP markers and 17 synonymous groups were identified in 114 cultivars. Chinese jujube (*Z. jujuba*) and Indian jujube (*Z. mauritiana*) are sister species that belong to different sections of genus *Ziziphus*. However, their interspecific crosses have been fertile, indicating cross compatibility and potential of interbreeding in both species (Sheng and Shen, 2011; Song et al., 2019). The development of marker-assisted germplasm characterization system for traits like self-compatibility need to be implemented in breeding programs. Fingerprinting using molecular markers has been quite useful in identification of the jujube species, different accessions, their parentage and cultivars (Xu et al., 2016).

6. Mitigation and Crop Improvement Approaches

Climate change resilience challenges could be effectively mitigated by more efficient utilization of the available wild relatives of jujube, elite plant resources and development of better varieties having novel traits for greater tolerance to abiotic stress factors. There is need to have deeper knowledge of plant growth patterns, different phenological stages and their alterations under changing climatic conditions. Being a tree fruit crop, it is difficult to study the impact of environmental cues under controlled conditions. Hence, better understanding of the crop adaptability patterns to diverse climatic conditions, greater integrated efforts and development of better simulation models are required to mitigate the impact of climate change, enhance climate resilience and improve plant growth and production. Some of these approaches are as follows:

a) Cultivation of *Z. mauritiana* can play an important role in mitigating the changing climate owing to its tree component as well as amassing of CO_2 in its biomass.
b) *Ziziphus mauritiana* fruit production under stress conditions indicates higher potential to be dealt as a candidate crop for agroforestry.
c) Heavy metal stress plays its role in stunting the plant growth. Recent studies confirmed the role of *Z. mauritiana* in reducing heavy metals, especially cadmium, using plant bark.
d) Abiotic stress tolerant species shall be preferred for interspecific breeding programs. Among Ziziphus species, *Z. rotundifolia* performed

better as compared with *Z. nummularia* at higher salinity levels and was more salt tolerant due to better physical attributes and osmolytic defense mechanism. Salt-tolerant (Banarsi Karaka, Narendra Ber Sel.-2, Ponda and Gola) and semi-tolerant (Narendra Ber Sel-1 and Gola) varieties may be planted in saline areas whereas Delhi White is recommended for cultivation in marginal lands.

e) Pruning, plant cover and candidate gene breeding approaches shall be used to alleviate drought stress in jujube.

f) There is need to enhance biodiversity, reforestation and conservation of other renewable resources, like fresh water and fertile soils. Use of mulching could help to conserve moisture and enhance soil microbial activity, particularly in the rain-fed areas. The pruned jujube branches may also be used as mulch in jujube species (Tang et al., 2021).

g) Perennial horticultural crops, like jujube also help in mitigation of climate change by absorbing excessive radiation.

h) Development of highly efficient hybrid breeding system is possible in Indian jujube by combining the male-sterile germplasm, utilization of embryo rescue technology, net controlled hybridization and molecular marker-based characterization. Male sterility has been reported in a few cultivars of Chinese jujube and it could help to overcome cross-breeding bearer and reduce frequency of embryo abortion.

i) Initiation of short-term and long-term breeding programs for the development of abiotic stress tolerant cultivars could be beneficial. Varieties having higher RUBISCO activity, less stomatal conductance, higher water use efficiency (WUE) and having genes for stress-related proteins, including chaperonins, dehydrins, aquaporins, enzymes regulating detoxification of ROS and dehydrins, shall be developed.

j) Thermographic tools could be used for higher throughput analysis of the heat, frost and drought-tolerant genotypes.

k) Existence of higher level of homonyms and synonyms in jujube germplasm collections in different parts of the world, including Pakistan, India, China and other countries, need to be addressed. There is a need for extensive genetic characterization of the candidate germplasm selection by DNA fingerprinting using genetic markers, like single nucleotide polymorphism (SNP) for genetic identification. The breeding strategy for Indian jujube and other related species may depend upon clonal selection from the progenies of inter-varietal crosses and different wild landraces (Wang et al., 2005), which usually present higher abiotic stress tolerance.

l) The development of a system of marker-assisted germplasm characterization for traits like self-compatibility need to be implemented in jujube breeding programs. Further, there is need to screen elite jujube germplasm for incompatibility as well as for future breeding and crop improvement programs.

m) Genome editing technologies including CRISPR/Cas9 tools could be utilized for creating site specific mutations and gene expression regulation with editing in transcription factors like WRKY, MAPK, DREB, etc.

n) Development of genomic tools in jujube is still limited; however, a draft genome (ID: 15586) of jujube species (*Z. jujuba*) and linkage maps have been developed (Liu et al., 2014). Sequencing of Chinese jujube cultivars has been quite useful and enabled identification of thousands of SNP markers widely used by breeders for varietal improvement (Liu et al., 2014; Huang et al., 2016). There is need to identify QTLs associated with abiotic stress tolerance across the jujube genome. Furthermore, next generation sequencing (NGS), whole genome sequencing (WGS) technologies may be used more extensively for screening of the available germplasm collections in *Z. mauritiana* and other species. The identification of novel stress-related candidate genes could be important for future crop improvement programs. Cultivation of abiotic stress-tolerant cultivars shall be enhanced in areas with more stress. Utilization of OMICs tools could be advantageous in identification of genes related to abiotic stress tolerance, including environmental contaminants.

o) There is need to develop international collaborations with leading horticulture, physiology and genetic research groups worldwide for the development of climate-resilient varieties.

7. Remote Sensing and Environmental Certification

Crop modelling systems enumerate crop development impacted by soil condition, climate and management situation and are considered a precise and mature method for yield prediction (de Wit et al., 2019). According to van Diepen et al. (1989), several crop models have been developed for different crops in the past few decades and one of them is World Food Studies (WOFOST). The remotely sensed information is incorporated into different crop models to estimate the obtained yield. Bai et al. (2019) integrated single-leaf area indices (LAI) to predict yield of jujube trees at field scale to maximum vegetative development stages derived from the satellite data into a calibrated WOFOST model. The research provided a strategy to employ remotely sensed state variables and a crop growth model to improve field-scale yield estimates for fruit tree crops. In addition, the simulation of growth dynamics of annual crop and yield prediction was carried out for *Z. jujuba* (Bai et al., 2020). Simulated growth dynamics of dry weights of various plant parts including stems, leaves, fruit, LAI and total biomass agreed with the measured values, showing Root Mean Square Error values. The proposed model was verified for prediction of accurate jujube yields. It also provided a fundamental strategy for simulating growth of other fruit trees. Remote sensing is a way forward towards environmental certification using this highly tolerant species to combat climate change.

8. Conclusion and Prospects

Drastic climatic changes have declined global fruit production and perennial fruit crops, like jujube species are also facing abrupt climate shifts. Jujube species, including *Z. mauritiana*, have been more tolerant to abiotic stress conditions, like salinity, drought, elevated temperatures, higher atmospheric CO_2 and pollutants like O_3. Jujube germplasm have wide genetic diversity for tolerance against such stress factors; hence, there is need to select climate-resilient strains followed by clonal multiplication and initiation of breeding programs for the development of improved hybrids having greater tolerance to climatic variability. Salt and drought-tolerant varieties shall be cultivated on marginal lands and could be helpful in land reforestation. Phenotypic studies shall be conducted more extensively under contrasting climatic and abiotic stress conditions to understand varietal responses to variable environmental factors. Utilization of microarray, NGS, WGS could help to underpin the candidate genes showing differential expression under contrasting climatic and abiotic stress conditions. Furthermore, system biology approaches are useful to understand complex multigenic traits, including abiotic stress tolerance. Integrated omic approaches may unravel interplay of the phenochemical, physiological, differential gene expression and transcriptional responses for future genetic improvement programs. The outcomes of basic research using conventional and modern breeding approaches could be translated into technologies and services for improving farming and development of improved varieties having better tolerance against abiotic and biotic stress factors. The integrated utilization of these strategies will enable mitigation of climate changes, increase productivity and further enhance resilience in the relatively climate-smart Indian jujube.

References

Aerts, R. and Chapin III, F.S. (2000). The mineral nutrition of wild plants revisited: A re-evaluation of processes and patterns. pp. 1–67. *In*: Fitter, A. and Raffaelli, D. (eds.). *Advances in Ecological Research*. Elsevier, London, UK.

Agrawal, R., Gupta, S., Gupta, N.K., Khandelwal, S.K. and Bhargava, R. (2013). Effect of sodium chloride on gas exchange, antioxidative defense mechanism and ion accumulation in different cultivars of Indian jujube (*Ziziphus mauritiana* L.). *Photosynthetica*, 51(1): 95–101.

Anjum, M.A., Rauf, A., Bashir, M.A. and Ahmad, R. (2018). The evaluation of biodiversity in some indigenous Indian jujube (*Ziziphus mauritiana*) germplasm through physico-chemical analysis. *Acta Scientiarum Polonorum Hortorum Cultus*, 17(4): 39–52.

Apel, K. and Hirt, H. (2004). Reactive oxygen species: metabolism, oxidative stress, and signal transduction. *Annual Review of Plant Biology*, 55: 373–399.

Arndt, S.K., Clifford, S.C., Wanek, W., Jones, H.G. and Popp, M. (2001). Physiological and morphological adaptations of the fruit tree *Ziziphus rotundifolia* in response to progressive drought stress. *Tree Physiology*, 21(11): 705–715.

Asatryan, A. and Tel-Zur, N. (2013). Pollen tube growth and self-incompatibility in three *Ziziphus* species (Rhamnaceae). *Flora-morphology, Distribution, Functional Ecology of Plants*, 208(5-6): 390–399.

Attipali, R.R., Kolluru, V.C. and Munusamy, V. (2004). Drought induced responses of photosynthesis and antioxidant metabolism in higher plants. *Journal of Plant Physiology*, 161: 1189–1202.

Austin, A.T., Yahdjian, L., Stark, J.M., Belnap, J., Porporato, A., Norton, U., Ravetta D.A. and Schaeffer, S.M. (2004). Water pulses and biogeochemical cycles in arid and semiarid ecosystems. *Oecologia*, 141(2): 221–235.

Awasthi, O.P. and More, T.A. (2008). Genetic diversity and status of *Ziziphus* in India. pp. 33–40. *In*: *Ist International Jujube Symposium*. 840, India.

Awasthi, O.P., More, T.A. and Liu, M.J. (2009). Genetic diversity and status of *Ziziphus* in India. *Acta Horticulture*, 840: 33–40.

Azam-Ali, S., Bonkoungou, E., Bowe, C., DeKock, C., Godara, A. and Williams, J. (2001). Fruits for the future *ber* and other jujubes. *Int. Centre Underutilized Crops*. Univ. of Southampton, Southampton, SO17 1BJ, UK.

Bagdi, D.L. and Bagri, G.K. (2016). Effect of saline irrigation water on gas exchange and proline metabolism in ber (*Ziziphus*). *Journal of Environmental Biology*, 37(5): 873.

Bai, T., Zhang, N., Mercatoris, B. and Chen, Y. (2019). Improving jujube fruit tree yield estimation at the field scale by assimilating a single landsat remotely sensed LAI into the WOFOST model. *Remote Sensing*, 11(9): 11–19.

Bai, T.C., Tao, W.A.N.G., Zhang, N.N., Chen, Y.Q. and Mercatoris, B. (2020). Growth simulation and yield prediction for perennial jujube fruit tree by integrating age into the WOFOST model. *Journal of Integrative Agriculture*, 19(3): 721–734.

Bal, J.S., Sharma, D.R. and Singh, P. (2011). Historic *Ber* (jujube) trees in golden temple Amritsar (Punjab)-twenty years of consistent care. pp. 37–41. *In*: *2nd International Jujube Symposium*. 993. India.

Bhatt, M.J., Patel, A.D., Bhatti, P.M. and Pandey, A.N. (2008). Effect of soil salinity on growth, water status and nutrient accumulation in seedlings of *Ziziphus mauritiana* (Rhamnaceae). *Journal of Fruit and Ornamental Plant Research*, 16(1): 383–401.

Bhattacherjee, T.N., Robbani, M., Ali, M. and Mehedi, M.N.H. (2019). Biodiversity of indigenous jujube germplasm available in Dumki Upazila. *Asian Journal of Plant Science and Research*, 9(1): 22–31.

Canter, L.W. (2018). *Environmental Impact of Agricultural Production Activities*. Broken Sound Parkway NW: CRC Press.

Chandra, D., Srivastava, R., Glick, B.R. and Sharma, A.K. (2018). Drought-tolerant Pseudomonas spp. improve the growth performance of finger millet [*Eleusine coracana* (L.) Gaertn.] under non-stressed and drought-stressed conditions. *Pedosphere*, 28(2): 227–240.

Chaturvedi, R.K., Joshi, J., Jayaraman, M., Bala, G. and Ravindranath, N.H. (2012). Multi-model climate change projections for India under representative concentration pathways. *Current Science*, 791–802.

Chen, D., Wang, Y., Wang, X., Nie, Z., Gao, Z. and Zhang, L. (2016). Effects of branch removal on water use of rain-fed jujube (*Ziziphus jujuba* Mill.) plantations in Chinese semiarid Loess Plateau region. *Agricultural Water Management*, 178: 258–270.

Cherry, M. (1985). The needs of the people. pp. 1–8. *In*: *Plants for Arid Lands*. Springer, Switzerland AG.

Choi, S.H., Ahn, J.B., Kozukue, N., Levin, C.E. and Friedman, M. (2011). Distribution of free amino acids, flavonoids, total phenolics, and antioxidative activities of jujube (*Ziziphus jujuba*) fruits and seeds harvested from plants grown in Korea. *Journal of Agricultural and Food Chemistry*, 59(12): 6594–6604.

Couto, R.R., Faversani, J., Ceretta, C.A., Ferreira, P.A.A., Marchezan, C., Facco, D.B., Garlet, L.P., Silva, J.S., Comin, J.J., Bizzi, C.A. and Flores, E.M.M. (2018). Health risk assessment and soil and plant heavy metal and bromine contents in field plots after ten years of organic and mineral fertilization. *Ecotoxicology and Environmental Safety*, 153: 142–150.

Craufurd, P.Q. and Wheeler, T.R. (2009). Climate change and the flowering time of annual crops. *Journal of Experimental Botany*, 60(9): 2529–2539.

Cruz, Z.N., Rodriguez, P., Galindo, A., Torrecillas, E., Ondono, S., Mellisho, C.D. and Torrecillas, A. (2012). Leaf mechanisms for drought resistance in *Ziziphus jujuba* trees. *Plant Sciencei* 197: 77–83.

Cui, N., Du, T., Kang, S., Li, F., Zhang, J., Wang, M. and Li, Z. (2008). Regulated deficit irrigation improved fruit quality and water use efficiency of pear-jujube trees. *Agricultural Water Management*, 95(4): 489–497.

Cui, X.M. (2011). *Acquirement and Molecular Identification of Hybrids between Chinese Jujube and Wild Jujube*. MS thesis, Agricultural University of Hebei, Baoding.

Dagar, J.C., Singh, G. and Singh, N.T. (2001). Evaluation of forest and fruit trees used for rehabilitation of semiarid alkali-sodic soils in India. *Arid Land Research and Management*, 15(2): 115–133.

Dagar, J.C., Sharma, P.C., Chaudhari, S.K., Jat, H.S. and Ahamad, S. (2016). Climate change vis-a-vis saline agriculture: impact and adaptation strategies. pp. 5–53. *In*: *Innovative Saline Agriculture*. Springer, New Delhi.

Dagar, J.C. and Yadav, R.K. (2017). Climate resilient approaches for enhancing productivity of saline agriculture. *Journal of Soil Salinity and Water Quality*, 9(1): 9–21.

de Wit, A., Boogaard, H., Fumagalli, D., Janssen, S., Knapen, R., van Kraalingen, D., Supit, I., van der Wijngaart, R. and van Diepen, K. (2019). 25 years of the WOFOST cropping systems model. *Agricultural Systems*, 168: 154–167.

Dichio, B., Xiloyannis, C., Angelopoulos, K., Nuzzo, V., Bufo, S.A. and Celano, G. (2003). Drought-induced variations of water relations parameters in Olea europaea. *Plant Soil*, 257: 381–389.

Dos Reis, S.P., Lima, A.M. and De Souza, C.R.B. (2012). Recent molecular advances on downstream plant responses to abiotic stress. *International Journal of Molecular Sciences*, 13(7): 8628–8647.

Ebert, G. (2000). Salinity problem in subtropical fruit production. *Acta Horticulture*, 531: 99–105.

Edwards, C.E., Parchman, T.L. and Weekley, C.W. (2012). Assembly, gene annotation and marker development using 454 floral transcriptome sequences in *Ziziphus celata* (Rhamnaceae), a highly endangered, Florida endemic plant. *DNA Research*, 19(1): 1–9.

Erwin, J.E. (1991). Diurnal variations in temperature affect carbon partitioning in New Guinea Impatiens and fuchsia shoots. *Horticulture Sciences*, 26: 117.

Farooq, M., Wahid, A., Kobayashi, N., Fujita, D.B.S.M.A. and Basra, S.M.A. (2009). Plant drought stress: effects, mechanisms and management. pp. 153–188. *In*: *Sustainable Agriculture*. Springer, Dordrecht.

Gao, Q.H., Wu, P.T., Liu, J.R., Wu, C.S., Parry, J.W. and Wang, M. (2011). Physico-chemical properties and antioxidant capacity of different jujube (*Ziziphus jujuba* Mill.) cultivars grown in loess plateau of China. *Scientia Horticulture*, 130(1): 67–72.

Godt, M.J.W., Race, T., and Hamrick, J.L. (1997). A population genetic analysis of *Ziziphus celata*, an endangered Florida shrub. *Journal of Heredity*, 88(6): 531–533.

Griffiths, H. (2003). *Effects of Air Pollution on Agricultural Crops*. Ministry of Agriculture, Food and Rural Affairs, Ontario, Canada.

Guo, Z., Ou, W., Lu, S.Y. and Zhong, Q. (2006). Differential responses of antioxidative system to chilling and drought in four rice cultivars differing in sensitivity. *Plant Physiology and Biochemistry*, 44(11-12): 828–836.

Gupta, A.K. (2004). Origin of agriculture and domestication of plants and animals linked to Early Holocene climate amelioration. *Current Science*, Bangalore, 87: 54–59.

Gusewell, S. (2004). N:P ratios in terrestrial plants: Variation and functional significance. *New Phytologist*., 164(2): 243–266.

Haldhar, S.M., Deshwal, H.L., Jat, G.C., Berwal, M.K. and Singh, D. (2016). Pest scenario of *ber* (*Ziziphus mauritiana* Lam.) in arid regions of Rajasthan. *Journal of Agriculture and Ecology*, 1: 10–21.

Haldhar, S.M., Bhargava, R., Krishna, H., Berwal, M.K. and Saroj, P.L. (2018). Bottom-up effects of different host plant resistance cultivars on ber (*Ziziphus mauritiana*)-fruit fly (*Carpomyia vesuviana*) interactions. *Crop Protection*, 106: 117–124.

Hatfield, J.L., Boote, K.J., Kimball, B.A., Ziska, L.H., Izaurralde, R.C., Ort, D., Thomson A.M. and Wolfe, D. (2011). Climate impacts on agriculture: Implications for crop production. *Agronomy Journal*, 3(2): 351–370.

Hauenschild, F., Matuszak, S., Muellner-Riehl, A.N. and Favre, A. (2016). Phylogenetic relationships within the cosmopolitan buckthorn family (Rhamnaceae) support the resurrection of Sarcomphalus and the description of *Pseudoziziphus* gen. nov. *Taxon*, 65(1): 47–64.

Hocking, D. (1993). *Trees for Drylands*. Oxford and IBH Publishing Co., New Delhi.

Hooda, P.S., Sindhu, S.S., Mehta, P.K. and Ahlawat, V.P. (1990). Growth, yield and quality of *ber* (*Ziziphus mauritiana* Lamk.) as affected by soil salinity. *Journal of Horticultural Science*, 65(5): 589–593.

Huang, J., Zhang, C., Zhao, X., Fei, Z., Wan, K., Zhang, Z., Pang, X., Yin, X., Bai, Y., Sun, X. and Gao, L. (2016). The jujube genome provides insights into genome evolution and the domestication of sweetness/acidity taste in fruit trees. *PLoS Genetics*, 12(12): p.e1006433.

Huang, Y.L., Yen, G.C., Sheu, F. and Chau, C.F. (2008). Effects of water-soluble carbohydrate concentrate from Chinese jujube on different intestinal and fecal indices. *Journal of Agricultural and Food Chemistry*, 56(5): 1734–1739.

Jackson, D.I., Looney, N.E., Morley-Buker, M. and Thiele, G.F. (2011). Miscellaneous fruit crops. pp. 280–293. *In*: Jackson D.I., Looney N.E. and Morley-Buker, M. (eds.). *Temperature and Subtropical Fruit Production*. Cambridge, UK: Cambridge University Press.

Jain, B.L. and Dass, H.C. (1988). Effect of saline water on performance of saplings of jujube (*Ziziphus-mauritiana*), Indian cherry (Cordia-Dichotoma var. Wallichii) and pomegranate (*Punica-granatum*) at nursery stage. *Indian Journal of Agricultural Sciences*, 58(5): 420–421.

Jat, L., Pareek, S. and Shukla, K.B. (2013). Physiological responses of Indian jujube (*Ziziphus mauritiana* Lamk.) fruit to storage temperature under modified atmosphere packaging. *Journal of Science of Food and Agriculture*, 93(8): 1940–1944.

Ji, Q., Wang, D.W., Zhou, J., Xu, Y.L., Shen, B.Q. and Zhou, F. (2019). Genome-wide characterization and expression analyses of the MYB superfamily genes during development stages in Chinese jujube. *Peer, J.*, 7: 1–28.

Kaaria, S. (1998). *Ziziphus mauritiania: A Valuable Tree for Arid and Semi-arid Lands*. Forest, Farm, and Community Tree Network, Winrock International. Arkansas, USA, 98–103.

Kaur, M., Sharma, A., Katnoria, J. and Nagpal, A. (2017). Air pollution tolerance index (APTI): An important determinant for the development of green space in and around industrial/urban area, bioremediation. pp. 422–449. *In*: Rathoure, A.K. (ed.). *Current Research and Applications*. IK International Publishing House Pvt. Ltd.: New Delhi, India.

Khalid, N., Hussain, M., Young, H.S., Boyce, Y., Aqeel, M. and Noman. A. (2018). Effects of road proximity on heavy metal concentrations in soils and common roadside plants in Southern California. *Environmental Science and Pollution Research*, 25: 35257–35265.

Khan, Z.I., Ugulu, I., Sahira, S., Ahmad, K., Ashfaq, A., Mehmood, N. and Dogan, Y. (2018). Determination of toxic metals in fruits of *Abelmoschus esculentus* grown in contaminated soils with different irrigation sources by spectroscopic method. *International Journal of Environmental Research*, 12(4): 503–511.

Kohandel, A., Pouladian, M. and Yadollahi, A. (2011). Method of reproduction and breeding *Ziziphus jujube* Mill. *In*: *Proceeding of the 1st National Conference on Barberry and Jujube*. Southern Khorasan province, Birjand, Iran (in Farsi).

Kulkarni, M., Schneider, B., Raveh, E. and Tel-Zur, N. (2010). Leaf anatomical characteristics and physiological responses to short-term drought in *Ziziphus mauritiana* (Lamk.). *Scientia Horticulturae*, 124(3): 316–322.

Lambers, H., Chapin, F.S. and Pons, T.L. (2008). *Plant Physiological Ecology*. New York, NY, USA: Springer.

Levitt, J. (1980). *Responses of Plants to Environmental Stresses*. 2nd ed., New York. Academic Press.

Li, X., Li, Y., Zhang, Z. and Li, X. (2015). Influences of environmental factors on leaf morphology of Chinese jujubes. *PloS One*, 10(5): p.e0127825.

Li, D., Niu, X and Tian, J. (2013). *The Illustrated Germplasm Resources of Chinese Jujube*. China Agriculture Press.

Li, D.K. (2006). Chinese jujube variety resource map. *Shanxi Fruits*, 3: 21.

Li, J.W., Fan, L.P., Ding, S.D. and Ding, X.L. (2007). Nutritional composition of five cultivars of Chinese jujube. *Food Chemistry*, 103(2): 454–460.

Lipiec, J., Doussan, C., Nosalewicz, A. and Kondracka, K. (2013). Effect of drought and heat stresses on plant growth and yield: A review. *International Agrophysics*, 27(4). 463–477.

Liu, D., Ye, X. and Jiang, Y. (2016). *Chinese Dates: A Traditional Functional Food*. CRC Press, Boca Raton, FL.

Liu, M.J. and Wang, M. (2008). *Germplasm Resources of Chinese Jujube*. Beijing: Chinese Forestry Publisher.

Liu, M., Wang, J., Wang, L., Liu, P., Zhao, J., Zhao, Z. and Ao, C. (2020). The historical and current research progress on jujube—A superfruit for the future. *Horticulture Research*, 7(1): 1–17.

Liu, M.J., Zhao, J. and Zhou, J.Y. (2010). *Jujube, Witches' Broom Disease*. China Agriculture Press, Beijing (China), 12: 50–51.

Liu, M.J., Zhao, J., Cai, Q.L., Liu, G.C., Wang, J.R., Zhao, Z.H. and Li, X.S. (2014). The complex jujube genome provides insights into fruit tree biology. *Nature Communications*, 5(1): 1–12.

Liu, M.J. (2018). The Challenges and countermeasures of jujube industry during transition period. *Chin. Fruits*, 1: 1–4.

Liu, P., Liu, M.J., Zhao, Z.H., Liu, X.Y., Wang, J.R. and Yan, C. (2008). Investigation on the characteristics of fruiting and seed development in Chinese jujube (*Ziziphus jujuba* Mill.). pp. 209–214. *In*: *1st International Jujube Symposium*, 840.

Liu, X.S., Chen, L., Wang, J.X., Li, L. and Peng, J.Y. (2013). Discovery and identification of natural triploid ploidy of Chinese jujube cultivar 'Pingguozao'. *Acta Horticulturae Sinica*, 40(3): 426–432.

Liu, Y., Feng, C., Yu, H., Wang, J. and Liu, M. (2016). Discovery and evaluation of natural mixoploid (2x+ 4x) variants in *Ziziphus jujuba* Mill. 'Dongzao'. *Acta Horticulturae Sinica*, 43: 966–974.

Lobell, D.B. and Field, C.B. (2007). Global scale climate-crop yield relationships and the impacts of recent warming. *Environmental Research Letters*, 2(1): 1002.

Lonergan, S. (1998). The role of environmental degradation in population displacement. *Environmental Change and Security Project Report*, 4(6): 5–15.

Mahajan, R.T.C.M. and Chopda, M. (2009). Phyto-pharmacology of *Ziziphus jujuba* Mill—A plant review. *Pharmacognosy Reviews*, 3(6): 320.

Mani, A. and Suresh, C.P. (2018). Climate Resilient Fruit Crops–Possible Solution to Ensure Nutritional Security in Changing Climate Scenario. Training Manual, p. 51.

Mantri, N., Patade, V., Penna, S., Ford, R. and Pang, E. (2012). Abiotic stress responses in plants: present and future. pp. 1–19. *In*: *Abiotic Stress Responses in Plants*. Springer, New York.

Maraghni, M., Gorai, M. and Neffati, M. (2011). The influence of water-deficit stress on growth, water relations and solute accumulation in wild jujube (*Ziziphus lotus*). *Journal of Ornamental Horticulture*, 1: 63–72.

Maruza, I.M., Musemwa, L., Mapurazi, S., Matsika, P., Munyati, V.T. and Ndhleve, S. (2017). Future prospects of *Ziziphus mauritiana* in alleviating household food insecurity and illnesses in arid and semi-arid areas: A review. *World Development Perspectives*, 5: 1–6.

Meghwal, P.R., Kumar, P. and Singh, D. (2018). Climate variability during flowering and fruiting reduces fruit yield of ber (*Ziziphus mauritiana*) in western Rajasthan. *Journal of Agriculture and Ecology*, 6: 31–38.

Memon, A., Memon, N., Luthria, D., Pitafi, A. and Bhanger, M. (2012). Phenolic compounds and seed oil composition of *Ziziphus mauritiana* L. fruit. *Polish Journal of Food and Nutrition Sciences*, 62(1): 15–21.

Mittler, R. (2002). Oxidative stress, antioxidants and stress tolerance. *Trends in Plant Science*, 7(9): 405–410.

Narendrula-Kotha, R., Theriault, G., Mehes-Smith, M., Kalubi, K. and Nkongolo, K. (2019). Metal toxicity and resistance in plants and microorganisms in terrestrial ecosystems. pp. 1–27. *In*: *Reviews of Environmental Contamination and Toxicology*. 249, Springer, Cham.

Nizamani, I.A., Rustamani, M.A., Nizamani, S.M., Nizamani, S.A. and Khaskheli, M.I. (2015). Population density of foliage insect pest on jujube, *Ziziphus mauritiana* Lam. ecosystem. *Journal of Basic and Applied Sciences*, 11: 304–313.

Pandey, S.D., Pathak, R.K. and Dwivedi, R. (1991). Effect of sodicity and salinity levels on chlorophyll free proline and amino acid content in *ber* leaves. *Journal of Horticulture Sciences*, 4: 33–36.

Pareek, O.P. (2001). *Ber*, International Center for Underutilized Crops, Southampton UK, pp. 248–266.

Pareek, S., Mukherjee, S. and Paliwal, R. (2007). Floral biology of Ber—A review. *Agricultural Reviews*, 28(4): 277–282.

Pasala, R.K., Minhas, P.S. and Wakchaure, G.C. (2017). Plant bioregulators: A stress mitigation strategy for resilient agriculture. pp. 235–259. *In*: Minhas, P.S., Rane, J. and Pasala, R.K. (eds.). *Abiotic Stress Management for Resilient Agriculture*. Springer, Singapore.

Pawlowska, A.M., Camangi, F., Bader, A. and Braca, A. (2009). Flavonoids of *Ziziphus jujuba* L. and *Ziziphus spina-christi* (L.) Wild (Rhamnaceae) fruits. *Food Chemistry*, 112(4): 858–862.

Pommer, C.V. (2012). Guava world-wide breeding: Major techniques and cultivars and future challenges. pp. 81–88. *In*: *3rd International Symposium on Guava and Other Myrtaceae*. 959.

Pradeep, T. and Jambhale, N.D. (2000). Ploidy level variations for stomata, chloroplast number, pollen size and sterility in ber (*Zizhyphus mauritiana* Lamk.). *Indian Journal of Genetics and Plant Breeding*, 60(4): 519–525.

Proshad, R., Kormoker, T., Mursheed, N., Islam, M.M., Bhuyan, M.I., Islam, M.S. and Mithu, T.N. (2018). Heavy metal toxicity in agricultural soil due to rapid industrialization in Bangladesh: A review. *International Journal of Advanced Geoscience*, 6(1): 83–88.

Qi, J., Dong, Z., Shen, L., Mao, Y., Li, Y., Liu, J. and Wang, X. (2009). Analysis of QTL for needle length in Chinese jujube. *Acta Horticulturae Sinica*, 36(6): 807–813.

Qu, Z.Z. and Wang, Y.H. (1993). *Chinese Fruit Trees Record – Chinese Jujube*. The Forestry Publ. House of China, Beijing, China.

Rao, M.V. and Dubey, P.S. (1992). Occurrence of heavy metals in air and their accumulation by tropical plants growing around an industrial area. *Science of Total Environment*, 126(1-2): 1–16.

Richardson, J.E., Fay, M.F., Cronk, Q.C.B., Bowman, D. and Chase, M.W. (2000). A molecular analysis of Rhamnaceae using rbcL and trnL-F plastid DNA sequences. *American Journal of Botany*, 87(9): 1309–1324.

Rouphael, Y., Cardarelli, M., Schwarz, D., Franken, P. and Colla, G. (2012). Effects of drought on nutrient uptake and assimilation in vegetable crops. pp. 171–195. *In*: *Plant Responses to Drought Stress*. Springer, Berlin, Heidelberg.

Rushforth, K. (1999). *Trees of Britain and Europe*. HarperCollins, London.

Sadio, S. and Negreros-Castillo, P. (2006). Trees outside forests: Facing smallholder challenges. pp. 169. *In*: *World Agroforestry into the Future*, Food and Agriculture Organization of the United Nations and Patricia Iowa State University, USA.

Sanaullah, M., Rumpel, C., Charrier, X., and Chabbi, A. (2012. How does drought stress influence the decomposition of plant litter with contrasting quality in a grassland ecosystem? *Plant and Soil*, 352(1-2): 277–288.

Sankhla, N., Gehlot, H.S., Choudhary, R., Joshi S. and Dinesh, R. (2008). Eco-physiological studies on Indian desert plants: effect of salt on antioxidant defense systems in *Ziziphus* spp. pp. 201–213. *In*: *Ecophysiology of High Salinity-tolerant Plants*. Springer, Dordrecht.

Sardans, J. and Penuelas, J. (2012). The role of plants in the effects of global change on nutrient availability and stoichiometry in the plant-soil system. *Plant Physiology*, 160(4): 1741–1761.

Serraj, R. and Sinclair, T.R. (2002). Osmolyte accumulation: Can it really help increase crop yield under drought conditions? *Plant Cell and Environment*, 25(2): 333–341.

Sharif, N., Jaskani, M.J., Naqvi, S.A. and Awan, F.S. (2019). Exploitation of diversity in domesticated and wild *ber* (*Ziziphus mauritiana* Lam.) germplasm for conservation and breeding in Pakistan. *Scientia Horticulturae*, 249: 228–239.

Sheng, J.P. and Shen, L. (2011). Chinese jujube (*Ziziphus jujuba* Mill.) and Indian jujube (*Ziziphus mauritiana* Lam.). pp. 299–326e. *In*: *Postharvest Biology and Technology of Tropical and Subtropical Fruits*. Woodhead Publishing.

Sherani, J., Jillani, M.S. and Ahmad, T. (2017). *Ber* (*Ziziphus mauritiana* L.) production and quality as influenced by different salinity levels in water. *Pakistan Journal of Agri. Sciences*, 54(3): 645–652.

Shukla, A.K., Singh, D., Shukla, A.K. and Dhandar, D.G. (2006). Genetic variability of *Ber* (*Ziziphus mauritiana* Lamk.) and its prospects for crop improvement. *Indian Journal of Plant Genetic Resources*, 19(1): 118–121.

Simpson, B.M. (2017). Preparing smallholder farm families to adapt to climate change. *Pocket Guide 2: Managing Crop Resources*. Catholic Relief Services: Baltimore, MD, USA.

Singh, A., Singh, R.K., Kumar, A., Kumar, A., Kumar, R., Kumar, N. and Sharma, D.K. (2020). Adaptation to social-ecological stressors: A case study with Indian jujube (*Ziziphus mauritiana* Lam.) growers of north-western India. *Environment, Development and Sustainability*, pp. 1–24.

Singh, S.K., Chhajer, S., Pathak, R., Bhatt, R.K. and Kalia, R.K. (2017). Genetic diversity of Indian jujube cultivars using SCoT, ISSR, and rDNA markers. *Tree Genetics and Genomes*, 13(1): 12.

Sivalingam, P.N., Singh, D., Chauhan, S., Changal, H.K., Bhan, C., Mohapatra, T. and Sharma, S.K. (2014). Establishment of the core collection of *Ziziphus mauritiana* Lam. from India. *Plant Genetic Resources*, 12(1): 40–142.

Sofo, A., Dichio, B., Xiloyannis, C. and Masia, A. (2004). Effects of different irradiance levels on some antioxidant enzymes and on malondialdehyde content during rewatering in olive tree. *Plant Science*, 166(2): 293–302.

Sofo, A., Manfreda, S., Fiorentino, M., Dichio, B. and Xiloyannis, C. (2008). The olive tree: A paradigm for drought tolerance in Mediterranean climates. *Plant Sciences*, 166: 293–302.

Soltani, A., Gholipoor, M. and Zeinali, E. (2006). Seed reserve utilization and seedling growth of wheat as affected by drought and salinity. *Environmental and Experimental Botany*, 55(1-2): 195–200.

Song, L., Meinhardt, L.W., Bailey, B. and Zhang, D. (2019). Genetic improvement of chinese jujube for disease resistances: status, knowledge gaps and research needs. *Crop Breeding, Genetics and Genomics*, 1(2): e190015.

Sudhir, H., Pralhad, R. and Mamata, L. (2020). Abatement of heavy metal from wastewater using surface modified adsorbent. *International Journal of Advanced Engineering, Management and Science (IJAEMS)*, 2: 280–284.

Sulusoglu, M., Cavusoglu, A., Dede, N. and Unver, H. (2014). Morphological, pomological and nutritional traits of jujube (*Ziziphus jujuba* Mill.). pp. 727–731. *In*: *49th Croatian and 9th International Symposium on Agriculture, Section: Pomology, Viticulture and Enology*. Dubrovnik, Croatia: Faculty of Agriculture, University of Zagreb.

Tang, M., Li, H., Zhang, C., Zhao, X., Gao, X. and Wu, P. (2021). Mulching measures improve soil moisture in rain-fed jujube (*Ziziphus jujuba* Mill.) orchards in the loess hilly region of China. *Sustainability*, 13(2): 610.

Tel-Zur, N. and Schneider, B. (2009). Floral biology of *Ziziphus mauritiana* (Rhamnaceae). *Sexual Plant Reproduction*, 22(2): 73–85.

Tian, H.Q. and Ma, D.Z. (1987). The embryological observation on a parthenocarpy of *Ziziphus jujuba* Mill. *Acta Botanica Sinica*, 29: 29–33.

Toppo, P. and Raj, A. (2018). Role of agroforestry in climate change mitigation. *Journal of Pharmacognosy and Phytochemistry*, 7(2): 241–243.

USGCRP. (2014). Hatfield, J., Takle, G., Grotjahn, R., Holden, P., Izaurralde, R.C., Mader, T., Marshall, E. and Liverman, D. 2014: Ch. 6: Agriculture. Climate Change Impacts in the United States: The Third National Climate Assessment, Melillo, J.M., Terese (T.C.) Richmond and Yohe, G.W. (eds.). U.S. Global Change Research Program, 150–174.

Van Diepen, C.V., Wolf, J., Van Keulen, H. and Rappoldt, C. (1989). WOFOST: A simulation model of crop production. *Soil Use and Management*, 5(1): 16–24.

Vaughan, M.M., Block, A., Christensen, S.A., Allen, L.H. and Schmelz, E.A. (2018). The effects of climate change associated abiotic stresses on maize phytochemical defenses. *Phytochemistry Reviews*, 17(1): 37–49.

Verma, S.S., Verma, R.P., Verma, S.K., Yadav, A.L. and Verma, A.K. (2018a). Responses of *ber* (*Ziziphus mauritiana* Lamk.) varieties to different level of salinity. *International Journal of Current Microbiology and Applied Sciences*, 7(7): 580–591.

Verma, S.S., Verma, R.S., Verma, S.K., Yadav, A.L. and Verma, A.K. (2018b). Impact of salt stress on plant establishment, chlorophyll and total free amino acid content of ber (*Ziziphus mauritiana* Lamk.) cultivars. *Journal of Pharmacognosy and Phytochemistry*, 7(2): 556–559.

Vurukonda, S.S.K.P., Vardharajula, S., Shrivastava, M. and Skz, A. (2016). Enhancement of drought stress tolerance in crops by plant growth promoting rhizobacteria. *Microbiological Research*, 184: 13–24.

Wang, D., Xiong, R.C., Wu, C.Y. and Gao, J.S. (2016). Jujube breeding research. *J. Tarim. Univ.*, 28: 109–18.

Wang, L., Luo, Z., Wang, L., Deng, W., Wei, H., Liu, P. and Liu, M. (2019). Morphological, cytological and nutritional changes of autotetraploid compared to its diploid counterpart in Chinese jujube (*Ziziphus jujuba* Mill.). *Scientia Horticulture*, 249: 263–270.

Wang, Y.P., Wang, G.X. and Li, Y.J. (2005). Advances in research on fruit diseases of Chinese jujube during storage. *Journal of Fruit Sciences*, 22: 376–380.

Waqas, M.A., Khan, I., Akhter, M.J., Noor, M.A. and Ashraf, U. (2017). Exogenous application of plant growth regulators (PGRs) induces chilling tolerance in short-duration hybrid maize. *Environmental Science and Pollution Research*, 24(12): 11459–11471.

Weekley, C.W. and Race, T. (2001). The breeding system of *Ziziphus celata*, a rare endemic plant of the Lake Wales Ridge, Florida: implications for recovery. *Biological Conservation*, 100: 207–213.

Xu, C., Gao, J., Du, Z., Li, D., Wang, Z., Li, Y. and Pang, X. (2016). Identifying the genetic diversity, genetic structure and a core collection of *Ziziphus jujuba* Mill. var. jujuba accessions using microsatellite markers. *Scientific Reports*, 6: 31503.

Yan, C., Liu, P., Liu, M.J., Wang, J.R., Dai, L., Li, D.K. and Kong, D.C. (2009). Characteristics of fruiting and fertility of different cultivars of *Ziziphus jujuba* Mill. *Journal of Plant and Genetic Resources*, 10: 121–125.

Yan, G. (1984). *Cytological Studies on Chinese Jujube*. Agricultural University of Hebei, Baoding.

Yao, S. (2013). Past, present, and future of jujubes—Chinese dates in the United States. *HortScience*, 48: 672–680.

Zafar, S.A., Noor, M.A., Waqas, M.A., Wang, X., Shaheen, T., Raza, M. and Rahman, M.U. (2018). Temperature extremes in cotton production and mitigation strategies. pp. 65–91. *In*: Rahman, M.R. and Zafar, Y. (eds.). *Past, Present and Future Trends in Cotton Breeding*. United Kingdom.

Zait, Y. and Schwartz, A. (2018). Climate-related limitations on photosynthesis and drought-resistance strategies of *Ziziphus spina-christi*. *Frontiers in Forests and Global Change*, 1: 3.

Zhang, M., Yang, L., Hao, R., Bai, X., Wang, Y. and Yu, X. (2020). Drought-tolerant plant growth-promoting rhizobacteria isolated from jujube (*Ziziphus jujuba*) and their potential to enhance drought tolerance. *Plant and Soil*, 452(1): 423–440.

Zietsman, P.C. and Botha, F.C. (1992). Flowering of *Ziziphus mucronata* subsp. *mucronata* (Rhamnaceae): Anthesis, pollination and protein synthesis. *Botanical Bulletin of Academia Sinica*, 33: 33–42.

Zörb, C., Geilfus, C.M. and Dietz, K.J. (2019). Salinity and crop yield. *Plant Biology*, 21: 31–38.

12

Environmental Friendly Snake Fruit (*Salacca* spp.) Cultivation for Agricultural Land Utilization and Conservation Supporting the Climate Change Resilience

Riry Prihatini,[1,*] *Tri Budiyanti*[1] and *Jameel M Al-Khayri*[2]

1. Introduction

Snake fruit plant (*Salacca* spp.) is an Indonesian indigenous commodity with an important commercial value. This tropical fruit belongs to the subfamily *Calamoideae* and Family *Arecaceae*. The number of species of *Salacca* is still debatable; however, several publications revealed 23 species of the genus that have been described, whereas many others have not been described yet. The species distribution is concentrated mainly in Sumatera, Borneo and Peninsular Malaysia. The fruit can be easily found growing in Indonesia, Malaysia, Thailand, the Philippines, Singapore, Myanmar, and India (Zumaidar and Miftahuddin, 2018).

Snake fruit plants are evergreen, thorny, usually dioecious palms developing in clusters shaped by progressive fanning at the basal stem.

[1] Indonesian Tropical Fruit Research Institute, Jalan Raya Solok-Arian km. 8 Solok 27301, Sumatera Barat, Indonesia.
[2] Department of Agricultural Biotechnology, College of Agriculture and Food Sciences, King Faisal University, Al-Ahsa 31982, Saudi Arabia.
Email: jkhayri@kfu.edu.sa
* Corresponding author: riry_silva@yahoo.com

Fig. 1. Snake fruit (*Salacca zalacca* var. *Zalacca*): (a) fruiting tree, (b) male inflorescences, (c) female inflorescence, (d) fruits, and (e) fruit segment with seeds.

The fruit is sub-globular to ellipsoid drupe, tapering near the base, round to taper at the upper part, with mostly brown to orange-brown peel, united imbricate scales which hold delicate prickles. The edible parts of the fruit are 2–8 mm thick, creamy to reddish creamy color pulp, beige brown to dark brown seeds (Ashari, 2002; Lim, 2012). The morphology of the snake fruit is presented in Fig. 1.

Snake fruit is cultivated mainly for the fruit which is rich in carbohydrates, protein, fat, fiber, ferrum, potassium, vitamins B, and C. In addition, the fruit is known for its antioxidant activity, such as proanthocyanin, epicatechin, polyphenols, and chlorogenic acid. It also shows anti-hyperuricemic activities as a bioactive compound that is used to manage gout disorder. Other potential uses of snake fruit as an anti-diabetic, anticancer, anti-aging, anti-inflammatory, anti-cholesterol and hyperglycemia therapeutic compound (Mazumdar et al., 2019; Zubaidah et al., 2018). The snake fruit's nutritional composition is given in Table 1.

Besides being consumed as a fresh fruit, snake fruits are also processed into products, such as chips, candied fruit, pickles, salad, juice, dodol (a sweet toffee-like confection), chips, wine, coffee (a drink with a similar colour to coffee) and jam. Snake fruit peels are also processed into crafts, such as bags, wallets, paintings, and photo frames. The rich health-beneficial phytochemicals in the fruit show promise to be developed either as fresh and processed fruits, or functional foods.

Applying appropriate cultivation technologies includes the use of a suitable variety, multiple cropping techniques and environment-friendly land management, both agricultural land productivity and conservation. Several reports on snake fruit genetic studies revealed a high diversity due to its natural cross-pollination reproductive system. Thus, this genetic

Table 1. Nutrition content of snake fruit (per 100 g fresh weight).

Substance	Content	References
Sucrose	7.6 mg	Chareoansiri and Kongkachuichai (2009)
Fructose	5.9 mg	
Glucose	3.9 mg	
Total sugar	17.4 mg	
Total carbohydrate	80.85%	Tun (2012)
Fiber	4.67%	
Protein	0.21%	
Vitamin C	91.72 mg	
Na	1.9 mg	Haruenkit et al. (2007)
K	191.2 mg	
Mg	7.16 mg	
Ca	6.11 mg	
Fe	30.17 µg	
Mn	249.9 µg	
Zn	35.1 µg	
Cu	8.4 µg	
Rb	1.24%	Tun (2012)
Br	0.18%	
S	0.9%	
Total phenolic compound	257.17 µL/mL	Cepkova et al. (2021)
Total antioxidants activities	10.56 M TE	
Flavonoid	5.35 mgGAE	Ilmiah et al. (2021)

diversity either intra- or inter-species of *Salacca* are precious resources to develop superior variety, such as high production and high fruit quality which are becoming a focus in mostly breeding programs, especially on the reduction of astringent taste of the fruit's characteristics. The high quality of fruit variety is also an interesting topic in terms of food security which is the biggest threat of climate change. In addition, other snake fruit cultivation methods could also minimize the environmental challenge, such as an upsurge in water requirement and reduction of water sources caused by climate change.

2. Snake Fruit Cultivation Requirement

Snake fruit plants require a high level of humidity. The plant grows well in the region with high and equally distributed rainfall per year—

1000–3000 mm per year, whereas in the moderate rainfall region, the fruit production gets minimized. On the contrary, regions with a long dry season (more than six months) are not suitable for snake fruit. Direct sunlight exposure should be avoided on snake fruit plantations. In addition, the temperature is another abiotic factor which limits snake fruit plant growth. The daily average optimum temperature for snake fruit is 20–30°C. By the end of 2019, Indonesian snake fruit production reached 955.763 tons from 27.050 ha production area or 35 ton/ha/year. Interestingly, the snake fruit may be found all year round even with the traditional cultivation practices, thus supporting food security in terms of continuous production.

Snake fruit plant is categorized into shallow rooting plant whose roots only penetrate the soil up to 100 cm below soil level. Thus, the appropriate soil type for snake fruit is organic-material-rich soil, with water-saving ability and not exposed to flood; last but not least, the fruit possesses a quantitatively adequate number of plant-required minerals. In general, snake fruit is well adapted to neutral soil, pH 6.0–7.0, though it adapts to acidic and base soil pH. Snake fruit is well adapted to a wide range of altitudes which should be 50–800 m above sea level. The best range for snake fruit plantation is up to 400 m; whereas beyond 900 m above sea level, snake fruit hardly produces any fruit.

The loss of biodiversity is another concern of climate change which impact the fruit due to the depletion of natural resources (Chapman, 2006). The natural cross-breeding of the snake fruit proves an advantage for the palm tree to face the upcoming climate change. Breeding is a time-consuming process. The traditional selection methods have dealt with plant features in varied environments over the past many years, thus, making it well suited for climate change adaptation. Modern breeding technologies, on the other hand, can aid in the gradual accumulation of positive alleles for 'invisible' physiological traits underlying climate adaptation (Snowdon et al., 2021).

Changing cultivated species is one of the short-term strategies recommended to face global warming (Debaeke et al., 2017). Because of the snake fruit plant's adaption ability to a wide range of abiotic conditions, the fruit is a decent candidate for the developmental region which is highly impacted by climate change.

2.1 *Agroecological-based Snake Fruit Development*

The world nowadays is facing a significant challenge regarding global climate change. As the earth's temperature is increasing, many crops are predicted to be unable to adjust to the new environment. The most typical effect of global climate change will be the decrease in crops productivity which will lead to an insufficient food supply. To overcome this, researchers around the globe have suggested some ideas, including the intensification and diversification of sustainable production.

The staple food is domineering in agricultural land which is also affected by climate change. The use of a monoculture system in the cultivation of those crops may lead to the plant's fragile adaptation to the environmental stress, thus intensification should be conducted such as managing crop variety's genetics, soil fertility, and pest and disease treatment.

To anticipate the loss of staple food production, other tolerable crops with health benefit value may be developed as one of the crops in the food estate program as launched by the Indonesian Government. Snake fruit which is has both a wide range of cultivation requirements and many beneficial phytochemicals is a suitable candidate for the crop diversification program. Most snake fruit plants have natural cross-pollination system, thus the genetic variety of the *Salacca* sp. is very broad. A wide genetic background of snake fruit also supports the breeding program which will maintain the plant adaptability to environmental stress (Budiyanti et al., 2015). Some characteristics of superior variety include big size fruit, high edible portion, sweet taste, minimum astringent taste, stable and high productivity (Sumantra and Martiningsih, 2016). For example in Indonesia, up to 50 snake fruit varieties have been released by the authorized government agency with the variety Pondoh as the most acknowledged one. Due to the natural cross breeding, the Pondoh variety then develop into many local varieties such as Pondoh Madu, Pondoh Hitam, Pondoh Gading, and Pondoh Manggala. Other snake fruit superior variety include Sari Intan (Table 2), Salak Bali Gula Pasir, and Salak Sidempuan.

Despite being tolerant, crop production is highly affected by the climate. Abiotic factors, namely average temperature, rainfall, climate fluctuation and other extreme weather states stimulate crop yield. As for the snake fruit, several studies on the effect of climate change revealed that climatic condition is primarily a dependent factor for phenology emergence. The humid condition will induce the appearance of male inflorescences; on the other hand, a relatively hot climate will encourage female inflorescences. This phenomenon is also found in oil palm (Adelina et al., 2021). In addition, a model plantation of the snake fruit showed that the fruit might be highly affected by climate change mainly due to the reduction in suitable plantation areas (Nuary et al., 2019).

To survive in the ecosystem changes, plants also need to change their strategy. Thus, to minimize the effect of environmental changes, suitable agro-ecological regions should be preferred for crops' commodity development. Agro-ecological is defined as the application of ecological theory and concept in the management of sustainable agricultural

Table 2. Morphological performance of snake fruit varieties Sari Intan 48, Sari Intan 541, and Sari Intan 295.

	Sari Intan 48	**Sari Intan 541**	**Sari Intan 295**
Breeding parents	Gula Pasir × Pondoh	Gondok × Pondoh	Pondoh × Mawar
Number of bunches per cob	2–3	2–4	2–3
Number of fruit per bunch	40–60	60–80	60–79
Fruit weight (g)	55.8–63.4	48.4–59,9	42–60
Fruit shape	Round to oval	Round to oval	Round to oval
Peel colour	Blackish brown	Dark brown	Blackish brown
Pulp colour	White yellow	Creamy yellow	White yellow
Number of fruit segment with seeds	1–3	1–3	1–3
Number of seedless fruit segment	0–2	0–2	0–2
Pulp thick (cm)	0.5–1.8	0.4–1.8	0.3–1.8
Sweet taste	Very sweet	Very sweet	Very sweet
Astringent taste	None	None	None
Sour taste	None	None	None
Pulp texture	Slightly firm	Crispy	Crispy
Aroma/fragrant	Yes	Yes	Yes
Soluble solid concentration (°Brix)	19.0–20.8	19.0–20.0	19–21
Water content (%)	78.46–81.26	75.90–76.07	77.3–79.0
Vitamin C (mg/100 g) content	31.88–85.42	29.82–44.96	21.03–35.79
Acid content (%)	0.62–0.82	0.34–1.25	0.51–1.23

ecosystems. The agro-ecological components which are related to crop development include land biophysical properties, i.e., shape, slope, region, altitude, latitude, climate and soil. Additionally, the plant mechanism to convert chemical energy into potential energy is a unique principle of agroecology.

Agro-ecological zone refers to the suitable crop system and land in agricultural activity. The understanding of ecological principles is important to find the most appropriate relationship between crops and agricultural biophysical factors to maximize sustainable crop production. Research on agro-ecology for mapping the agricultural commodity region is significant with respect to natural and environmental resources. Research usually recommends land use and fertilizer application methods. The agro-ecological principles for snake fruit commodity development are:

a) Isothermic soil temperature type, with 15–22°C temperature as an annual average at 700–1500 m above sea-level altitude. This type of soil is mostly found in tropical latitude with an average annual temperature of 22°C or higher, seldom reaching a freezing condition (Buol, 2005);
b) Alfisol soil type with gray to a brown upper layer, medium pH and haplic luvisol horizon. Alfisol soil taxonomy is defined as the type of soil in which the volcanic residues are altered to aluminous and or ferruginous soil. Alfisol is moderately-leached fertile soil. This type of soil is mostly found in the temperate, humid and sub-humid regions of the world (Bockheim and Hartemink, 2017);
c) Udic soil moisture regime with dry surface sections is less than 90 cumulative days per year. This type of soil-moisture regime is common to the soil of humid region which gets evenly distributed rainfall.

Some studies on the snake fruit agro-ecology reveal that land altitude influences the fruit's sugar content in the cultivar, Sidimpuan. Biophysical environment, such as altitude, soil type, rainfall and temperature significantly influence the snake fruit production. In addition, it is reported that snake fruit cultivar, Bali Gula Pasir's pulp thickness, vitamin C content, sugar-acid ratio are also affected by the altitude of the land (Sumantra et al., 2012). The cultivation practice strongly affects the snake fruit quality, including the taste, which is the main consumer quality preference. Another study on the effect of climate change on snake fruit quality revealed that flood simulation affects the quality of the fruit as is shown by the decrease in the physical appearance, the hardness and the total acid content of the fruit (Nasrudin and Elizani, 2019).

2.2 *Snake Fruit Multiple Cropping and Cultivation Practices*

The current definition of cropping means cultural practices applied to crop sequences at the field level. Cropping techniques provide a variety of alternatives that can be combined to create climate-smart agriculture involving the technology of genetic and machinery engineering. Cropping systems influence adaptation and mitigation strategies and impact food production (Debaeke et al., 2017). Some of the well-known cropping systems are mono-cropping, ratooning, sole crop, sequential cropping, intercropping, mixed intercrop, row intercrop, strip intercrop and multiple cropping.

Multiple cropping or multi-cropping is one form of polyculture and is defined as the practice of planting more than one type of crop on the same agricultural land during one growing season. The choice species of multiple cropping vary with geographical location, level of input and local preferences (Azam-Ali, 2003). Multiple cropping is suggested to be the oldest method and still can be practiced in many agricultural systems, including on the Indonesian snake fruit plantations. There is also double cropping, in which the second type of crop is planted after the first crop has been harvested. Another variety of multiple cropping is relay cropping, in which the second type of crop is sowed before the harvest of the first type of crop (Petrie and Bates, 2017).

Multiple cropping systems may be applied as guides for bestowing numerous ecosystem benefits, namely food, feed and fibers (Gaba et al., 2015). In addition, multiple cropping systems influence the soil's organic carbon pools, microbial activity, crop yield and quality (Sanchez-Navarro et al., 2019). Thus, multiple cropping is an alternative technique to help meet the world demand for food, especially in facing global warming (Petrie and Bates, 2017).

The farmers usually choose multiple cropping in snake fruit cultivation primarily due to economic reasons, that is, to maximize agricultural land use to earn more income. Another justification is the ecological reason which entails reduction in the use of fertilizers and pesticides, increase in efficiency of water use and land conservation. In many regions, snake fruits are cultivated in multiple cropping systems, both with annual, such as vegetable crops, and other perennial crops. Additionally, multiple cropping is adopted to provide shade to snake fruit seedlings. In older plants, the shady plants modify the microclimate on the plantation by regenerating lower light interception and temperature. Heat in the final stage affects the snake fruit production (Sumantra et al., 2012).

Setting up snake fruit plantations should be initiated by clearing the land of weeds, wood chops and other pests which may perturb the cultivation process. Since snake fruit requires avoidance of direct sunlight exposure, some other plants should be planted to serve as shade trees. Banana plants are mostly chosen as shade trees due to their fast-growing characteristics and the highly economic value of their fruits. The shade trees should be planted at 6 × 2.5 m distance before planting the snake fruit seedlings. The snake fruit seedlings should be planted at a 3 × 2.5 m planting distance.

The planting holes for snake fruit seedlings should be prepared in 60 × 60 × 60 cm dimensions. The upper soil should be removed from the lower layer and mixed with approximately 10 kg of compost manure before being replaced into the initial position. This setting is allowed to remain free for two weeks to improve the soil texture and fertility before filling

Fig. 2. The multiple copping system in snake fruit cultivation: (a) snake-fruit-banana multiple cropping and (b) snake-fruit-chili multiple cropping.

up with snake fruit seedlings. After planting, some care and maintenance will be called for. Some recommended snake fruit planning care and maintenance for high fruit production and quality are listed below:

2.2.1 *Weeding*

Weeding involves taking care of the land by rooting out or removing weeds every four weeks. Unlike mechanical equipment, manual weeding will not destroy the soil layer but only has a minimal effect on the soil. This method is employed to improve the plants' nutritional uptake and prevent pest and disease life cycles. Manual weeding should be applied to snake fruit plantations instead of tillage since the root systems of snake fruit penetrate the soil only up to 100 cm depth with roots density up to 1,440 cm/cm^3 (Nasution, 2017).

Since the snake fruit roots usually emerge above the soil level, the roots need to be re-buried by adding more soil on the emerging parts, thus preventing plants from collapsing. This technique is also conducted to improve soil texture surrounding the plants and is usually done during weeding.

2.2.2 *Watering*

Snake fruit plant dislikes both droughts and flooded land; thereby water management is important to optimal growth. Heavy watering should be applied at the seedling stage, after pruning of fronds and fruit maturation. In addition, light watering is advisable on initiation of flowering and a week before the harvest period. The optimized 20 liter per plant for every 20 days increases both snake fruit cultivar Suwaru's production and the farmer's financial income three-fold as compared to the control plants (Sudaryono, 2017). Thus, water is very crucial for optimum growth of snake fruit plants, though water supply availability is a serious problem in many agricultural plantations.

Several studies on the efficiency of water used for optimizing snake fruit production showed that water is the main component to determine productivity. As mentioned, in the Palmae family description, *Salacca* spp., similar to coconut, flowers throughout the year. In practice, the plants flower every three months. Thus there will be four flowering periods per year. However, snake fruit production fluctuates throughout the year due to the varied fruit set percentage when many flowers are unable to develop into fruits. Thus, snake fruit maximum harvest occurs only two times a year or in many cases, one harvest per year on many snake fruit plantations. A study case reports that the fluctuating snake fruit production is caused by water and rainfall limitations; thus a drip irrigation system is recommended to optimize the snake fruit yield (Rai et al., 2014).

2.2.3 Fertilization

Fertilization includes the application of a mixture of fertilizers which is composed of 300 g ZA + 37.5 g urea + TSP (triple superphosphate) 75 g + 175 g KCL + 200 g Dolomit per plant; half-dose should be applied after harvest period, whereas the rest is applied for 30 days after the first application. The chemical fertilizer can be substituted with 10 kg/plant/year of cow manure to increase the plant's productivity and financial income (Adijaya and Yasa, 2015). In addition, the use of 150 g per plant of arbuscular mycorrhiza (*Acaulospora* sp.) is also recommended (Amnah and Friska, 2018).

2.2.4 Pest and Disease Management

Management should be conducted before the symptoms of infestation occur. Pests commonly found in snake fruit plants are borer insects (*Silphidae*), whitefly, and mice, whereas some diseases that commonly infect the snake fruits are leaf spot, flower rot and fruit rot. Steps adopted for prevention of pest and disease include early eradication, balanced fertilization, watering and maintaining good sanitation.

2.2.5 Trimming Midrib and Thinning Young Shoots

Maintaining the number of fonds at 10 per plant in snake fruit is crucial. The older and lower fonds should be pruned. In addition, each snake fruit plant should only have two young shoots to facilitate optimum growth. This practice is aimed at reducing humidity and optimizing the generative character of growth and development.

2.2.6 Fruit Thinning

This method aims to improve fruit maturation, maximum fruit growth and homogeny in size. Fruit thinning is conducted in fruits that are abnormal, ailing or in squeezed position with other fruits until only 50% of fruits from the initial number are left. The practice is followed two to three months after the first flowers are bloom. On an average, by maintaining 20 fruits per bunch, up to 14 kg of fruits per plant can be harvested of similar size and quality.

2.2.7 Pollination

Two types of pollination occur on snake fruit plants–human-made and insect pollination, which is mostly facilitated by the Curculionidae beetle (Apriniarti et al., 2019). Human-made pollination increases fruit set up to 60% compared to insect pollination. Human-made pollination is conducted two days after the inflorescences begin to bloom by brushing the pollen into female inflorescence; nevertheless, the percentage of snake fruit is very low.

A study regarding the fruit set percentage of the snake fruit revealed that failure in the fruit set is associated with the low level of endogenous IAA (Indole Acetic Acid), sugar content on leaves and flowers (Rai et al., 2015). Another research suggests the application of 30% (v/v) chitosan as an anti-transpirant agent to increase the fruit set of snake fruit up to 84% in the off-season (Sunarka et al., 2015). A deeper study implied that heat units on two different altitudes impacted phenology, inducing the development of midrib, the appearance of a spatha and also the harvest time. The heat unit also influences the number of fruits, the edible portions of the fruit, pulp thickness and TSS (total soluble acid) (Sumantra et al., 2014).

3. Snake Fruit on Soil Conservation

The main objectives of soil conservation are to prevent the loss of the top layer of soil caused by erosion and to avert soil fertility reduction due to overexploitation and chemical contamination. In the agricultural ecosystem, downgraded organic carbon in the soil will also deplete the yield potential. Soil erosion devastates 10 million ha of agricultural land annually, and 40% of the loss is caused by tillage erosion (Pimental, 2006). Soil with significant organic carbon loss is difficult to mend by using chemical fertilizers due to the high amount of carbon dioxide released during the fertilizer application (Purwanto and Alam, 2020).

Most chemicals are applied to the agricultural land to increase soil fertility as it will have a positive impact on crop production. Nonetheless, the overuse of chemical fertilizers has proven to be toxic to the soil as they destroy the microbes and weeds, thus disturbing the agroecosystem. In addition, the fertilizer residues usually pollute the underground water; thus an inappropriate fertilizer technique may be harmful to the environment.

On the other hand, soil erosion may also occur as a result of natural disasters, namely landslides, floods and other extreme weather condition, like strong winds. Additionally, the land slope is increasing the risk of soil erosion. Thus soil conservation is a crucial area needing attention. Soil conservation technologies include contour mowing, terrace farming, windbreaks, mineralization, soil organism, salinity management, cover crop or crop rotation and soil-conservation farming.

Plant rooting systems influence the ecosystem resilience to abiotic stress, for example, drought. Roots link soil, underground water and atmosphere, thus affecting the cycle of hydrology and climate. Information on the plant rooting system is crucial to study plant-mediated global change. While deep-rooting plants are commonly found in well-drained upland, the shallow rooting system may be found in waterlogged lowlands (Fan et al., 2017).

A study of rooting characteristics suggests that this parameter may be applied as a significant element in selecting a potential plant for a specific land slope. Tensile strength is reduced, revealing that the diameter of lower roots plays a major role in determining high tensile strength. Thus, it is suggested that shallow root types should be planted on the top and base of the slope, whereas the middle part of the slope should contain deep root-system plants to prevent slope slide (Saifuddin and Normaniza, 2016). Another research suggests that shallow-root plants are crucial in reducing soil mass loss and landslide mostly because of their biomass. In addition, it is also revealed that plant biomass is the main factor to determine soil-root reinforcement level and eco-engineering potential of certain plant species (Gonzalez-Ollauri and Mickovski, 2016).

Snake fruit is a perennial plant with continuously growing branches, shoots, strong roots and does not overload the land. The plant is suitable to be cultivated on moderate to steep land slope. The snake fruit roots spread in the vertical and horizontal directions beyond the soil surface, with the highest density located in the upper layer. In addition, snake fruit plants are also characterized by a wide canopy, relatively light biomass and have water-retention capacity. Thus, plants with this intense rooting on the upper soil layer are effective in reducing soil erosion (Nasution, 2017). A further study revealed the application of snake fruit plants in soil conservation as part of agroforestry. Soil erosion on the catchment area of the snake fruit is significantly lower than mixed watering, implying that snake fruit plants can uphold much more (12 times more) soil sediment than mixed agroforestry. Thus, snake fruit is recommended as a useful plant in soil conservation (Nasution et al., 2019).

4. Conclusion and Prospects

Facing the upcoming climate change is a concern of many people. The aspects covered by climate change resilience are both the ability to minimize the negative effects of the phenomenon and adaptation to improve the sustainability of the ecosystem. The risk of biodiversity loss and overexploitation of several major crops due to climate change might be overcome by empowering the underutilized crops, such as snake fruit.

The high nutrient and antioxidant content and phytochemical properties of snake fruit prove beneficial in meeting people's nutritional requirements. Continuous fruit production is also a promising aspect of these tropical fruits, thus supporting food security and sustainability which are the major concerns of climate-change impact. From the economic point of view, the snake fruit agricultural industry may be developed by adopting good agricultural practices in farm management. Last but not least, land conservation encouraged by snake fruit plantation is also a major contribution of these plants in fostering climate-change resilience.

References

Adelina, R., Suliansyah, I., Syarif, A. and Warnita, W. (2021). Phenology of flowering and fruit set in snake fruit (*Salacca sumatrana* Becc.). *Acta Agrobot.*, 74: 1–12. https://doi.org/10.5586/AA.742.

Adijaya, I.N. and Yasa, I.M.R. (2015). *Pengaruh pupuk organik dan penjarangan buah terhadap produktivitas salak Gula Pasir. J. Pengkaj. dan Pengemb, Teknol. Pertan.*, 18: 195–206.

Amnah, R. and Friska, M. (2018). The usage of arbuscular mychorriza on the growth of salak Sidimpua (*Salacca sumatrana* Becc.) seedling. *J. Pertan. Trop.*, 5: 455–461.

Apriniarti, M.S., Suryobroto, B., Atmowidi, T., Kahono, S. and Manalu, W. (2019). The pollen load of snake fruit fowers. *Int. J. Ecosyst. Ecol. Sci.*, 9. https://doi.org/https://doi.org/10.31407/ijees9322.

Ashari, S. (2002). *On the Agronomy and Botany of Salak (Salacca zalacca)*. Wageningen University.

Azam-Ali, S.N. (2003). Production systems and agronomy: Multicropping. pp. 978–984. *In*: Thomas, B., Murphy, D.J. and Murray, B.G. (eds.). *Encyclopedia of Applied Science*. Academic Press, New York.

Bockheim, J.G. and Hartemink, A.E. (2017). *The Soils of Wisconsin*. Springer International Publishing, Cham.

Budiyanti, T., Hadiati, S., Prihatini, R. and Sobir. (2015). Genetic diversity of indonesian snake fruits as food diversification resources. *Int. J. Adv. Sci. Eng. Inf. Technol.*, 5: 41–44. https://doi.org/10.18517/ijaseit.5.3.513.

Buol, S.W. (2005). Tropical soils: Humid tropical. *Encycl. Soils Environ.*, 187–195.

Cepkova, R.H., Jagr, M., Janovska, D., Dvoracek, V., Kozak, A.K. and Viehmannova, I. (2021). Comprehensive mass spectrometric analysis of snake fruit: Salak (*Salacca zalacca*). *J. Food Qual.*, 6621811. https://doi.org/10.1155/2021/6621811.

Chapman, H. (2006). WRAMS, sustainable water recycling. *Desalination*, 188: 105–111.

Chareoansiri, R.I.N. and Kongkachuichai, R. (2009). *Sugar Profiles and Soluble and Insoluble Dietary Fiber Contents of Fruits in Thailand Markets*, 60. https://doi.org/10.1080/09637480802609376.

Debaeke, P., Pellerin, S. and Scopel, P. (2017). Climate-smart cropping systems for temperate and tropical agriculture: Mitigation, adaptation and trade-offs. *Cah. Agric.*, 26. https://doi.org/10.1051/cagri/2017028.

Fan, Y., Miguez-Macho, G., Jobbagy, E.G., Jackson, R.B. and Otero-Casal, C. (2017). Hydrologic regulation of plant rooting depth. *PNAS*, 114: 10572–10577.

Gaba, S., Lescourret, F., Boudsocq, S., Enjalbert, J., Hinsinger, P., Journet, E.P., Louarn, G., Malezieux, E., Pelzer, E., Prudent, M. and Ozier-lafontaine, H. (2015). Multiple cropping systems as drivers for providing multiple ecosystems services: from concepts to design. *Agron. Sustain. Dev.*, 35: 607–623. https://doi.org/10.1007/s13593-014-0272-z.

Gonzalez-Ollauri, A. and Mickovski, S.B. (2016). Using root spread information of pioneer plants to quantify their mitigation potential against shallow landslides and erosions in temperate humid climate. *Ecol. Eng.*, 95: 302–315. https://doi.org/10.1016/j.ecoleng.2016.06.0280925-8574.

Haruenkit, R.A.H., Oovarodom, S.U.P., Eontowicz, H.A.L., Eontowicz, M.A.L., Ajewicz, M.I.S., Owalska, T.E.K., Icon, E.F.D.E., Uzma, N.U.E.R.O. and Orinstein, S.H.G. (2007). Comparative Study of Health Properties and Nutritional Value of Durian, Mangosteen, and Snake Fruit: Experiments *in vitro* and *in vivo*. *J. Agric. Food Chem.*, 55: 5842–5849.

Ilmiah, H.H., Sulistyaningsih, E. and Joko, T. (2021). *Fruit Morphology, Antioxidant Activity, Total Phenolic and Flavonoid Contents of Salacca zalacca (Gaertner) Voss by Applications of Goat Manures and Bacillus velezensis B-27*, 36: 270–282.

Lim, T.K. (2012). Salacca glaberescens. pp. 432–437. *In*: Lim, T.K. (ed.). *Edible Medicine and Non-medicine Plants*. Springer, Dordecht. https://doi.org/10.1007/978-90-481-8661-7_54.

Mazumdar, P., Pratama, H., Lau, S.E., Teo, C.H. and Harikrishna, J.A. (2019). Biology, phytochemical profile and prospects for snake fruit: An antioxidant-rich fruit of Southeast Asia. *Trends Food Sci. Technol.*, 91: 147–158. https://doi.org/10.1016/j.tifs.2019.06.017.

Nasrudin, Elizani, P. (2019). Study of the Impact of La Nina on the quality of snakefruits (*Salacca edulis* Reinw.) during the storage at room temperature. *J. Galung Trop.*, 8: 103–111.

Nasution, Y. (2017). *Hubungan kerapatan panjang akar tanaman salak Sidimpuan dengan kedalaman efektif di Kecamatan Angkola. J. LPPM UGN*, 8: 44–48.

Nasution, Y., Rasyidin, A.Y. and Saidi, A. (2019). Evaluation of *Salacca sumatrana* as a soil conservation crop in South Tapanuli, North Sumatera, Indonesia. *Biodiversitas*, 20: 664–670. https://doi.org/10.13057/biodiv/d200307.

Nuary, R.B., Sukartiko, A.C. and Machfoedz, M.M. (2019). Modeling the plantation area of geographical indication product under climate change: Salak Pondoh Sleman (*Salacca edulis* cv Reinw). *IOP Conf. Ser. Earth Environ. Sci.*, 365: 0–9. https://doi.org/10.1088/1755-1315/365/1/012020.

Petrie, C.A. and Bates, J. (2017). 'Multicropping', intercropping and adaptation to variable environment in Indus South Asia. *J. World Prehystory*, 30: 81–130.

Pimental, D. (2006). Soil erosion: A food and environment threats. *Environ. Dev. Sustain.*, 8: 119–137.

Purwanto, B.H. and Alam, S. (2020). Impact of intensive agriculture management on carbon and nitrogen dynamics in the humid tropics. *Soil Sci. Plant Nutr.*, 66: 50–59. https://doi.org/10.1080/00380768.2019.1705182.

Rai, I.N., Wiraatmaja, I.W., Semarjaya, C.G.A. and Astiari, K.A. (2014). Application of irrigation for producing fruit of salak gula pasir (*Salacca zalacca* var. Gula Pasir) off season on dry land. *J. Degrad. Min. L. Manag.*, 2(1): 219–222. Doi://https://doi.org/10.115243/jdmlm.2014.021.219.

Rai, I.N., Semarjaya, C.G.A., Wiraatmaja, I.W. and Astiari, K.A. (2015). Relationship between IAA, sugar content, and fruit-set in snake fruit (*Salacca zalacca*). *J. Appl. Hortic.*, 18: 213–216.

Saifuddin, M. and Normaniza, O. (2016). Rooting characteristics of some tropical plants for slope protection. *J. Trop. For. Sci.*, 28: 469–478.

Sanchez-Navarro, V., Zornoza, R., Faz, A. and Fernandez, J.A. (2019). Comparing legumes for use in multiple cropping to enhance soil organic carbon, soil fertility, aggregates stability and vegetables yields under semi-arid conditions. *Scientia Horticulturae*, 246: 835–841. https://doi.org/10.1016/j.scienta.2018.11.065.

Snowdon, R.J., Wittkop, B., Chen, T.W. and Stahl, A. (2021). Crop adaptation to climate change as a consequence of long-term breeding. *Theor. Appl. Genet.*, 134: 1613–1623. https://doi.org/10.1007/s00122-020-03729-3.

Sudaryono, T. (2017). *Peningkatan produksi dan pendapatan usahatani salak Suwaru. J. Hijau Cendekia*, 2: 78–82.

Sumantra, I.K., Ashari, S., Wardiyati, T. and Suryanto, A. (2012). Diversity of shed trees and their influence on the microclimate of agro ecosystem and fruit production of Gula Pasir salak (*Salacca zalacca* var. *Amboniensis*) fruit. *Int. J. Basic Appl. Sci.*, 12: 214–221.

Sumantra, I.K., Suyasdi Pura, I.N.L. and Ashari, S. (2014). Heat unit, phenology and fruit quality of salak (*Salacca zalacca* var. *Amboinensis*) cv. Gula Pasir on different elevation in Tabanan regency-Bali. *Agric. For. Fish.*, 3: 102–107. https://doi.org/10.11648/j.aff.20140302.18.

Sumantra, I.K. and Martiningsih, N.G.A.G.E. (2016). Evaluation of superior characters of salak gula pasir cultivars in two harvest season at the new development area in Bali. *Int. J. Basic Appl. Sci.*, 16: 19–22.

Sunarka, I.K., Rai, I.N. and Kartini, N.L. (2015). *Pengaruh konsentrasi antitranspiran chitosan terhadap pembuahan dan produksi salak Gula Pasir di luar musim. Agrotrop*, 5: 30–36.

Tun, T. (2012). Study on some chemical analysis of *Salacca wallichiana* Mart. (yengan) fruits. *Univ. Reseach J.*, 5: 275–288.

Zubaidah, E., Ifadah, R.A., Kalsum, U., Lyrawati, D., Dwi, W., Putri, R., Blanc, P.J., Zubaidah, E., Ifadah, R.A., Kalsum, U., Lyrawati, D., Dwi, W., Putri, R., Zubaidah, E. and Ifadah, R.A. (2018). Anti-diabetes activity of Kombucha prepared from different snake fruit cultivars. https://doi.org/10.1108/NFS-07-2018-0201.

Zumaidar, Miftahuddin. (2018). Species distribution of genus Salacca. *In*: *Proceeding of IOP Conference, Journal of Physics: Cof. Series*, p. 1116: 052083. https://doi.org/10.1088/1742-6596/1116/5/052083.

Index

M

N

O

P

R

S

T

V

W

Z